Harmonic Analysis in China

Mathematics and Its Applications

Volume 327

Harmonic Analysis in China

edited by

Minde Cheng
Department of Mathematics,
Peking University,
Beijing, China

Dong-gao Deng
Department of Mathematics,
Zhongshan University,
Guangzhou, China

Sheng Gong
Department of Mathematics,
University of Science and Technology,
Hefei, China

and

Chung-Chun Yang
Department of Mathematics,
The Hong Kong University of Science and Technology,
Clear Water Bay,
Kowloon, Hong Kong

SPRINGER-SCIENCE+BUSINESS MEDIA, B.V.

A C.I.P. Catalogue record for this book is available from the Library of Congress.

ISBN 978-94-010-4064-8 ISBN 978-94-011-0141-7 (eBook)
DOI 10.1007/978-94-011-0141-7

Printed on acid-free paper

TABLE OF CONTENTS

PREFACE

In August 1986, while attending the International Congress of Mathematicians held at University of Berkeley, U.S.A., Prof. C.C. Yang suggested that he and I co-edit a book on harmonic analysis in China. At that time he had already edited a book on complex analysis of one complex variable functions in China for Contemporary Mathematics Series of American Mathematical Society. I was very interested in his suggestion, but I asked him to postpone the project for a number of reasons. Previously, I had organized a Summer Symposium on Analysis in China in 1984 at Peking University, and seven expository lectures were delivered by Prof. E.M. Stein, Prof. R.R. Coifman and their colleagues. Our intention was to encourage further scientific exchanges between the mathematicians of USA and PRC and to expose students at the summer school to the level of current research in those important fields covered by the series of seven expository lectures. A book entitled "Beijing Lecturers in Harmonic Analysis" was edited by Prof. E.M. Stein and published as "Annals of Mathematics Studies No. 12" by Princeton University Press in 1986. At that time I believed that this book would have a great influence on the development of harmonic analysis in China a few years later. Moreover, a course on harmonic analysis was already scheduled to be held at the Nankai Institute of Mathematics from March 1 to June 30, 1988. Therefore , I thought that the project proposed by Prof. Yang would be more fruitful if it were started after the meeting at the Nankai Institute. The proceedings of that meeting was published by Springer Verlag as "Lecture Notes in Mathematics" No. 1494 in the year 1991. Since then, Prof. C.C. Yang of the Hong Kong University of Science and Technology, Prof. D.G. Deng of Zhongshan University, Prof. S. Gong of Chinese University of Science and Technology, and myself started to edit the present book in late of 1992.

Professor C.C. Yang played an important role in editing this book and soliciting contributions outside of China, while Prof. D.G. Deng and Prof. S. Gong both paid great attention to recommending Chinese authors and their representative works. In fact, many authors of this book were participants of the two above-mentioned meetings held in Peking University and the Nankai Institute respectively. Thus the contents of this book in a certain sense represent current research on and progress in harmonic analysis in China. However, owing to the limitations of space, we cannot include all recent studies on harmonic analysis in China within this book. Furthermore, there is no space in our book to deal with early research work in this field. We have already held a meeting, "The 1993 International conference on Analysis, Hangzhou", in memory of late Prof. K.K. Chen on the anniversary of his 100th birthday. Professor Chen was a pioneer of harmonic analysis not only in China but also in Japan. In fact he wrote a book entitled "Theory of Trigonometric Series" in Japanese which was published in Japan in 1931. This book had a major influence on the development of harmonic analysis in Japan. I would like to mention it here as one example of the early research works on harmonic analysis by a Chinese mathematician. We, the editors would like to dedicate this book to the pioneer of harmonic analysis in China the late Prof. K.K. Chen. We would also like to express our heartfelt thanks to Prof. Hazewinkel, the editor-in-chief of the Mathematics and its Applications Book Series of Kluwer Academic Publishers, and his staff in making this endeavour possible.

M. Cheng

Peking University

Contributors

Der-Chen Chang	張德健	Department of Mathematics University of Maryland College Park, MD, 20742, USA
Han-Lin Chen	陳翰麟	Institute of Mathematics Academia Sinica, Beijing, China
Jiecheng Chen	陳杰誠	Department of Mathematics Hangzhou University Hangzhou, 310028, China
Dong-Gao Deng	鄧東皋	Department of Mathematics Zhongshan University Guangzhou, 510275, China
Dashan Fan	范大山	Department of Mathematics University of Wisconsin-Milwaukee Milwaukee, WI, 53201, USA
Charles Fefferman		Department of Mathematics Princeton University Princeton, NJ, 08544, USA
Yongsheng Han	韓永生	Department of Mathematics Auburn University Auburn University, Alabama, 36849-5310, USA
Jing-Song Huang	黃勁松	Department of Mathematics The H.K. University of Science & Technology Clear Water Bay, Kowloon, Hong Kong
Qing-Tang Jiang	蔣慶堂	Department of Mathematics Peking University Beijing, 100871, China
Yinsheng Jiang	江寅生	Department of Mathematics Xinjiang University Urumuchi, 830046, China
Guang-Ri Jin	金光日	Department of Mathematics Jilin University Jilin, China
Jian-Shu Li	厲建書	Department of Mathematics University of Maryland College Park, MD, 20742, USA
Shi-Xiong Li	李世雄	Department of Mathematics Anhui University Hefei, Anhui 230039, China

Xue-Zhang Liang	梁學章	Department of Mathematics Jilin University Changchun, Jilin, China
Rui-Lin Long	龍瑞麟	Institute of Mathematics Academia Sinica Beijing 100080, China
Shanzhen Lu	陸善鎮	Department of Mathematics Beijing Normal University Beijing, 100875, China
Yibiao Pan	潘翼彪	Department of Mathematics Pittsburg University Pittsburg, PA, USA
Li-Zhong Peng	彭立中	Department of Mathematics Peking University Beijing, 100871, China
Weiyi Su	蘇維宜	Department of Mathematics Nanjing University Nanjing, 210008, China
Silei Wang	王斯雷	Department of Mathematics Hangzhou University Hangzhou, 310028, China
Weixing Zheng	鄭維行	Department of Mathematics Nanjing University Nanjing, 21008, China
Xuean Zheng	鄭學安	Department of Mathematics Anhui University Hefei, Anhui 230039, China
Kehe Zhu	朱克和	Department of Mathematics State Unviersity of New York Albany, New York, USA

ON L^p ESTIMATES OF THE CAUCHY-RIEMANN EQUATION

DER-CHEN CHANG* CHARLES FEFFERMAN*

University of Maryland Princeton University

Dedicated to Professor James Hummel on the occasion of his retirement

ABSTRACT. Let $\Omega \subset\subset \mathbb{C}^{n+1}$ be a smooth, bounded pseudo-convex domain of finite type. We assume that the Levi form of $\partial\Omega$ is diagonalizable. In this paper, we obtain the L^p estimates for solving operators of the Cauchy-Riemann equation and the sub-Laplacian $\Box_b$.

§1. Introduction to the $\bar\partial$ problem.

Let $\Omega \subset\subset \mathbb{C}^{n+1}$ be a bounded domain with smooth boundary, *i.e.*,

$$\partial\Omega = \{z \in \mathbb{C}^{n+1} : \rho(z) = 0\}$$

with $d\rho(z) \neq 0$, $\forall z \in \partial\Omega$.

One of the basic problem in the theory of several complex variables is to solve the inhomogeneous Cauchy-Riemann equations:

$$\bar\partial\mu = f \tag{1.1}$$

with good bounds on Ω, where

$$f = \sum_{j=1}^{n+1} f_j \bar\omega_j$$

is a given $(0,1)$-form on Ω.

The system (1.1) is overdetermined when $n \geq 1$. In order to solve the problem, f has to satisfy the consistency condition:

$$\bar\partial f = 0.$$

However, the solution for the system (1.1) is highly non-unique. We always can add holomorphic functions to a solution of (1.1) to produce other solutions.

For the problem (1.1), there are essentially three aspects to this matter:

* Both authors are supported by a grant from the National Science Foundation

M. Cheng et al. (eds.), Harmonic Analysis in China, 1–21.

(i) Existence of solutions;

(ii) Choosing a *good solution*, where "good" means smooth or bounded;

(iii) Estimates and regularity.

If we introduce the Hilbert space $L^2(\Omega)$, we can define a **"canonical"** solution (or a Kohn solution) μ by requiring that .

$$\mu \perp \{\text{holomorphic functions}\}. \tag{1.2}$$

For $\mu, \nu \in C^\infty(\overline{\Omega})$ and D^* the formal adjoint of a first order differential operator D, we have

$$\int_\Omega (D\mu)\overline{\nu} dV = \int_\Omega \mu\overline{(D^*\nu)} dV + \int_{\partial\Omega} \mu\overline{(A^{\#}\nu)} d\sigma,$$

where $A^{\#}$ is a zeroth order operator on the boundary. The domain of the adjoint D^* consists of those $\nu \in C^\infty(\overline{\Omega})$ such that the integral over the boundary equals 0.

In our case $D = \overline{\partial}$ and the domain of the adjoint is

$$\mathrm{dom}(\overline{\partial}^*) = \{\nu \in C^\infty(\overline{\Omega}) : A^{\#}\nu = 0 \text{ on } \partial\Omega\}.$$

Note that with $\mu = \overline{\partial}^* u$, then for F holomorphic

$$< \overline{\partial}^* u, F > = < u, \overline{\partial} F > = 0,$$

whenever the $(0,1)$ form $u \in \mathrm{dom}(\overline{\partial}^*)$. This means that if we solve the equation

$$\overline{\partial}\,\overline{\partial}^* u = f, \quad u \in \mathrm{dom}(\overline{\partial}^*),$$

then we can solve our original problem (1.1) by putting $\mu = \overline{\partial}^* u$.

In fact, problem (1.1) is equivalent to the case $\overline{\partial} f = 0$ of the system

$$\Box u = (\overline{\partial}\,\overline{\partial}^* + \overline{\partial}^*\overline{\partial})u = f, \tag{1.3}$$

$$u \in \mathrm{dom}(\overline{\partial}^*), \quad \overline{\partial} u \in \mathrm{dom}(\overline{\partial}^*). \tag{1.4}$$

To see this, note that

$$0 = \overline{\partial} f = \overline{\partial}(\overline{\partial}\,\overline{\partial}^* + \overline{\partial}^*\overline{\partial})u = \overline{\partial}\,\overline{\partial}^*\overline{\partial} u,$$

and so

$$0 = < \overline{\partial} u, \overline{\partial}\,\overline{\partial}^*\overline{\partial} u > = < \overline{\partial}^*\overline{\partial} u, \overline{\partial}^*\overline{\partial} u >$$

showing that $\overline{\partial}^*\overline{\partial} u = 0$. Comparing this with (1.3) gives $\overline{\partial}\,\overline{\partial}^* u = f$, which by our earlier calculation is equivalent to (1.1). Note that the system (1.3) and (1.4) is not over determined, which maps the space of all smooth $(0,q)$ forms $\mathcal{B}^{(0,q)}(\Omega)$ to itself for $1 \le q \le n+1$. For general u, the system (1.3) and (1.4) is called the

"$\overline{\partial}$-Neumann problem".

The formalism of the $\overline{\partial}$-Neumann problem was introducing by D.C.Spencer in the early 1950's. The existence and regularity properties of the solving operator $\mathbf{N}$ of the $\overline{\partial}$-Neumann problem and hence the solving operator $\overline{\partial}^*\mathbf{N}$ of the equation (1.1) are based on the geometric properties of Ω.

§2. A parametrix for the $\overline{\partial}$-Neumann problem.

Let us first review some basic properties of complex geometry of a domain in $\mathbb{C}^{n+1}$ (see also [BFG]).

Definition 2.1. *Let $\mathcal{M}$ be a $(2n+1)$-dimensional manifold. Then a CR structure on $\mathcal{M}$ is given by a subbundle $T^{1,0}(\mathcal{M})$ of the complex tangential bundle $\mathbb{C}T(\mathcal{M})$ satisfying the following properties:*

(1) $T^{1,0}(\mathcal{M}) \cap \overline{T^{1,0}(\mathcal{M})} = \{0\}$.

(2) The fiber dimension of $T^{1,0}(\mathcal{M})$ is n.

(3) If Z and Z' are local vector fields with values in $T^{1,0}(\mathcal{M})$, then the commutator $[Z, Z'] = ZZ' - Z'Z$ also has values in $T^{1,0}(\mathcal{M})$.

In general, a manifold $\mathcal{M}$ with a fixed CR structure is called a CR manifold.

Definition 2.2. *Let $Z_1, \ldots, Z_n$ be C^∞ vector fields on an open set $U \subset \mathcal{M}$ which are a local basis of sections of $T^{1,0}(\mathcal{M})$ on U. Let T be a real vector field on U such that $Z_1,\ldots,Z_n$, $\overline{Z}_1,\ldots,\overline{Z}_n$, T is a basis of the complex vector fields. The vector field $[Z_j, \overline{Z}_k]$ in terms of this basis is given by*

$$[Z_j, \overline{Z}_k] = c_{jk}\sqrt{-1}T + \sum_{\ell=1}^{n} a_{jk}^{\ell} Z_\ell + \sum_{\ell=1}^{n} b_{jk}^{\ell} \overline{Z}_\ell. \tag{2.1}$$

The Hermitian form (c_{jk}) is called the Levi-form. $\mathcal{M}$ is called pseudoconvex if each point of $\mathcal{M}$ has a neighborhood on which the vector field T can be chosen so that $(c_{jk}) \geq 0$. The Levi form is said to be diagonalizable on U if the local basis $Z_1, \ldots, Z_n$ can be chosen so that

$$c_{jk} = \delta_{jk}\lambda_j \qquad \text{on} \quad U. \tag{2.2}$$

Remarks.

(1) When the Levi-form (c_{jk}) is positive definite, we call $\mathcal{M}$ strongly pseudoconvex.

(2) If $\mathcal{M}$ is a hypersurface in $\mathbb{C}^{n+1}$ then it has the CR sturcture induces by $\mathbb{C}^{n+1}$ where

$$T^{1,0}(\mathcal{M}) = T^{1,0}(\mathbb{C}^{n+1}) \cap \mathbb{C}T(\mathcal{M});$$

that is, the fiber $T_p^{1,0}(\mathcal{M})$ consists of vectors of the form

$$\sum_{j=1}^{n+1} a_j \frac{\partial}{\partial z_j}$$

which are tangent to $\mathcal{M}$ at p.

(3) Let $p_0 \in \mathcal{M} \subset \mathbb{C}^{n+1}$, ρ a local defining function of $\mathcal{M}$ near p_0, that is, $\rho = 0$ on $\mathcal{M}$ and $d\rho \neq 0$, and let $z_1, \dots, z_{n+1}$ be coordinates with origin at p_0 such that $\frac{\partial \rho}{\partial z_{n+1}}(0) \neq 0$. Define Z_j and T by

$$Z_j = \frac{\partial}{\partial z_j} - \frac{\partial \rho / \partial z_j}{\partial \rho / \partial z_{n+1}} \frac{\partial}{\partial z_{n+1}}, \qquad j = 1, \dots, n$$
$$T = -\sqrt{-1}\left(\frac{\partial \rho}{\partial \overline{z}_{n+1}} \frac{\partial}{\partial z_{n+1}} - \frac{\partial \rho}{\partial z_{n+1}} \frac{\partial}{\partial \overline{z}_{n+1}}\right).$$

Then in (2.1) we have $a^{\ell}_{jk} = b^{\ell}_{jk} = 0$. If the Levi form is diagonalizable then in general a basis which diagonalizes the Levi form does not have $a^{\ell}_{jk} = b^{\ell}_{jk} = 0$.

In the rest of this paper, we always assume that Ω be a bounded domain in $\mathbb{C}^{n+1}$ with smooth boundary $\partial\Omega$ whose Levi form is diagonalizable. Observe that the condition diagonalizable Levi form is automatically satisfied if dim($\partial\Omega$)=3 and also if all of the eigenvalues of the Levi form are positive, *i.e.*, Ω is strongly pseudoconvex.

Now let us go back to the $\overline{\partial}$-Neumann problem. First of all, we can rewrite the system (1.3), (1.4) as the following. Let g be a smooth Hermitian metric on $\mathbb{C}^{n+1}$. Then there is an open neighborhood U of $\partial\Omega$ such that if ρ denotes a signed geodesic distance in the metric g to $\partial\Omega$, then

$$\Omega^+ = \Omega \cap U = \{z \in U : \rho(z) > 0\};$$
$$\nabla \rho(z) \neq 0 \qquad \text{for all} \quad z \in U.$$

We choose a smooth orthonomal basis for $(0,1)$-form on U, given by $\overline{\omega}_1, \dots, \overline{\omega}_{n+1}$, where

$$\overline{\omega}_{n+1} = \sqrt{2}\overline{\partial}\rho.$$

We let $\overline{Z}_1, \dots, \overline{Z}_{n+1}$ be the dual basis of antiholomorphic vector fields on U. Then $Z_1,\dots,Z_n$, $\overline{Z}_1,\dots,\overline{Z}_n$ are tangential on $\partial\Omega$, and in fact on the set U we have

$$Z_1(\rho) = \cdots = Z_n(\rho) = \overline{Z}_1(\rho) = \cdots = \overline{Z}_n(\rho) = 0$$

$$Z_{n+1}(\rho) = \overline{Z}_{n+1}(\rho) = \frac{1}{\sqrt{2}}.$$

Hence if we define a real vector field T by

$$T = \frac{1}{2i}(Z_{n+1} - \overline{Z}_{n+1})$$

then on the set U we have

$$T(\rho) = 0$$

so T is also tangential on $\partial\Omega$, and the vector fields

$$\{\mathrm{Re}(Z_1), \dots, \mathrm{Re}(Z_n), \mathrm{Im}(Z_1), \dots, \mathrm{Im}(Z_n), T\}$$

spans the real tangent space to $\partial\Omega$ at every point of $\partial\Omega$. If $\frac{\partial}{\partial\rho}$ is the vector field dual to the one form $d\rho$ then it is easy to see that

$$Z_{n+1} = \frac{1}{\sqrt{2}}\frac{\partial}{\partial\rho} + iT.$$

Because the vector fields split into "tangential" and "normal" part, we may consider a (0,1)-form u as follows:

$$u = \sum_{j=1}^{n+1} u_j\overline{\omega}_j = \sum_{j=1}^{n} u_j\overline{\omega}_j + u_{n+1}\overline{\omega}_{n+1} = u_{(t)} + u_{(n)}.$$

Then the $\overline{\partial}$-Neumann problem is the following boundary value problem:

$$\begin{aligned} \Box u &= f \qquad \text{on} \quad \Omega; \\ u_{n+1} &= 0 \qquad \text{on} \quad \partial\Omega; \\ \overline{Z}_{n+1}(u_j) - [S(u)]_{j,n+1} &= 0 \qquad \text{on} \quad \partial\Omega \end{aligned}$$

with

$$[S(u)]_{j,n+1} = \sum_{\ell=1}^{n} \overline{s}^{\ell}_{j,n+1} u_j, \qquad j = 1, \dots, n.$$

Here the matrix S is defined by the equations

$$\overline{\partial}\overline{\omega}_\ell = \sum_{j<k} \overline{s}^{\ell}_{jk}\overline{\omega}_j \wedge \overline{\omega}_k.$$

The computation of $\Box$ is an algebraic exercise. In the coordinates given by $\overline{\omega}_1,\dots,\overline{\omega}_{n+1}$, $\Box$ is a matrix valued differential operator.

$$\begin{aligned} \Box u = &\begin{bmatrix} \Box_1 & 0 & \cdots & 0 & 0 \\ 0 & \Box_2 & \cdots & 0 & 0 \\ \cdots & & & & \\ 0 & 0 & \cdots & \Box_n & 0 \\ 0 & 0 & \cdots & 0 & \Box_{n+1} \end{bmatrix} u \\ &+ (h_{n+1}\mathbf{I}_{n+1} + S^t)(\overline{Z}_{n+1}u) - (\overline{S}(Z_{n+1}u)) \\ &+ \varepsilon(Z, \overline{Z})u + \varepsilon(u) \end{aligned}$$

with

$$\Box_\ell = \left(-\frac{1}{2}\right)\sum_{j=1}^{n}(Z_j\overline{Z}_j + \overline{Z}_jZ_j) - Z_{n+1}\overline{Z}_{n+1} + \left(\sum_{j=1}^{n}\lambda_j - 2\lambda_\ell\right) iT$$

for $\ell = 1, \dots, n$, and

$$\Box_{n+1} = \left(-\frac{1}{2}\right)\sum_{j=1}^{n}(Z_j\overline{Z}_j + \overline{Z}_jZ_j) - Z_{n+1}\overline{Z}_{n+1} + \left(\sum_{j=1}^{n}\lambda_j\right) iT.$$

The function h_{n+1} comes from the volume elements of the metric. Here $\varepsilon(u)$ and $\varepsilon(Z,\overline{Z})u$ represent an expression that depends on u and the first order derivatives $Z_j u$ and $\overline{Z}_j u$, $j = 1, 2, \dots, n$, with smooth coefficients (which are tangential to the boundary, but does not depend on higher derivatives or on the Z_{n+1} and $\overline{Z}_{n+1}$ derivatives of u).

In fact, for the "normal" part $u_{(n)}$ of u, we are solving a standard coercive boundary value problem:

$$\begin{aligned} \Box u_{n+1} &= f_{n+1} \qquad \text{on} \qquad \Omega \\ u_{n+1} &= 0 \qquad\quad\ \text{on} \qquad \partial\Omega. \end{aligned}$$

It is easy to see that the $(n+1)$-component $\mathbf{N}_{n+1}$ of solving operator $\mathbf{N}$ for the $\overline{\partial}$-Neumann problem will gain two in all directions, *i.e.*,

$$\mathbf{N}_{n+1} : L^p_k(\Omega) \to L^p_{k+2}(\Omega)$$

for all $k \in \mathbb{Z}_+$ and $1 < p < \infty$. Now let us turn to the "tangential" part $u_{(t)}$. Because of the second boundary condition, we need to deal with a non-coercive boundary value problem for $u_{(t)}$. We are using a method to reduce the problem of inverting a certain pseudodifferential operator on $\partial\Omega$. The reduction to the boundary is accomplished through the use of two operators associated to the domain Ω and the operator $\Box$, namely a Poisson operator $\mathbf{P}$ and a Green's operator $\mathbf{G}$. We let $\mathbf{R}$ denote the operator of restriction to the boundary, then the operator $\mathbf{P}$ maps $(0,1)$-forms on the boundary $\partial\Omega$ to $(0,1)$-forms on Ω and has the property, that modulo C^∞ errors,

$$\begin{aligned} \Box \circ \mathbf{P} &= 0 \ \text{ on } \Omega; \\ \mathbf{R} \circ \mathbf{P} &= \mathbf{I} \ \text{ on } \partial\Omega. \end{aligned}$$

The Green's operator $\mathbf{G}$ has the property that modulo C^∞ errors,

$$\begin{aligned} \Box \circ \mathbf{G} &= \mathbf{I} \ \text{on } \Omega; \\ \mathbf{R} \circ \mathbf{G} &= 0 \ \text{ on } \partial\Omega. \end{aligned}$$

We now try to find a solution to problem (1.3) of the form

$$u_{(t)} = \mathbf{P}(u^b_{(t)}) + \mathbf{G}(f_{(t)}) \tag{2.3}$$

where $u^b_{(t)}$ is a $(0,1)$-form on the boundary to be detemined. It follows from the defining properties of $\mathbf{P}$ and $\mathbf{G}$ that, modulo C^∞ errors,

$$\Box(u_{(t)}) = f_{(t)} \tag{2.4}$$

for **any** $u_{(t)}$ of the form given in (2.3). Thus we want to determine $u^b_{(t)}$ so that the second $\overline{\partial}$-Neumann condition is satisfied. Now let us consider

$$\overline{Z}_{n+1}(u_{(t)}) = \overline{Z}_{n+1}\left(\mathbf{P}(u^b_{(t)})\right) + \overline{Z}_{n+1}\left(\mathbf{G}(f_{(t)})\right)$$

and so the second $\overline{\partial}$-Neumann condition is equivalent to

$$\mathbf{R}\left(\overline{Z}_{n+1}\left(\mathbf{P}(u^b_{(t)})\right)\right)=\mathbf{R}\left(\frac{1}{\sqrt{2}}\left(\frac{\partial}{\partial\rho}-iT\right)\left(\mathbf{P}(u^b_{(t)})\right)\right) \tag{2.5}$$
$$=-\mathbf{R}\left(\overline{Z}_{n+1}\left(\mathbf{G}(f_{(t)})\right)\right).$$

But we know that the Dirichlet to Neumann operator N^+ associated with the domain Ω and the operator $\Box$ is

$$\mathbf{R}\frac{\partial}{\partial\rho}\mathbf{P}(u^b_{(t)})=N^+(u^b_{(t)}).$$

We now make the following definition:

Definition 2.3. *The Calderón operator $\Box^+$ of the $\overline{\partial}$-Neumann problem is*

$$\Box^+=\frac{1}{\sqrt{2}}N^+-\frac{i}{\sqrt{2}}T.$$

The proceeding discussion then shows that we have:

Proposition 2.4. *The problem (1.3) is equivalent (modulo C^∞ error terms) to the problem of solving*

$$\Box^+(u^b_{(t)})=-\mathbf{R}\left(\overline{Z}_{n+1}\left(\mathbf{G}(f_{(t)})\right)\right),$$

or

$$\Box^+(u^b_\ell\overline{\omega}_\ell)=\left[\left(\frac{1}{\sqrt{2}}\left(-\frac{\partial}{\partial\rho}+iT\right)\right)(\mathbf{G}f_\ell)\overline{\omega}_\ell\right]_{\partial\Omega},$$

for $\ell=1,\cdots,n$; i.e., the problem (1.3) is equivalent to the problem of inverting the operator $\Box^+$.

From the symbolic calculation, we know that the principal symbol of the Calderón operator of the $\overline{\partial}$-Neumann problem on $\overline{\Omega}^+$ is

$$\sigma(\Box^+)=\frac{1}{\sqrt{2}}(\sigma(T)-\Delta)-\frac{1}{2\sqrt{2}}\left(2\lambda_j-\sum_{k=1}^n\lambda_k\right)\frac{\sigma(T)}{\Delta}$$

where

$$\Delta=\sqrt{2\sum_{j=1}^n|\sigma(Z_j)|^2+\sigma(T)^2}.$$

It is easy to see that $\Box^+$ is a first order pseudodifferential operator doubly characteristic on half of the line bundle

$$\Sigma^+=\{(z,t;\xi,\tau);\ \sigma(T)>\Delta\}$$

on the cotangent bundle $T^*(\partial\Omega)$. On the other hand, we also can contruct the Calderón operator $\Box^-$ of the $\overline{\partial}$-Neumann problem on

$$\overline{\Omega}^- = \{z \in U : \rho(z) \leq 0\}.$$

The principal symbol of $\Box^-$ is

$$\sigma(\Box^-) = \frac{1}{\sqrt{2}}(\sigma(T) + \Delta) + \frac{1}{2\sqrt{2}}\left(2\lambda_j - \sum_{k=1}^{n} \lambda_k\right)\frac{\sigma(T)}{\Delta}.$$

This is also a first order pseudodifferential operator characteristic on the other half of the line bundle

$$\Sigma^- = \{(z, t; \xi, \tau);\ \sigma(T) < -\Delta\}$$

but elliptic on the characteristics of $\Box^+$. Now we shall need to consider the compositions $\Box^+ \circ \Box^-$ and $\Box^- \circ \Box^+$.

The important phenomenon is that

$$\Box^+ \circ \Box^- = -\Box_b + \text{zero order terms}$$

and

$$\Box^- \circ \Box^+ = -\Box_b + \text{zero order terms.}$$

Here $\Box_b$ is the complex sub-Laplacian on $(0,1)$-forms defined on the boundary $\partial\Omega$. More precisely,

$$\Box_b = \begin{bmatrix} \Box_1' & 0 & \cdots & 0 \\ 0 & \Box_2' & \cdots & 0 \\ \cdots & & & \\ 0 & 0 & \cdots & \Box_n' \end{bmatrix} + \mathfrak{L}$$

with

$$\Box_\ell' = -\left(\frac{1}{2}\right)\sum_{j=1}^{n}(Z_j\overline{Z}_j + \overline{Z}_jZ_j) + i\left(\sum_{k=1}^{n}\lambda_k - 2\lambda_\ell\right)T$$

for $\ell = 1, \ldots, n$ and $\mathfrak{L} = [\mathfrak{L}_{jk}]$ of the form

$$\mathfrak{L}_{jk} = \sum_{k=1}^{n} a_{jk}^k Z_k + \sum_{k=1}^{n} b_{jk}^k \overline{Z}_k + e_{jk}.$$

Now we can write down a parametrix for the $\overline{\partial}$-Neumann problem.

Case (1): $n \geq 2$ and Ω is strongly pseudo-convex.

By a result of Folland-Stein [FoS], $\Box_b$ has an inverse $\mathbf{K}$ such that

$$\Box_b \circ \mathbf{K} = \mathbf{I} + (\text{smoothing operators}).$$

Putting these result together, for $n \geq 2$ and $j = 1, \cdots, n$ we have

$$\mathbf{N}_j = \mathbf{G} + \mathbf{P}(-\mathbf{K}\Box^- \mathbf{R}\overline{Z}_{n+1}\mathbf{G}) + S_{-\infty}$$

where $S_{-\infty}$ is a smoothing operator.

Case (2): $n = 1$ **and** Ω **is finite type** m[1]

$$\begin{aligned}\Box_b &= -\frac{1}{2}(Z_1\overline{Z}_1 + \overline{Z}_1 Z_1) - i\lambda T \\ &= -\overline{Z}_1 Z_1.\end{aligned}$$

We consider the Lewy's equation

$$Z_1 u = f.$$

It is **not** in general locally solvable. However, by a theorem of Greiner-Kohn-Stein [GKS] (strongly pseudo-convex) and results obtained by Fefferman-Kohn [FK], Christ [Chr], Nagel-Rosay-Stein-Wainger [NRSW] (finite type), we know that there exist an operator $\mathbf{K}$ such that

$$(-\overline{Z}_1 Z_1)\mathbf{K} = \mathbf{K}(-\overline{Z}_1 Z_1) = \mathbf{I} - \mathbf{S}.$$

Here $\mathbf{S}$ is the Szegö projection from $L^2(\Omega)$ to the Hardy space $\mathcal{H}^2(\Omega)$.

Let us consider pseudodifferential operators Γ^+ and Γ^- with symbols in the class $S^0_{1,0}$ such that the principal symbol of Γ^+ equals 1 on the set

$$\{\Delta < \frac{1}{4}\sigma(T)\}$$

and whose principal symbol equal 0 on the set

$$\{\Delta > \frac{1}{2}\sigma(T)\}.$$

We define $\Gamma^- = \mathbf{I} - \Gamma^+$. Now we define

$$\mathbf{K}^+ = Q_{\Gamma^-} + \Gamma^+\mathbf{K}\Box^-,$$

where Q_{Γ^-} denote the parametrix for $\Box^+$ in the support of Γ^-, *i.e.*,

$$Q_{\Gamma^-}\Box^+ = \mathbf{I} - \Gamma^+ + S_{-\infty}.$$

Then we have

$$\begin{aligned}\mathbf{K}^+\Box^+ &= Q_{\Gamma^-}\Box^+ + \Gamma^+\mathbf{K}\Box^-\Box^+ \\ &= \mathbf{I} - \Gamma^+ + \Gamma^+\mathbf{K}(\Box_b + T_0) \\ &= \mathbf{I} - \Gamma^+ + \Gamma^+(\mathbf{I} - \mathbf{S} + S_{-\infty}) \\ &= \mathbf{I} - \Gamma^+\mathbf{S} + S_{-\infty}.\end{aligned}$$

Now we have to use the microlocal analysis to show that $\Gamma^+\mathbf{S}$ is an infinitely smoothing operator. Once we can do that, we have for the $\mathbb{C}^2$ case

$$\mathbf{N}_1 = \mathbf{G} + \mathbf{P}\left((-Q_{\Gamma^-} + \Gamma^+\mathbf{K}\Box^-)\mathbf{R}\overline{Z}_2\mathbf{G}\right) + S_{-\infty}.$$

[1] We will define "finite type" in section 3

Theorem A. *The Neumann operator*

$$\mathbf{N} = (\mathbf{N}_1, \cdots, \mathbf{N}_n, \mathbf{N}_{n+1})$$

has a unique extension so that the indicated mappings are bounded

(1)

$$\mathcal{P}(Z, \overline{Z})\mathbf{N} : L^p_k(\overline{\Omega}) \to L^p_k(\overline{\Omega}),$$

for $1 < p < \infty$, $k \in \mathbb{Z}^+$, *where* $\mathcal{P}$ *is any second degree polynomial in "good" vector fields. (*$\mathbf{N}$ *gains two in "good" directions);*

(2) *As a consequence of* (1),

$$\mathbf{N} : L^p_k(\overline{\Omega}) \to L^p_{k+\frac{2}{m}}(\overline{\Omega}),$$

for $1 < p < \infty$, $k \in \mathbb{Z}^+$.

(3)

$$\overline{\partial}\mathbf{N} \lrcorner \overline{\partial}\rho : L^p_k(\overline{\Omega}) \to L^p_{k+1}(\overline{\Omega}),$$

for $1 < p < \infty$, $k \in \mathbb{Z}^+$.

(4) *Suppose* $\alpha > 0$ *and if* $f \in \Lambda_\alpha(\overline{\Omega})$, *then*

$$\mathbf{N}(f) \in \Lambda_{\alpha+\frac{2}{m}}(\overline{\Omega}) \cap \Gamma_{\alpha+2}(\overline{\Omega}).$$

Here Λ_α *is the standard Lipschitz space and* Γ_α *is the non-isotropic Lipschitz space defined by Folland and Stein* [FS] *(strongly pseudo-convex case) and defined by Nagel, Rosay, Stein, and Wainger* [NRSW] *(finite type in* $\mathbb{C}^2$ *).*

This theorem was proved by Greiner-Stein [GS], Beals-Greiner-Stanton [BGS], Chang [Ch], Henkin,[H], Phong-Stein [PS1], [PS2], Lieb-Range [LR],... when Ω is strongly pseudo-convex. When Ω is finite type in $\mathbb{C}^2$, the theorem was proved by Chang-Nagel-Stein in [CNS1], [CNS2].

From Theorem A, it is easy to get the following results for the Kohn soultion of the Cauchy-Riemann equation.

Corollary. *The solving operator* $\overline{\partial}^*\mathbf{N}$ *of the Cauchy-Riemann equation has a unique extension so that the indicated mappings are bounded*

(1)

$$Z_j(\overline{\partial}^*\mathbf{N}) : L^p_k(\overline{\Omega}) \to L^p_k(\overline{\Omega}), \qquad \text{and} \qquad \overline{Z}_j(\overline{\partial}^*\mathbf{N}) : L^p_k(\overline{\Omega}) \to L^p_k(\overline{\Omega}),$$

for $1 < p < \infty$, $k \in \mathbb{Z}_+$, *and* $j = 1, \ldots, n$.

(2) *As a consequence of* (1),

$$\overline{\partial}^*\mathbf{N} : L^p_k(\overline{\Omega}) \to L^p_{k+\frac{1}{m}}(\overline{\Omega}),$$

for $1 < p < \infty$, $k \in \mathbb{Z}^+$.

(3) *Suppose* $\alpha > 0$ *and if* $f \in \Lambda_\alpha(\overline{\Omega})$, *then*

$$\overline{\partial}^*\mathbf{N}(f) \in \Lambda_{\alpha+\frac{1}{m}}(\overline{\Omega}) \cap \Gamma_{\alpha+1}(\overline{\Omega}).$$

On of the application of $\overline{\partial}$ is to find conditions on a zero set $\mathcal{Z}$ in a given domain Ω that ensures that it is defined by a function f in some given function class. We say a holomorphic function f in the Nevanlinna class $\mathcal{N}(\Omega)$ if and only if

$$f \quad \text{holomorphic in } \Omega \text{ and } \quad \sup_{\varepsilon>0} \int_{\partial\Omega_\varepsilon} \log^+ |f| d\sigma < \infty.$$

We say a holomorphic function f in Hardy space $\mathcal{H}^p(\Omega)$ for some $0 < p < \infty$ if and only if

$$f \quad \text{holomorphic in } \Omega \text{ and } \quad \sup_{\varepsilon>0} \int_{\partial\Omega_\varepsilon} |f|^p d\sigma < \infty.$$

It is well known, in a strongly pseudoconvex domain Ω in $\mathbb{C}^{n+1}$ as well as in a pseudoconvex domain of finite type in $\mathbb{C}^2$ (see [H], [Sk], [BC1], [BC2], [CNS1], and [CNS2]) that a zero variety in Ω is the zero variety of a function f in the Nevanlinna class $\mathcal{N}(\Omega)$ if and only if the zero variety $\mathcal{Z}$ satisfies the Blaschke condition

$$\int_{\mathcal{Z}} \delta(z) d\sigma(z) < \infty$$

where $\delta(z)$ denotes the distance to the boundary; or if the corresponding (1,1)-current $\theta_{\mathcal{Z}}$ representing $\mathcal{Z}$ (see [L]) satisfies

$$\int_\Omega \delta(z) trace(\theta_{\mathcal{Z}}) < \infty.$$

This result is usually referred as the Henkin-Skoda Theorem. It is also natural to wonder whether $\mathcal{Z}$ is the zero set of an $\mathcal{H}^p$ function. In the one dimensional case, it is well known that the same Blaschke condition is necessary and sufficient. In higher dimension, it is much more complicated. In [V], Varopoulos gave a general condition on an analytic set in a strongly pseudoconvex domain Ω to be the zero set of a $\mathcal{H}^p$ function (*i.e.*, for some p), the so-called uniform Blaschke condition. Let us first introduce some notations about the non-isotropic geometry. For every k-tuple of integers define smooth functions $\lambda_{i_1,\dots,i_k}$, $\alpha^1_{i_1,\dots,i_k}$, $\alpha^2_{i_1,\dots,i_k}$ on $\partial\Omega$ by the equation

$$[X_{i_k}, [\dots, [X_{i_2}, X_{i_1}]\dots]] = \lambda_{i_1,\dots,i_k} T + \alpha^1_{i_1,\dots,i_k} Z_1 + \alpha^2_{i_1,\dots,i_k} \overline{Z}_1$$

with $X_{i_j} = Z_1$ or $\overline{Z}_1$ for $1 \le j \le k$. For each integer $\ell \ge 2$ define a smooth function Λ_ℓ on $\partial\Omega$ by the equation

$$\Lambda_\ell(x) = \left[\sum |\lambda_{i_1,\dots,i_k}(x)|^2\right]^{\frac{1}{2}},$$

where the sum is taken over all k-tuples with $2 \le k \le \ell$. Finally set

$$\Lambda(x,\delta) = \sum_{j=2}^m \Lambda_j(x)\delta^j.$$

Let $h \to \mu(x,h)$ be the function inverse to $\delta \to \Lambda(x,\delta)$. Thus clearly

$$\mu(x,h) \approx \min_{2\le j\le m} \left(\frac{h}{\Lambda_j(x)}\right)^{\frac{1}{j}}.$$

If $\rho(x)$ denotes the distance of $x \in \Omega$ from the boundary, then we let $\mu(x)$ be defined by

$$\mu(x) = \mu(\pi(x), \rho(x)).$$

Thus $\rho(x)$ is essentially the radius of the largest "normal" disc in Ω centered at x, while $\mu(x)$ is essentially the radius of the largest "tangential" disc in Ω centered at x.

Now let us identify a small neighborhood of $\partial\Omega$ in $\overline{\Omega}$ with $\partial\Omega \times [0,\varepsilon_0)$. More precisely, we choose a map $\Phi : \partial\Omega \times [0,\varepsilon_0) \to \overline{\Omega}$ such that Φ is a diffeomorphism of $\partial\Omega \times [0,\varepsilon_0)$ onto a neighborhood $\overline{\Omega} \cap U$ of $\partial\Omega$ in $\overline{\Omega}$ such that

- for any $x \in \partial\Omega$, $\Phi(x,0) = x$;
- for any $x \in \partial\Omega$ and any $0 < t \le \varepsilon_0$, $\delta(\Phi(x,t)) \simeq t$.

For $z = \Phi(x,t)$, let us write $x = \pi(z)$. For any $x \in \partial\Omega$ and any $0 < \varepsilon \le \varepsilon_0$, we define the *"tent"* over $\mathcal{B}(x,\varepsilon)$ as the set

$$\mathcal{B}^{\#}(x,\varepsilon) = \Phi\left(\{\mathcal{B}(x,\varepsilon) \times (0,\varepsilon)\}\right).$$

Then $\mathrm{Vol}[\mathcal{B}^{\#}(x,\varepsilon)] \simeq \mu(x,\varepsilon)^2 \times \varepsilon^2$.

For any $z, w \in \overline{\Omega} \cap U$, let us note

$$D(z,w) = \delta(z) + \delta(w) + d(\pi(z), \pi(w))$$

where $d(x,y)$ is given by: $d(x,y) = \inf\{\delta > 0;$ there exists Φ piecewise smooth from $[0,1]$ to $\partial\Omega$ such that $\Phi(0) = x$, $\Phi(1) = y$, and if $\beta_1(t)$, $\beta_2(t)$, $\gamma(t)$ are the coordinates of $\Phi(t)$ in the basis X_1, X_2, T, $|\beta_j(t)| \le \mu(\Phi(t),\delta)$, $|\gamma(t)| \le \delta\}$. Here $Z = X_1 + iX_2$.

Then, z, w belong to the tent $\mathcal{B}^{\#}(\pi(z), CD(z,w))$, and conversely, if w and z belong to some tent $\mathcal{B}^{\#}(x', \varepsilon)$, then $\varepsilon \ge cD(z,w)$.

We will call a positive measure ν defined on Ω a Carleson measure (*i.e.*, $\nu \in \mathcal{C}(\Omega)$) if and only if there exists a constant C such that $\nu(\Omega) \le C$ and if, for any $x \in \partial\Omega$, any $0 < \varepsilon \le \varepsilon_0$,

$$\nu(\mathcal{B}^{\#}(x,\varepsilon)) \le C\sigma(\mathcal{B}(x,\varepsilon)).$$

The smallest constant C satisfying the preceding property is called the Carleson norm of ν and is denoted by $\|\nu\|_{\mathcal{C}}$. Now the uniform Blaschke condition can be expressed as follows.

We say that a positive (1,1)-current θ satisfies the uniform Blaschke condition (which will be denoted by (UB)) if and only if

$$\rho|\theta| + \mu(|\partial\rho \wedge \theta| + |\overline{\partial}\rho \wedge \theta|) + \frac{\mu^2}{\rho}|\partial\rho \wedge \overline{\partial}\rho \wedge \theta|$$

is a Carleson measure with respect to the non-isotropic structure on $\partial\Omega$.

Then, we have the following result.

Theorem B. *Let $\Omega = \{\rho > 0\}$ be a bounded pseudoconvex domain of finite type in $\mathbb{C}^2$ and $\mathcal{Z}$ be a zero-variety of a holomorphic function f in Ω so that the (1,1) current $\theta = \frac{i}{\pi}\partial\overline{\partial}\log|f|$ satisfies*

(1) *the Blaschke condition then, $\mathcal{Z}$ is defined by a holomorphic function in the Nevanlinna class.*
(2) *the uniform Blaschke condition then, $\mathcal{Z}$ is defined by an $\mathcal{H}^p$ function for some $p > 0$.*

The scheme of the proof is well known. We have to find real valued solutions of $i\partial\overline{\partial}u = \theta_{\mathcal{Z}}$ with control on u since any such solutions is $\log|f|$ for an f which defines $\mathcal{Z}$. The equation $i\partial\overline{\partial}u = \theta_{\mathcal{Z}}$ is solved in two steps. First, we solve $idw = \theta_{\mathcal{Z}}$ with w satisfying $w = -\overline{w}$. Then we decompose w into

$$w = w_{1,0} + w_{0,1}$$

where $w_{1,0}$ and $w_{0,1}$ are bidegree $(1,0)$ and $(0,1)$ forms. Note that $w_{0,1}$ is $\overline{\partial}$-closed since θ is a $(1,1)$ form. If we let v be the solution to

$$\overline{\partial}v = w_{0,1},$$

it follows that

$$\begin{aligned} i\partial\overline{\partial}(2\operatorname{Re}(v)) &= i\partial\overline{\partial}(v+\overline{v}) = i(\partial\overline{\partial}v - \overline{\partial}\partial\overline{v}) \\ &= i(\partial w_{0,1} - \overline{\partial}\overline{w}_{0,1}) = i\{(\partial+\overline{\partial})w_{0,1} + (\partial+\overline{\partial})w_{1,0}) \\ &= idw = \theta. \end{aligned}$$

Thus if we put $u = 2\operatorname{Re}(v)$, the function u satisfies $i\partial\overline{\partial}u = \theta_{\mathcal{Z}}$. The first step is stated in Theorem 2.5 below.

Theorem 2.5.

Let $\Omega = \{\rho > 0\}$ be a pseudoconvex domain of finite type in $\mathbb{C}^2$. Assume that $\theta = \frac{i}{\pi}\partial\overline{\partial}\log|f|$ is a closed (1,1) positive current that satisfies the (UB) condition. Then, there is a real solution w to $dw = \theta$ such that

$$|w| + \frac{\mu}{\rho}|\overline{\partial}\rho \wedge w_{0,1}|$$

is a Carleson measure with respect to the nonisotropic structure on $\partial\Omega$.

In order to prove Theorem B, we need the following estimates on the solution of $\overline{\partial}$-equation.

Theorem C. *Suppose f is a smooth $(0,1)$ form. Then we have the a priori inequality*

$$\text{(HS)} \qquad \|\bar{\partial}^* \mathbf{N} f\|_{L^1(\partial\Omega)} \leq C \cdot \left(\|f\|_{L^1(\Omega)} + \left\| \left(\frac{\mu}{\rho}\right) f \wedge \bar{\partial}\rho \right\|_{L^1(\Omega)} \right).$$

The estimate (HS) was proved by Henkin and Skoda independently when Ω is strongly pseudo-convex (see Henkin [H] and Skoda [Sk]). When Ω is finite type in $\mathbb{C}^2$, the theorem was proved in [CNS2].

Theorem D. *Suppose f is a smooth $(0,1)$ form. Then we have the a priori inequality*

$$\text{(V)} \qquad \|\exp(\bar{\partial}^* \mathbf{N} f)\|_{L^p(\partial\Omega)} \leq C \cdot \left(\|f\|_{\mathfrak{C}(\Omega)} + \left\| \left(\frac{\mu}{\rho}\right) f \wedge \bar{\partial}\rho \right\|_{\mathfrak{C}(\Omega)} \right).$$

Here $\|f\|_{\mathfrak{C}(\Omega)}$ is the Carleson norm of the function (form) f.

The estimate (V) was proved by Varopoulos in [V] when Ω is strongly pseudo-convex. When Ω is finite type in $\mathbb{C}^2$, the theorem was proved in [CF1].

§3. Weakly pseudo-convex domains in $\mathbb{C}^n$, $n \geq 3$.

To get similar results for weakly pseudo-convex domains is not so easy. Sibony [S1], [S2] constructed a family of domains such that the sup-norm estimate (with no gain) for $\bar{\partial}$ is fail. More precisely, for any fixed δ, $0 \leq \delta \leq 1$, there is a pseudoconvex domain $\Omega_\delta \subset\subset \mathbb{C}^3$, with smooth boundary, strongly pseudoconvex except at one point, and for every $k \in \mathbb{Z}_+$ there exists a form $f \in C^{k,\delta}_{(0,1)}(\overline{\Omega}_\delta)$, such that $\bar{\partial} u = f$ has no solution in $C^{k,\delta}(\overline{\Omega}_\delta)$. Here $C^{k,\delta}(\overline{\Omega}_\delta)$ denotes the space of all C^k functions in $\overline{\Omega}_\delta$, such that for all α, $|\alpha| \leq k$,

$$\sup_{x,y \in \Omega_\delta} \frac{|D^\alpha f(x) - D^\alpha f(y)|}{|x-y|^\delta} < \infty.$$

In fact, Catlin [C1], [C2] obtained remarkable results in 1983 and 1987: the $\bar{\partial}$-Neumann problem on a pseudo-convex domain satisfies the sub-elliptic estimate at a point $P \in \partial\Omega$ if and only if P is of finite type (*i.e.*, every complex analytic curve through the boundary $\partial\Omega$ has bounded order of contact with $\partial\Omega$.) Here sub-elliptic estimate means

$$\|u\|_{L^2_{k+\varepsilon}(\Omega)} \leq C \cdot \left(\|\Box u\|_{L^2_k(\Omega)} + \|u\|_{L^2_k(\Omega)} \right).$$

for some $\varepsilon > 0$.

Remark. When the Levi form is diagonalizable, the finiteness condition is equivalent to the following:

"For each $j = 1,\dots,n$, the Lie algebra generated by Z_j and $\overline{Z}_j$ contains T modulo $Z_1,\dots,Z_n,\overline{Z}_1,\dots,\overline{Z}_n$."

Equivalently,

"For each j there exists a monomial $\mathcal{P}_j$ in Z_j, $\overline{Z}_j$, $j = 1,\dots,n$, such that

$$\mathcal{P}_j(Z_j,\overline{Z}_j)\lambda_j \neq 0. \tag{3.1}$$

The type of Z_j denoted by m_j is given by

$$m_j = \deg(\mathcal{P}_j) + 2,$$

where $\mathcal{P}_j$ is the monomial of the smallest degree for which (3.1) holds. We define m, the type of $\partial\Omega$, by $m = \max\{m_j\}$."

We also would like to point out that the pseudo-metric associated to the Szegö projection for strongly pseudo-convex domains in $\mathbb{C}^n$ or finite type domains in $\mathbb{C}^2$ is the same as the pseudo-metric associated to the sum of square of vector fields and the operator $\Box_b$ (see *e.g.*, [NSW], [NRSW], [CNS2]). Unlike the cases of strong pseudo-convexity and finite type in $\mathbb{C}^2$, the parametrix $\mathbf{K}$ of $\Box_b$ for decoupled domains of finite type in $\mathbb{C}^3$ is controlled by the above pseudo-metric ρ and another nonequivalent metric $\tilde{\rho}$. This striking result was first discovered by Machedon [M]. For example, let us consider the domain $\{(z_1,z_2,z_3) : \operatorname{Im} z_3 > \mathcal{P}(z_1) + |z_2|^2\}$ in $\mathbb{C}^3$ with $\mathcal{P}$ a subharmonic but not harmonic polynomial then the kernel $K(x,y)$ of the solving operator $\mathbf{K}$ for $\Box_b$ satisfying the size estimate

$$|K(x,y)| \leq C\frac{\tilde{\rho}^2(x,y)}{\mathrm{Vol}(B^{\tilde{\rho}}(y,\tilde{\rho}(x,y)))}\log\left(2+\frac{\tilde{\rho}(x,y)}{\rho(x,y)}\right)$$

and the log cannot be removed. We also know that

$$|Z^J K(x,)| \leq C\frac{\tilde{\rho}^2(x,y)}{\rho^{|J|}(x,y)\mathrm{Vol}(B^{\tilde{\rho}}(y,\tilde{\rho}(x,y)))}\log\left(2+\frac{\tilde{\rho}(x,y)}{\rho(x,y)}\right)$$

for Z^J a $|J|$th order monomial in Z_1 and $\overline{Z}_1$ with $|J| > 0$. Here

$$\rho = \text{metric associated to } \sum_{j=1}^{2}(X_j^2+Y_j^2)$$

and B^ρ the associated ball

$$\tilde{\rho} = \text{metric associated to } (1-\lambda)^{-1}\left(X_1^2+Y_1^2+\lambda(X_2^2+Y_2^2)\right)$$

and $B^{\tilde{\rho}}$ the associated ball

where $\lambda = \frac{1}{4}\triangle\mathcal{P}$. Hence it is not so surprising that the solving operator $\mathbf{K}$ for $\Box_b$ fails to achieve the expected result, *i.e.*,

$$\mathcal{P}(Z,\overline{Z})\mathbf{K} : L^2(\partial\Omega) \nrightarrow L^2(\partial\Omega).$$

Indeed, this was first proved by Rothschild [R]. However, we still have some positive results. Recently, Chang-Fefferman [CF] obtained the L^p, $1 < p < \infty$, estimates for $\mathbf{K}$ when Ω is finite type with diagonalizable Levi form.

Theorem E. *Let $\Omega \subset\subset \mathbb{C}^{n+1}$ be a bounded weakly pseudoconvex domain of finite type m. Suppose the Levi form of $\partial\Omega$ is diagonalizable. Suppose further that the range of $\overline{\partial}_b$ is closed in $L^2(\partial\Omega)$, then*

(1) *For all $\varepsilon > 0$ the operator*

$$\mathbf{K} : L^p_k(\partial\Omega) \to L^p_{k+\frac{2}{m}-\varepsilon}(\partial\Omega)$$

boundedly. Here $L^p_k(\partial\Omega)$ denotes the L^p Sobolev spaces of forms (of degree $(0,q)$ with $1 \le q \le n$) of order k, $k \ge 0$.

(2) *For all $\varepsilon > 0$ the operators*

$$\overline{\partial}_b\mathbf{K},\ \overline{\partial}_b^*\mathbf{K},\ \mathbf{K}\overline{\partial}_b,\ \mathbf{K}\overline{\partial}_b^*$$

map $L^p_k(\partial\Omega)$ into $L^p_{k+\frac{1}{m}-\varepsilon}(\partial\Omega)$.

(3) *For all $\varepsilon > 0$ the operators*

$$\overline{\partial}_b\overline{\partial}_b^*\mathbf{K},\ \overline{\partial}_b^*\overline{\partial}_b\mathbf{K},\ \mathbf{K}\overline{\partial}_b\overline{\partial}_b^*,\ \mathbf{K}\overline{\partial}_b^*\overline{\partial}_b,\ \overline{\partial}_b\mathbf{K}\overline{\partial}_b^*,\ \overline{\partial}_b^*\mathbf{K}\overline{\partial}_b$$

map $L^p_k(\partial\Omega)$ into $L^p_{k-\varepsilon}(\partial\Omega)$. Here again we assume $\mathbf{K}$ acts on forms of type $(0,q)$ with $1 \le q \le n$.

(4) *If $\overline{\partial}_b u = f$, if f is orthogonal to nullspace of $\overline{\partial}_b$, and if f is a $(0,1)$-form in $L^p_k(\partial\Omega)$ then $u \in L^p_{k+\frac{1}{m}-\varepsilon}(\partial\Omega)$ for all $\varepsilon > 0$.*

(5) *If $f \in L^p_k(\partial\Omega)$ then $\mathbf{S}_b(f) \in L^p_{k-\varepsilon}(\partial\Omega)$ for all $\varepsilon > 0$ where $\mathbf{S}_b$ is the orthogonal projection on square-integrable CR functions.*

We only sketch the main idea of the proof of Theorem E here. The details will appear in somewhere else (see [CF]). Once again, we are using microlocal analysis. We first choose coordinates $x_1, \dots, x_{2n}, t$ on a small neighborhood U of $p_0 \in \partial\Omega$ such that $T = \frac{\partial}{\partial t}$. Let

$$\mathcal{R}^+ = \left\{(\xi,\tau) \in \mathbb{R}^{2n+1} : (|\xi|^2 + \tau^2)^{\frac{1}{2}} > \frac{1}{2} \quad \text{and} \quad \tau > c|\xi|\right\}$$

with τ the dual variable of t. We also let

$$\mathcal{R}^- = \left\{(\xi,\tau) \in \mathbb{R}^{2n+1} : (|\xi|^2 + \tau^2)^{\frac{1}{2}} > \frac{1}{2} \quad \text{and} \quad \tau < -c|\xi|\right\}$$

and $\mathcal{R}^0$ be the set of $(\xi,\tau) \in \mathbb{R}^{2n+1}$ such that either $|\tau| < c|\xi|$ or $(|\xi|^2 + \tau^2)^{\frac{1}{2}} < 1$. Let $\chi^+, \chi^-, \chi^0 \in C^\infty(\mathbb{R}^{2n+1})$ be a partition of unity, *i.e.*, $\chi^+ + \chi^- + \chi^0 = 1$ and $\mathrm{supp}(\chi^+) \subset \mathcal{R}^+$, $\mathrm{supp}(\chi^-) \subset \mathcal{R}^-$, $\mathrm{supp}(\chi^0) \subset \mathcal{R}^0$. Suppose further these functions are homogeneous of order 0 for $(|\xi|^2 + \tau^2)^{\frac{1}{2}} \ge 1$, that is, $\chi(\delta(\xi,\tau)) = \chi(\xi,\tau)$ whenever $(|\xi|^2 + \tau^2)^{\frac{1}{2}} \ge 1$ and $\delta \ge 1$. We define projection operators P^+, P^-, and P^0 by

$$\widehat{P^+u}(\xi,\tau) = \chi^+(\xi,\tau)\hat{u}(\xi,\tau), \qquad \widehat{P^-u}(\xi,\tau) = \chi^-(\xi,\tau)\hat{u}(\xi,\tau),$$

and

$$\widehat{P^0 u}(\xi,\tau) = \chi^0(\xi,\tau)\hat{u}(\xi,\tau).$$

Hence for $\zeta = 1$ on supp(u), we have

$$\zeta P^+ u + \zeta P^- u + \zeta P^0 u = u.$$

Since $\Box_b$ is elliptic on the support of χ^0, we have

$$\|\zeta P^0 u\|^2_{L^p_1(\overline{\Omega})} \leq c\left\{< \zeta P^0 \Box_b u, \zeta P^0 u > + \|u\|^2_{L^p(\overline{\Omega})}\right\}.$$

Denoting by $\mathcal{C}^+$ and $\mathcal{C}^-$ the spaces of functions of the form $\zeta P^+ u$ and $\zeta P^- u$ respectively, we will say that $\Box_b^+$ (resp. $\Box_b^-$) is the restriction of $\Box_b$ to $\mathcal{C}^+$ (resp. $\mathcal{C}^-$) and $\mathbf{K}^+$ (resp. $\mathbf{K}^-$) is the restriction of $\mathbf{K}$ to $\mathcal{C}^+$ (resp. $\mathcal{C}^-$). Now the crucial step to prove Theorem E reduces to study operators of the form

$$A = -\sum_{j=1}^{n} X_j \circ \phi_j \circ X_j - \sum_{j=1}^{n} Y_j \circ \psi_j \circ Y_j + \lambda |T|$$

where $\lambda \geq 0$, and $|T|$ is a pseudodifferential operator of order one whose symbol is smooth and on the region $\mathcal{R}^+ \cup \mathcal{R}^-$ is equal to $|\tau|$. The $X_j = \mathrm{Re}(Z_j)$ and $Y_j = \mathrm{Im}(Z_j)$ are real vector fields, the functions $\phi_j \geq 0$, $\psi_j \geq 0$ are defined on a neighborhood of p_0.

We know that $\sum_{j=1}^{n} X_j \circ \phi_j \circ X_j + \sum_{j=1}^{n} Y_j \circ \psi_j \circ Y_j$ is subelliptic (see [RS] and [FS]). In fact,

$$\|u\|^2_{L^p_{\frac{1}{m}}(\overline{\Omega})} \leq C\left\{\sum_{j=1}^{n} < \phi_j X_j u +, X_j u > + \sum_{j=1}^{n} < \psi_j Y_j u, Y_j u > + \|u\|^2_{L^p(\overline{\Omega})}\right\}.$$

The problem now is showing that, under suitable restrictions on λ, the operator A^{-1} and the operator

$$\left(-\sum_{j=1}^{n} X_j \circ \phi_j \circ X_j - \sum_{j=1}^{n} Y_j \circ \psi_j \circ Y_j\right)^{-1}$$

have approximately the same mapping properties. We will restrict ourselves to $\mathbf{K}^+$. The corresponding result for $\mathbf{K}^-$ then will follow by changes of notation. Let us introduce pseudodifferential operators of order zero Γ_δ, $\tilde{\Gamma}_\delta$ on $\partial\Omega$, with symbol $\tilde{\Gamma}_\delta = 1$ on the support of the symbol of Γ_δ, and with Γ_δ, $\tilde{\Gamma}_\delta$ having symbols supported in the cotangent bundle $\{(x,t;\xi,\tau) \in T^*(\partial\Omega) : (|\xi|^2 + \tau^2)^{\frac{1}{2}} \sim \frac{1}{\delta}\}$, $(0 < \delta < \frac{1}{2})$. More precisely, let $\chi \in C_0^\infty(\{(\xi,\tau) \in \mathbb{R}^{2n+1} : 0 < a < (|\xi|^2 + \tau^2)^{\frac{1}{2}} < b\})$. For $u \in \bigcup_{k_0} L^p_{-k_0}$, $\delta > 0$, $R > 0$, the operator $\Gamma_\delta u$ is defined by

$$\widehat{\Gamma_\delta u}(\xi,\tau) = \chi\left(\frac{\delta}{R}(\xi,\tau)\right)\hat{u}(\xi,\tau).$$

We first prove this theorem for the case $p = 2$. Indeed, we have the following proposition.

Proposition 3.1. *Let $\Omega \subset\subset \mathbb{C}^{n+1}$ be a bounded weakly pseudoconvex domain of finite type m. Suppose the Levi form of $\partial\Omega$ is diagonalizable. Suppose further that the range of $\overline{\partial}_b$ is closed in $L^2(\partial\Omega)$, then*

$$\mathbf{K} : L^2_k(\partial\Omega) \to L^2_{k+\frac{2}{m}}(\partial\Omega)$$

boundedly. More precisely, the following a priori estimates hold:

$$\|\phi\|^2_{L^2_{k+\frac{1}{m}}(\partial\Omega)} \leq C \cdot \left\{ \langle \Box_b \phi, \phi \rangle + \|\phi\|^2_{L^2(\partial\Omega)} \right\}$$

$$\|\phi\|^2_{L^2_{k+\frac{2}{m}}(\partial\Omega)} \leq C \cdot \left\{ \|\Box_b \phi\|^2_{L^2_k(\partial\Omega)} + \|\phi\|^2_{L^2(\partial\Omega)} \right\}$$

for all $\phi \in C_0^\infty(\partial\Omega)$.

Using Proposition 3.1 and L^∞-estimates (see [FKM]), it is not hard to obtain the following results:

Proposition 3.2. *Let Z and Z' be smooth vector fields of the form*

$$\sum_{j=1}^n g_j X_j + \sum_{j=1}^n h_j Y_j$$

with $|g_j|^2 \leq \phi_j$ and $|h_j|^2 \leq \psi_j$ on $\partial\Omega$. For $k \in \mathbb{Z}^+$, $1 < p < \infty$, and all $\varepsilon > 0$, we have the following estimates:

(1) $\|\Gamma_\delta A^{-1} u\|_{L^p_{k+\frac{2}{m}-\varepsilon}} \leq C \left\{ \|\tilde{\Gamma}_\delta u\|_{L^p_k} + C\delta^M \|u\|_{L^p_{-k_0}} \right\}$;

(2) $\|\Gamma_\delta Z A^{-1} u\|_{L^p_{k+\frac{1}{m}-\varepsilon}} \leq C \left\{ \|\tilde{\Gamma}_\delta u\|_{L^p_k} + C\delta^M \|u\|_{L^p_{-k_0}} \right\}$;

(3) $\|\Gamma_\delta A^{-1} Z u\|_{L^p_{k+\frac{1}{m}-\varepsilon}} \leq C \left\{ \|\tilde{\Gamma}_\delta u\|_{L^p_k} + C\delta^M \|u\|_{L^p_{-k_0}} \right\}$;

(4) $\|\Gamma_\delta Z Z' A^{-1} u\|_{L^p_{k-\varepsilon}} \leq C \left\{ \|\tilde{\Gamma}_\delta u\|_{L^p_k} + C\delta^M \|u\|_{L^p_{-k_0}} \right\}$;

(5) $\|\Gamma_\delta Z A^{-1} Z' u\|_{L^p_{k-\varepsilon}} \leq C \left\{ \|\tilde{\Gamma}_\delta u\|_{L^p_k} + C\delta^M \|u\|_{L^p_{-k_0}} \right\}$;

(6) $\|\Gamma_\delta A^{-1} Z Z' u\|_{L^p_{k-\varepsilon}} \leq C \left\{ \|\tilde{\Gamma}_\delta u\|_{L^p_k} + C\delta^M \|u\|_{L^p_{-k_0}} \right\}$.

In these estimates, M and k_0 are arbitrary large. The constants C depend on M, k_0, but not on u.

Theorem E is now easily to obtain by using Littlewood-Paley theory.

Following the same method as in Phong [P], it is easy to get the following corollary from Theorem E.

Corollary 3.3. *Let $\Omega \subset \mathbb{C}^{n+1}$ be a bounded pseudoconvex domain with smooth boundary $\partial\Omega$. Suppose $\partial\Omega$ is of finite type m and the Levi form of $\partial\Omega$ is diagonalizable. Given a form f on Ω, we will denote its restriction to the boundary $\partial\Omega$ by f_b. Then we have if $\overline{\partial} u = f$ with f a $(0,1)$-form and $f \in L^p(\Omega)$, u_b, $f_b \in L^p(\partial\Omega)$ and u_b orthogonal to the nullspace of $\overline{\partial}_b$, then $u \in L^p_{\frac{1}{m}-\varepsilon}(\Omega)$ for all $\varepsilon > 0$.*

§4. Concluding remarks.

We indicate two further directions of research here:

(1) In Theorem E(b), we just used the identity

$$\mathbf{S}_b = \mathbf{I} - \overline{\partial}_b^* \mathbf{K} \overline{\partial}_b$$

to obtain the result. Hence the projection operator $\mathbf{S}_b$ has ε loss. It is very interesting to know that when can we get the optimal estimates for the "type zero" operator, *e.g.,* the Szegö projection, etc. However, there are some partial results of this question by doing deeper analysis.

We say that Ω is decoupled of finite type near $\zeta \in \partial\Omega$ if there exists a holomorphic coordinate system $(z_1, ..., z_{n+1})$ mapping ζ onto 0 and a neighborhood U_ζ of ζ onto a neighborhood U of 0 and smooth, sub-harmonic but not harmonic functions $\{f_k\}_{\{k=1,\dots,n\}}$, $f_k : \mathbb{C} \to \mathbb{R}$ with $f_k(0) = 0$, and each f_k vanishing to finite order at 0, such that

$$\left\{ z \in U; r(z) = 2\,\mathrm{Re}(z_{n+1}) + \sum_{k=1}^{n} f_k(z_k) < 0 \right\} \simeq \Omega \cap U_\zeta .$$

In this case, the Szegö projection maps $L^p_k(\partial\Omega)$ to itself for $k \in \mathbb{Z}^+$ and $1 < p < \infty$ and maps $\Lambda_\alpha(\partial\Omega)$ to itself for $\alpha > 0$. This result was obtained by Chang and Grellier [CG]. Recently, Stein and McNeal [SM] obtain the similar result for bounded, smooth, convex domains in $\mathbb{C}^{n+1}$.

(2) As we pointed it out in Theorem E, the solving operator $\mathbf{K}$ of $\Box_b$ doesnot have the optimal gain along **all** good directions. However, the operator $\mathbf{K}$ still gain two along some particular directions. For example, let us consider the domain $\{(z_1, z_2, z_3) : \mathrm{Im}\, z_3 > \mathcal{P}(z_1) + |z_2|^2\}$ in $\mathbb{C}^3$ again. It can be shown that

$$|Z^J \overline{Z}_2 \mathbf{K}(x,y)| \leq C \frac{\rho^{1-|J|}(x,y)}{\mathrm{Vol}(B^\rho(y, \rho(x,y)))}$$

for Z^J a $|J|$th order monomial in Z_j, $\overline{Z}_j$, $j = 1, 2$, taken in x, y. This tells us that $Z_j \overline{Z}_2 \mathbf{K}$, $\overline{Z}_j \overline{Z}_2 \mathbf{K}$, $j = 1, 2$, are standard Calderón-Zygmund operators with respect to the metric $\rho(x,y)$. It is interesting to know along what direction that two derivatives of $\mathbf{K}$ will have optimal gain.

REFERENCES

[BC1] A. Bonami & P. Charpentier, *Solutions de l'équation $\bar{\partial}$ et zéros de la classe de Nevanlinna dans certains domaines faiblement pseudoconvexes*, Ann. Inst. Fourier **32** (1982), 53-89.

[BC2] ———, *Estimations des (1,1) courants positifs fermés dans les domaines de $\mathbb{C}^2$*, Lecture Notes in Mathematics, Springer-Verlag, Berlin **1094** (1984), 44-52.

[BFG] M. Beals, C.L. Fefferman, & R. Grossman, *Strictly pseudo-convex domains in $\mathbb{C}^n$*, Bull. Amer. Math. Soc. **8** (1983), 125-322.

[BGS] R. Beals, P.C. Greiner, & N. Stanton, *L^p and Lipschitz estimates for the $\bar{\partial}$-equation and the $\bar{\partial}$-Neumann problem*, Math. Ann. **277** (1987), 185-196.

[C1] D. Catlin, *Necessary conditions for subellipticity of the $\bar{\partial}$-Neumann problem*, Ann. of Math. **117** (1983), 147-171.

[C2] ———, *Subelliptic estimates for the $\bar{\partial}$-Neumann problem on pseudo-convex domains*, Ann. of Math. **126** (1987), 131-191.

[Ch] D.C. Chang, *On L^p and Hölder estimates for the $\bar{\partial}$-Neumann problem on strongly pseudo-convex domains*, Math. Ann. **282** (1988), 267-297.

[CF1] D.C. Chang & C.L. Fefferman, *Characterization of the zero sets of $\mathcal{H}^p$ functions for pseudoconvex domains in $\mathbb{C}^2$ of finite type*, (in preparation) (1994).

[CF2] ———, *L^p estimates on CR manifolds with a diagonalizable Levi form*, preprint (1994).

[CG] D.C. Chang & S. Grellier, *Régularité de la projection de Szegö dans les domaines découplés de type fini de $\mathbb{C}^n$*, C. R. Acad. Sci. Paris **315** (1992), 1365-1370.

[CNS1] D.C. Chang, A. Nagel & E.M. Stein, *Estimates for the $\bar{\partial}$-Neumann problem for pseudoconvex domains in $\mathbb{C}^2$ of finite type*, Proc. Natl. Acad. Sci. USA **85** (1988), 8771-8774.

[CNS2] ———, *Estimates for the $\bar{\partial}$-Neumann problem in pseudoconvex domains of finite type in $\mathbb{C}^2$*, Acta Math. **169** (1992), 153-228.

[Chr] M. Christ, *Regularity properties of the $\bar{\partial}_b$-equation on weakly pseudo-convex CR manifolds of dimension three*, J. Amer. Math. Soc..

[FK] C.L. Fefferman & J.J. Kohn, *Hölder estimates on domains of complex dimension two and on three dimensional CR manifolds*, Adv. in Math. **69** (1988), 223-303.

[FKM] C.L. Fefferman, J.J. Kohn, & M. Machedon, *Hölder estimates on CR manifolds with a diagonalizable Levi form*, Adv. in Math. **84** (1990), 1-90.

[FS] C.L. Fefferman & A. Sanchez-Calle, *Fundamental solutions for second order subelliptic operators*, Ann. of Math. **14** (1986), 247-272.

[FoS] G.B. Folland & E.M. Stein, *Estimates for the $\bar{\partial}_b$ complex and analysis on the Heisenberg group*, Comm. Pure Appl. Math. **27** (1974), 429-522.

[GS] P.C. Greiner & E.M.Stein, *Estimates for the $\bar{\partial}$-Neumann problem (Math. Notes #19)*, Princeton University Press, Princeton, New Jersey, 1977.

[H] G.M. Henkin, *Solutions with estimates of the H. Lévy and Poincaré-Lelong equations. Constructions of functions of the Nevanlinna class with prescribed zeros in strictly pseudoconvex domains*, Dokl. Akad. Nauk. SSSR **224** (1975).

[K1] J.J.Kohn, *Boundaries of complex manifolds*, Proceedings, Conference on Complex Analysis, Minneapolis, 1964 (1965), 81-94.

[K2] ———, *Boundary behavior of $\bar{\partial}$ on weakly pseudo-convex manifolds of dimension two*, J. Differentail Geom. **6** (1972), 523-542.

[K3] ———, *Estimates for $\bar{\partial}_b$ on pseudo-convex CR manifolds*, Proc. Symposia in Pure Math. **43** (1985), 207-217.

[K4] ———, *Subellipticity of the $\bar{\partial}$-Neumann problem on pseudo-convex domains: Sufficient conditions*, Acta Math. **142** (1979), 79-122.

[L] P. Lelong, *Fonctionnelles analytiques et fonctions entières (n variables)*, Montréal, les Presses de l'Univ. de Montréal (1968).

[LR] I. Lieb & R.M. Range, *Integral representations and estimates in the theory of the $\bar{\partial}$-Neumann problem*, Ann. of Math. **123** (1986), 265-301.

[M] M. Machedon, *Estimates for the parametrix of the Kohn Laplacian on certain domains*, Invent. Math. **91** (1988), 339-364.

[MS] J. McNeal & E.M. Stein, *Mapping properties of the Bergman projection on convex domains of finite type*, (preprint) (1993).

[NRSW] A. Nagel, J.P. Rosay, E.M. Stein, & S. Wainger, *Estimates for the Bergman and Szegö kernels in* $\mathbb{C}^2$, Ann. of Math. **129** (1989), 113-149.

[NSW] A. Nagel, E.M. Stein, & S. Wainger, *Balls and metrics defined by vector fileds I: Basic properties*, Acta Math. **155** (1985), 103-147.

[P] D.H. Phong, *On* L^p *and Hölder estimates for the* $\overline{\partial}$ *equations on strongly pseudo-convex domains*, Ph. D. thesis, Princeton University, 1977.

[PS1] D.H. Phong & E.M. Stein, *Hilbert integrals, singular integrals and Radon transforms I*, Acta Math. **157** (1986), 99-157.

[PS2] ———, *Hilbert integrals, singular integrals and Radon transforms II*, Invent. Math. **86** (1986), 75-113.

[RS] L.P. Rothschild & E.M. Stein, *Hypoelliptic differential operators and nilpotent groups*, Acta Math. **137** (1976), 247-320.

[S1] N. Sibony, *Un exemple de domaine pseudoconvexe régulier où l'équation* $\overline{\partial}u = f$ *n'admet pas de solution bournée pour* f *bournée*, Invent. Math. **62** (1980), 235-242.

[S2] ———, *On Hölder estimates for* $\overline{\partial}$, Several complex variables (Stockholm, 1987/1988), Princeton Univ. Math. Notes **38** (1993), 587-599.

[Sk] H. Skoda, *Valeurs au bord pour les solutions de l'opérateur d'' et caractérisation de zéros des fonctions de la classe de Nevanlinna*, Bull. Soc. Math. France (1976), 225-299.

[V] N. Varopoulos, *Zeros of* $\mathcal{H}^p$ *functions in several complex variables*, Pac. J. Math. **88** (1980), 189-246.

RECENT PROGRESS IN HARDY SPACES ON MANIFOLDS

Jiecheng CHEN* and Silei WANG*

Department of Mathematics, Hangzhou University
Hangzhou 310028, P.R.China

Abstract: In this paper, we introduce a kind of Hardy spaces $H_\Phi(M)$ on a connected complete Riemannian manifold M with nonnegative Ricci curvature, set up two characterizations of $H_\Phi(M)$ for H-increasing function Φ, and, make some brief discussions about Lipschitz spaces $\Lambda_\alpha(M)$ ($\alpha > 0$) and dual of $H_\Phi(M)$ for $\Phi(x) = x^p$ ($0 < p \leq 1$). At same time, we set up some distribution inequalities about radial maximal function, nontangential maximal function and square function.

Key words: Hardy space, characterization, dual space, manifold.

§1. Introduction and definition of $H_\Phi(M)$

As we know, the theory of Hardy spaces on R^n is one of the main progresses of harmonic analysis in 1970s. In recent years, it has been extended to some more general geometric bodies, such as compact Lie groups, R^n with nonisotropic metrics, nilpotent Lie groups, Lipschitz domains in R^n, product spaces, etc. For example, we may refer to [2-5, 15-16, 18-20, etc.]. In this paper, we shall continue to try to set up a corresponding theory on manifolds, and some earlier trials were made in [11, 13]. Now, let us first mention some notations used in this paper.

Let M be an n-dimensional connected complete Riemannian manifold with nonnegative

*) Supported by NNSF of P.R.China

M. Cheng et al. (eds.), Harmonic Analysis in China, 22–54.

Ricci curvature, Δ its Laplace-Beltrami operator, ∇ its gradient operator, P_t its Poisson semi-group, $B_x(t)$ a geodesic ball with center x and radius t, d(x,y) the geodesic distance between x and y. Also, $p_t(x,y)$ denotes the Poisson kernel, $V_x(t)$ the volume of $B_x(t)$, $(B_x(t))^\wedge := B_x(t)\times(0,t]$, $aB_x(t) := B_x(at)$, $\Delta^\perp := \Delta + (\frac{\partial}{\partial t})^2$, $\nabla^\perp := (\nabla, \frac{\partial}{\partial t})$, $M^\perp := M\times R^1_+$, $\Gamma_\alpha(x) := \{(y,t) \in M^\perp: d(x,y) < \alpha t\}$ $(\alpha > 0)$, $|E| := \int_E dx$ where $E \subset M$ and dx denotes the Riemannian volume element. For a measurable or differentiable function u on $M^\perp$, set

$$(1)\qquad \begin{cases} N_+(u)(x) = \sup_{t>0} |u(x,t)| \\ N_\alpha(u)(x) = \sup_{(y,t)\in\Gamma_\alpha(x)} |u(y,t)| \\ S_\alpha(u)(x) = \left(\int_{\Gamma_\alpha(x)} |\nabla^\perp u(y,t)|^2 V_x^{-1}(\alpha t)\, tdtdy\right)^{1/2} \end{cases}$$

where $\alpha > 0$. Also, for $f \in L^2_c(M)$ (i.e., $f \in L^2(M)$ and supp (f) is compact). Set $N_{P,+}(f) = N_+(u_f)$, $N_{P,\alpha}(f) = N_\alpha(u_f)$, $S_{P,\alpha}(f) = S_\alpha(u_f)$ where $u_f(x,t) := P_t(f)(x)$. To define Hardy spaces, we need the notation of H-increasing function.

We call Φ an H-increasing function if Φ is an increasing function on $[0,+\infty)$ such that $\Phi(+0) = 0$, $\Phi(+\infty) = +\infty$, and

$$(2)\qquad \phi(\gamma) := \sup_{t>0} \Phi(t/\gamma) / \Phi(t) \to 0 \quad (\gamma\to 0).$$

we say $f \in L^*_\Phi(M)$ if f is measurable function on M and

$$\|f\|_\Phi := \inf \{\delta: \int_M \Phi(|f(x)|/\delta)\, dx \le \Phi(1)\} < +\infty.$$

Now, we claim that $\|.\|_\Phi$ is a quasi-norm, i.e.

$$(3)\qquad \begin{cases} \|f\|_\Phi = 0 \iff f = 0 \\ \|\alpha f\|_\Phi = |\alpha|\, \|f\|_\Phi \text{ for any number } \alpha \\ \|f + g\|_\Phi \le B_\Phi (\|f\|_\Phi + \|g\|_\Phi). \end{cases}$$

Actually, the first two properties are obvious. For the last one, let $\alpha > \|f\|_\Phi$, $\beta > \|g\|_\Phi$, $A = \inf \{\gamma : \phi(\gamma) \le 1/2\}$, then, by monotonicity of Φ

$$\Phi((|f|+|g|)/(2A(\alpha+\beta))) \le \Phi(|f|/(A\alpha)) + \Phi(|g|/(A\beta)) \le (1/2)\, (\Phi(|f|/\alpha) + \Phi(|g|/\beta)).$$

By integrating the left and the right sides of the above inequalities over M, we get

$$\int_M \Phi((|f|+|g|)/(2A(\alpha+\beta))) \le \Phi(1).$$

So, $\|f+g\|_\Phi \le 2A(\alpha+\beta)$. Let $\alpha \to \|f\|_\Phi$, $\beta \to \|g\|_\Phi$, we get desired inequality with $B_\Phi = 2\alpha$. Of course, if Φ is convex, $\|\,.\,\|_\Phi$ is a norm (i.e., $B_\Phi = 1$), and, if $\Phi(t) = t^p$ $(0 < p < \infty)$, $\|\,.\,\|_\Phi$ is a norm for $1 \le p < \infty$ and a quasi-norm with best $B_\Phi = 2^{1/p-1}$ for $0 < p < 1$. In Lemma 19, we shall give two kinds of special H-increasing functions.

Now, we may introduce a kind of Hardy spaces $H_\Phi(M)$ as follows. For an H-increasing function Φ, let $\|f\|_{H_\Phi} = \|N_{P,+}(f)\|_\Phi$. The space $H_\Phi(M)$ is defined as the completion of $L^2_c(M)$ in the H_Φ-norm, $\|\,.\,\|_{H_\Phi}$ (actually, it is a quasi-norm except that Φ is convex). For $\Phi(t) = t^p$ $(0<p<\infty)$, $H_\Phi(M)$ will be denoted by $H^p(M)$, and it is obvious that $H^p(M) = L^p(M)$ for $1 < p < \infty$.

In [10, 13], we have made some efforts on $H^p(M)$. Here, we shall set up some basic characterizations of $H_\Phi(M)$ (see Theorem 9) and give some brief discussions about $Lip_\alpha(M)$ and dual of $H^p(M)$ (see Theorems 1 and 2). At same time, we get some distribution inequalities about radial maximal function, non-tangential maximal function and square function of a given nice function.

In §2, we briefly discuss about Lip_α-classes and the dual of $H^p(M)$. The main result of this paper is Theorem 9 which appears in §6. In §3 — §5, we set up some good-λ inequalities which are necessary for the proof of the main theorem. Our work is a kind of generalizations of the corresponding theory on the Euclidean spaces, and some differences can be found in §3 — §5 especially when M is compact or noncompact. A different generalization of Hardy spaces were made by Anderson and Schoen[1] when the manifold M is negatively curved, by viewing M as a generalized geometric model of the open unit disc not the unit circle (or noncompact R^1). In whole paper, $C_{\alpha,\beta,n,\dots}$ always denotes a positive number depending only on α, β, n, ... It may be different in different occurances.

§2. $\Lambda_\alpha(M)$ AND THE DUAL OF $H^p(M)$

We know that $(H^p(M))^* = H^q(M)$ $(1/q + 1/p = 1)$ for $1 < p < \infty$, because $H^p(M) = L^p(M)$ in that case. For p = 1, we proved[11]

Theorem A. *$(H^1(M))^* = BMO(M)$ with equivalent norms.*

Here, we shall only consider the case when $0 < p < 1$. At first, we introduce Lip_α-spaces on M as follows. For $\alpha > 0$, let k_α be the maximal integer which is less than α. Then, we say $f \in \Lambda_\alpha(M)$ if $f \in L^\infty(M)$ and

$$\|f\|_{\Lambda_\alpha} := \|f\|_\infty + \|(-\Delta)^{k_\alpha/2}(f)\|_{\Lambda_{\alpha-k_\alpha/2}} \qquad \text{for } \alpha > 1$$

$$\|f\|_{\Lambda_\alpha} := \|f\|_\infty + \sup_{x,y\in M} |f(x) - f(y)| / d(x,y)^\alpha \qquad \text{for } 0 < \alpha \leq 1$$

where $(-\Delta)^{k_\alpha/2}(f)$ is defined in the following distribution sense

$$\int_M h(x)\,(-\Delta)^{k_\alpha/2}(f)(x)\,dx := \int_M f(x)\,(-\Delta)^{k_\alpha/2}(h)(x)\,dx$$

for $h \in \{g \in C^\infty(M)\colon \Delta^k(g) \in L^1(M) \text{ for } k = 0, 1, 2, \ldots\}$.This definition is somewhat different from the known definition of Lip_α for $M = R^n$ (see [24]) because the derivatives of higher orders will, generally speaking, involve more curvature properties of M. Now, we give a characterization of $\Lambda_\alpha(M)$ which is useful when we consider dual space of $H^p(M)$.

Theorem 1. *$f \in \Lambda_\alpha(M)$ iff there is a harmonic function u on $M^\perp$, such that $u = P(f)$ and $u \in \Lambda_\alpha^\perp(M)$ i.e.*

$$\|u\|_{\Lambda_\alpha^\perp} := \|u\|_\infty + \|t^{\dot{k}_\alpha-\alpha} (\tfrac{\partial}{\partial t})^{\dot{k}_\alpha} u(x,t)\|_\infty < \infty$$

where $\dot{k}_\alpha$ is the smallest integer which is larger than α. Furthermore

$$C_{n,\alpha}^{-1} \leq \|f\|_{\Lambda_\alpha} / \|u\|_{\Lambda_\alpha^\perp} \leq C_{n,\alpha}.$$

In an earlier paper [10], we have set up a similar characterization for BMO(M)-functions.

Lemma 1(*gradient estimates*). *For function u on $M^\perp$ and ball $B \subset M^\perp$, if u is harmonic, then*

$$\sup_{(x,t)\in(1/2)B} |\nabla^\perp u(x,t)| \leq C_n\, r_B^{-1} \sup_{(x,t)\in(1/2)B} |u(x,t)|$$

where r_B is the radius of B.

This lemma can be followed from [22] since $M^\perp$ has non-negative Ricci curvature under the usual product metric $ds^2_{M^\perp} = ds^2_M + dt^2$.

Lemma 2. *For any harmonic function u on $M^{\perp}$, there is a bounded function f on M such that $u = P(f)$.*

This lemma is essentially set up in [13] where the technical condition $\inf_{x\in M} V_x(1) > 0$ has been moved in [29] by the method of gradient estimates.

Proof of Theorm 1. At first, we consider the case when $0 < \alpha < 1$. From Lemma 2, it is enough to prove $\|f\|_{\Lambda_\alpha} \sim \|u\|_{\Lambda_\alpha^\perp}$ for $f \in L^\infty(M)$ and $u = P(f)$. Now, if $u \in \Lambda_\alpha^\perp(M)$, we have

$$f(x) - f(y) = \int_0^t \frac{\partial}{\partial s}u(x,s)\, ds - \int_0^t \frac{\partial}{\partial s}u(y,s)\, ds + u(y,t) - u(x,t)$$

$$= O(1)\,(t^\alpha \|u\|_{\Lambda_\alpha^\perp} + \|\nabla u(.,t)\|_{L^\infty(M)}\, d(x,y)) = O(1)\, d(x,y)^\alpha$$

for $t = d(x,y)$, because by Lemma 1,

$$|\nabla^\perp u(z,t)| \le \int_t^\infty |\frac{\partial}{\partial s} \nabla^\perp u(z,s)|\, ds = O(1) \int_t^\infty s^{-1} \sup\{|\frac{\partial}{\partial r}u(y,r)| : y\in B_z(s), |r-s|<s/2\}\, ds$$

$$= O(1) \int_t^\infty s^{-1} \|u\|_{\Lambda_\alpha^\perp} s^{-1+\alpha}\, ds \le C_{n,\alpha} \|u\|_{\Lambda_\alpha^\perp} \cdot t^{-1+\alpha}.$$

So, $\|f\|_{\Lambda_\alpha} \le C_{n,\alpha} \|u\|_{\Lambda_\alpha^\perp}$ because $\|f\|_\infty \le \|u\|_\infty$ is obvious. Conversely, if $f\in \Lambda_\alpha(M)$, we have

$$\frac{\partial}{\partial t}u(x,t) = (\frac{\partial}{\partial t}P_t)(f) = \int_M \frac{\partial}{\partial t}p_t(x,y)\,(f(y) - f(x))dy = O(1)\int_M d(x,y)^\alpha V_x^{-1}(t+d(x,y))\,(t+d(x,y))^{-1}\, dx$$

$$= O(1) \int_{B_x(t)} d(x,y)^\alpha V_x^{-1}(t+d(x,y))\,(t+d(x,y))^{-1}\, dx$$

$$+ O(1) \int_{B_x(t)^c} d(x,y)^{\alpha-1} V_x^{-1}(d(x,y))\, dx = O(1)\, t^{\alpha-1}.$$

So, $\|u\|_{\Lambda_\alpha^\perp} \le C_{n,\alpha} \|f\|_{\Lambda_\alpha}$ because $\|u\|_\infty \le \|f\|_\infty$ is obvious. For general $\alpha \ge 1$, it is not hard to show

$$(\frac{\partial}{\partial t})^{k_\alpha} u(x,t) = ((\frac{\partial}{\partial t})^{k_\alpha} P_t)(f) = ((-\Delta)^{k_\alpha/2} P_t)(f) = P_t((-\Delta)^{k_\alpha/2}(f))$$

whenever $u \in \Lambda_\alpha^\perp(M)$ or $f \in \Lambda_\alpha(M)$. So

$$\|f\|_{\Lambda_\alpha} = \|f\|_\infty + \|(-\Delta)^{k_\alpha/2}(f)\|_{\Lambda_{\alpha-k_\alpha}} \sim \|u\|_\infty + \|(\tfrac{\partial}{\partial t})^{k_\alpha/2}(u)\|_{\Lambda^\perp_{\alpha-k_\alpha}}$$

$$\sim \|u\|_\infty + \|(\tfrac{\partial}{\partial t})^{k_\alpha}(u)\|_\infty + \|t^{k'_\alpha-\alpha}(\tfrac{\partial}{\partial t})^{k'_\alpha}(u)\|_\infty \sim \|u\|_{\Lambda^\perp_\alpha}$$

because for $t \geq 1$

$$|(\tfrac{\partial}{\partial t})^{k_\alpha} u(x,t)| \leq C_{n,\alpha} \sup \{|u(y,s)|: y \in B_x(t/2), |t-s| < t/2\} \leq C_{n,\alpha} \|u\|_{\Lambda^\perp_\alpha} \qquad \text{(by Lemma 1)}$$

and, for $0 < t \leq 1$, if α is a non-integer (in this case, $k'_\alpha = k_\alpha+1$), we have

$$|(\tfrac{\partial}{\partial t})^{k_\alpha} u(x,t)| \leq |\int_t^1 (\tfrac{\partial}{\partial t})^{k_\alpha} u(x,s)\, ds| + |(\tfrac{\partial}{\partial t})^{k_\alpha} u(x,t)\big|_{t=1}|$$

$$\leq \int_t^1 \|u\|_{\Lambda^\perp_\alpha} s^{-(k_\alpha-\alpha)} ds + \|u\|_{\Lambda^\perp_\alpha} \leq C_{n,\alpha} \|u\|_{\Lambda^\perp_\alpha} .$$

if α is an integer (in this case $k'_\alpha = k_\alpha+2$), we have

$$|(\tfrac{\partial}{\partial t})^{k_\alpha} u(x,t)| \leq |\int_t^1 \int_r^1 (\tfrac{\partial}{\partial s})^{k'_\alpha} u(x,s)\, ds dr| + |(\tfrac{\partial}{\partial t})^{k'_\alpha-1}(u)(x,1)\,(1-t)| + |(\tfrac{\partial}{\partial t})^{k'_\alpha-2}(u)(x,1)|$$

$$\leq C_{n,\alpha} \Big(\int_t^1 \int_r^1 s^{-(k'_\alpha-\alpha)} ds dr + 1\Big) \|u\|_{\Lambda^\perp_\alpha} \leq C_{n,\alpha} \|u\|_{\Lambda^\perp_\alpha} .$$

Q.E.D.

For $f \in L^2(M)$ and harmonic function u on $M^\perp$, we may introduce following notations

$$\|f\|_{X_\alpha} = \|((-\Delta)^{k_\alpha/2}(f)(x) - (-\Delta)^{k_\alpha/2}(f)(y)) / d(x,y)^{\alpha-k_\alpha}\|_\infty$$

$$\|u\|_{X^\perp_\alpha} = \|t^{\alpha-k'_\alpha} (\tfrac{\partial}{\partial t})^{k'_\alpha} u(x,t)\|_\infty$$

then, by checking the above proof, we can easily find

Corollary 1. *For $f \in L^2(M)$, $\|f\|_{X_\alpha} \sim \|u\|_{X^\perp_\alpha}$ where $u = P_t(f)$.*

In the case when $M = R^n$, we know that the dual of $H^p(R^n)$ is essentially $\Lambda_\alpha(R^n)$ for $\alpha = n(p^{-1}-1)$. It is natural to conjecture that the dual of $H^p(M)$ is also $\Lambda_\alpha(M)$ essentially. But, the problem is rather different now. For example, we don't know if $|(\nabla^{\perp})^k u(x,t)|^p$ is subharmonic for $p > (n-1)/(n-1+k)$, $k > 1$, harmonic function u on $M^{\perp}$ (answer is positive for $k = 1$, see [12]); and, generally speaking, Riesz potential $(-\Delta)^{-\alpha/2}$ may be not (L^p,L^q)-bounded for any pair (p,q) (see [9]). So, we don't know if following Lemma 3 holds in general (see [16, Lemma 6] for $M = R^n$).

Lemma 3. *If $\inf_{x\in M} V_x(1) > 0$, then for any $g \in H^p(M) \cap L^2(M)$*

$$\int_{M\times R^1_+} t^{\alpha-1} |P_t(g)(x)|\, dxdt \le C_{n,\alpha} \|g\|_{H^p}$$

for $0 < p < 1$, $\alpha = n(p^{-1}-1)$.

Lemma 4 (mean value inequality) [13]. *For a domain $D \subset M^{\perp}$, if $A \ge R_D / r_D \ge A^{-1}$ where $R_D := \inf\{R: B_R \supset D$ where B_R is a geodesic ball in $M^{\perp}$ with radius $R\}$ and $r_D := \sup\{r: B_r \subset D\}$, then for any nonnegative subharmonic function w on $M^{\perp}$, $R > \varepsilon > 0$ and $0 < p < \infty$, following inequality holds*

$$\sup\{w(z)^p : z\in D_\varepsilon\} \le C_{n,p,\varepsilon,A} |D|^{-1} \int_D w(x,t)^p\, dxdt$$

where $D_\varepsilon := \{z\in M^{\perp}: d(z,D^c) > \varepsilon\}$ and $|D| := \int_D dxdt$.

Proof of Lemma 3. At first, we note that inequality $A \le a(Bt^{\alpha} + Ct^{-n})$ for all $t \in R^1_+$ implies

(4) $$A \le 2aB^{n/(\alpha+n)}C^{\alpha/(n+\alpha)}.$$

Now

$$G(x) := \int_0^\infty t^{\alpha-1} |P_t(g)(x)|\, dt \le \inf_{T>0} \Big(\sup_{t>0} |P_t(g)(x)| \int_0^T t^{\alpha-1} dt + C_{n,p} \int_T^\infty t^{\alpha-1} t^{-n/p} dt\Big)$$

$$\le \inf_{T>0} C_{n,p} \big(\sup_{t>0} |P_t(g)(x)| T^{\alpha} + \|g\|_{H^p} T^{-n}\big) \le C_{n,p} \sup_{t>0} |P_t(g)(x)|^p \|g\|_{H^p}^{1-p}$$

by (4), where the first inequality is true because, by Lemma 4, we have

$$|P_t(g)(x)| \le C_{n,p} \big(|B_x(t/2)\times(t/2,2t)|^{-1} \int_{B_x(t/2)\times(t/2,2t)} |P_s(g)(y)|^p\, dxdt\big)^{1/p}$$

$$\le C_{n,p} \|g\|_{H^p} V_x(t/2)^{-1/p} \le C_{n,p} \|g\|_{H^p} t^{-1/p}$$

since $V_x(t) \ge V_x(1)\, t^n \ge c\, t^n$. Therefore

$$\int_M G(x)\,dx \le C_{n,p} \|g\|_{H^p}^{1-p} \int_M \sup_{t>0} |P_t(g)(x)|^p\,dx = C_{n,p} \|g\|_{H^p}.$$

Q.E.D.

Now, we can prove that, under the condition of Lemma 3, the dual of $H^p(M)$ is essentially $\Lambda_\alpha(M)$. We have

Theorem 2. *If $\inf_{x\in M} V_x(1) > 0$, $0 < p < 1$, $\alpha = n(p^{-1}-1)$, then for $f \in L^2(M) \cap \Lambda_\alpha(M)$ and $g \in H^p(M) \cap L^2(M)$, we have*

$$\left|\int_M f(x)g(x)\,dx\right| \le C_{n,p} \|f\|_{\tilde{\Lambda}_\alpha} \|g\|_{H^p}.$$

Proof. By spectral decomposition theorem and Theorem 1, it is easy to see that

$$\left|\int_M f(x)g(x)\,dx\right| = C_{n,p} \left| \int_{M^\perp} t^{2k'_\alpha-1} \left(\frac{\partial}{\partial t}\right)^{k'_\alpha} P_t(g) \left(\frac{\partial}{\partial t}\right)^{k'_\alpha} P_t(f)\,dx\,dt \right|$$

$$\le C_{n,p} \|f\|_{\tilde{\Lambda}_\alpha} \int_{M^\perp} t^{k'_\alpha-1+\alpha} \left|\left(\frac{\partial}{\partial t}\right)^{k'_\alpha} P_t(g)\right| dx\,dt \le C_{n,p} \|f\|_{\tilde{\Lambda}_\alpha} \|g\|_{H^p}$$

because

$$\int_{M^\perp} t^{k'_\alpha-1+\alpha} \left\|\left(\frac{\partial}{\partial t}\right)^{k'_\alpha} P_t(g)\right\|_1 dt \le \int_0^\infty t^{k'_\alpha-1+\alpha} \|P_{t/2}(g)\|_1 \sup_{z\in M} \left\|\left(\frac{\partial}{\partial s}\right)^{k'_\alpha} p_s(z,\cdot)\big|_{s=t/2}\right\|_1 dt$$

$$\le C_{n,\alpha} \int_0^\infty t^{-1+\alpha} \|P_{t/2}(g)\|_1\,dt \le C_{n,\alpha} \|g\|_{H^p}$$

by Lemma 3. Q.E.D.

In a sequent paper, we shall make some more detailed discussions about $\Lambda_\alpha(M)$ and the dual of $H^p(M)$. We find that, generally speaking, $(H^p(M))^* \ne \Lambda_\alpha(M)$ $(\alpha = n(p^{-1} - 1)$, $0 < p < 1)$. The idea of the proof of this fact is to get a kind of atomic decomposition of H^p (which bases on the characterizations of H^p we set up in Theorem 9) and consider duals of the atomic Hardy spaces.

§3. Square Function Controls Nontangential Maximal Function

IN DISTRIBUTION SENSE

We know that for $M = R^n$, C.Fefferman and E.M.Stein set up following distribution inequality in [16] (also see [2] for $M = T^1$, the unit circle)

Theorem B. *For harmonic function u on R_+^{n+1} and $0 < \alpha < \beta < \infty$, $\tau > 0$, following inequality holds*

$$|\{N_\alpha(u) > \tau\}| \le C_{n,\alpha,\beta}(\tau^{-2} \int_{S_\beta(u)\le\tau} S_\beta(u)^2(x)\,dx + |\{S_\alpha(u) > \tau\}|).$$

Since then, this kind of inequalities was studied extensively, for example, we may refer to [4, 13, 14, 17, 19, 23, etc.]. Here, we shall set up some similar inequalities on manifold M. Our results are Theorem 3 and Theorem 4 which show some differences between the cases when M is compact and non-compact.

Theorem 3. *For harmonic function u on $M^\perp$ (if M is compact, we assume that $\int_M u(x,t_0)\,dx = 0$ for some $t_0 > 0$), $0 < \alpha < \beta < \infty$, $\tau > 0$, γ and $\delta > c_M$ where c_M is a sufficiently large positive number depending only on M, following inequality holds*

$$|\{N_\alpha(u) > \tau\gamma,\ S_\beta(u) \le \tau/\delta\}| \le C_n(\tau\gamma)^{-2}\, |\{N_\alpha(u) > \tau\}|$$

Without loss of generalities, we may assume $\tau = 1$, and, for simplicity, we only consider the case when $\alpha = 1$ and $\beta > 1$. Let $B_0 = B_{x_0}(r_0)$, $\Gamma(x) = \Gamma_1(x)$, $\Gamma(x,r) = \{(x',t)\in\Gamma(x): 0 < t < r\}$, $\Gamma'(x,r) = \{(x',t)\in\Gamma(x): t \ge r\}$, $E = \{x\in M: S_\beta(u) \le \delta^{-1}\}$, $W = \{(x,t)\in M^\perp: d(x,E) < t\}$, $\sigma(y) = d(y,E) := \inf_{z\in M} d(z,y)$, $N(u) = N_1(u)$, $S(u) = S_1(u)$.

Lemma 5. *Suppose $W \cap \Gamma(x,r_0) \ne \emptyset$ for some $x \in B_0$, then $(x_0,3r_0) \in W$.*

Proof. By assumption, there is y ($\in$ E) such that $\Gamma(y) \cap \Gamma(x,r_0) \ni$ some (x_*,t_*), so

$$d(x_0,y) \le d(x_*,y) + d(x_*,x_0) \le t_* + d(x_*,x) + d(x,x_0) \le 3r_0.$$

Lemma 6. *For $A^{-1} \le s/t \le A$, $d(x,y) \le Bt$, we have $C_{n,A,B}^{-1} \le V_x(s)/V_x(t) \le C_{n,A,B}$.*

This lemma can be followed from [6 or 7].

Lemma 7. *For $(y,t) \in W$, $|\nabla^\perp u(y,t)| \le C_{n,\beta}(\delta t)^{-1}$.*

Proof. For $(y,t) \in W$, $\exists\ x_0 \in E$ such that $(y,t) \in \Gamma(x_0)$, so, by the subharmonicity of $|\nabla^\perp u(y,t)|^2$ and Lemma 4, we have

$$|\nabla^{\perp}u(y,t)|^2 \le \frac{C_n}{t\,|B_y(t/4)|} \int_{d(z,y)\le t/4,\ |s-t|\le t/4} |\nabla^{\perp}u(z,s)|^2\,dzds$$

$$\le C_n t^{-2} \int_{\Gamma_\beta(x_0)} s\,|\nabla^{\perp}u(z,s)|^2\, V_x^{-1}(s)\,dzds \le \frac{C_n}{(\delta t)^2}$$

because $\{(z,s): d(z,y) \le t/4, |s-t| \le t/4\} \subset \Gamma_\beta(x_0)$.

Definition. For a non-negative measure $d\mu$ on $M^{\perp}$, if

$$\|d\mu\|_{CM} := \sup\{\mu(B_x(r)) / V_x(r): x \in M, r > 0\} < +\infty,$$

we call $d\mu$ a Carleson measure on $M^{\perp}$, where $\mu(B_x(r)) := \int_{B_x(r)} d\mu$.

Lemma 8. *For $\beta > 1$, $d\mu := |\nabla^{\perp}u(y,t)|^2 \chi_W(y,t)\,tdtdy$ is a Carleson measure on $M^{\perp}$ with norm $\le C_{n,\beta}\,\delta^{-2}$, where χ_W denotes the characteristic funvtion of the set W.*

Proof. At first, we notice that $\Gamma_{(\beta-1)/2}(x) \cap W \subset \Gamma_\beta(z_x)$ for z_x ($\in E$) nearest to x because $\Gamma_\beta(z_x) \supset \Gamma_{(\beta-1)/2}(x) \cap \{(y,t): t > t_x\}$ where t_x is the smallest which satisfies $(z_x,t_x) \in \Gamma_{(\beta-1)/2}(x) \cap W$. So

$$\int_{W\cap \hat{B}_0} |\nabla^{\perp}u(y,t)|^2\,tdtdy \le C_{n,\beta}\int_{B_0}\int_{\Gamma_{(\beta-1)/2}(x)\cap W} |\nabla^{\perp}u(y,t)|^2\,tV_x(t)\,dtdydx$$

$$\le C_{n,\beta}\int_{B_0} S^2(u)(z_x)\,dx \le C_{n,\beta}\,\delta^{-2}\,|B_0|.$$

Q.E.D.

Lemma 9. *Let $u_{B_0}(y,t) = u(y,t) - u(x_0,3r_0)$, $n_{B_0}(u)(x) = \sup\{|u_{B_0}(y,t)|: (y,t) \in \Gamma(x,r_0) \cap W\}$, then, there is a finite sequence of balls $\{B_i\}_1^m$, where B_i has radius r_i ($< r_0/3$) and center x_i ($\in B_0$), such that*

1). $\sup_{y\in B_i} \sigma(y) < 3r_i$ and $\inf_{y\in B_i} |u_{B_0}(y,3r_i)| > \gamma - c_*$

2). $\{2B_i\}_1^m$ are disjoint mutually

3). $|\{x \in B_0: n_{B_0}(u)(x) > \gamma\}| \le C_{n,\beta}\,|\cup_1^m (B_0 \cap B_i)|$

where $\gamma \ge 2c_*$ and c_* depending only on n and β.

Proof. By a standard selection proceedure of ball-covering of a set (e.g., see [4]), it is enough to show that for any $x \in \{x \in B_0: n_{B_0}(u)(x) > \gamma\}$, there is a ball B_x with radius $r_x < r_0/3$

and center x such that 1) holds. Now, for such an x, by definition of $n_{B_0}(u)$, $\Gamma(x,r_0) \cap W = \emptyset$ which implies $(x_0,3r_0) \in W$ by Lemma 1, and $|u_{B_0}(z_x,r_x)| > \gamma$ for some $(z_x,r_x) \in \Gamma(x,r_0) \cap W$. So, for some $\xi \in$ the geodesic line joining z_x and x_0

$$\gamma < |u_{B_0}(z_x,r_x)| \le |u(z_x,r_x) - u(z_x,3r_0)| + |u(z_x,3r_0) - u(x_0,3r_0)|.$$

$$\le \int_{r_x}^{3r_0} |\tfrac{\partial}{\partial t}u(z_x,t)|\, dt + |\nabla u(\xi,3r_0)|\, d(z_x,x_0) \le C_{n,\beta}\, \delta^{-1} (\ln(r_0/s) + 1) \quad \text{(by Lemma 7)}$$

so, $r_0 \ge \exp(\delta\gamma / C_{n,\beta} - 1)\cdot r_x > 3r_x$ (if $\gamma > 3C_{n,\beta}/\delta$) since the geodesic lines joining $(z_x,3r_0)$ and $(x_0,3r_0)$, (z_x,r_x) and $(z_x,3r_0)$ are in W. Now, we can conclude that the ball $B_x := B_x(r_x)$ is what we are searching for. Actually, we have $B_x \times \{3r_x\} \subset W$ because for any $y \in B_x$, we may first take $\Gamma(y') \subset W$ such that $(z_x,r_x) \in \Gamma(y') \cap \Gamma(x,r_0)$, then $d(y,y') \le d(y,x) + d(y',x) < r_x + d(y',z_x) + d(z_x,x) < 3r_x$, which means that $(y,3r_x) \in \Gamma(y') \subset W$. By this fact and the definition of W, we get $\sigma(y) < 3r_x$ for all $y \in B_x$. On the other hand, the geodesic line joining (z_x,r_x) and any point of $B_x \times \{3r_x\}$ are contained in W, so, for $y \in B_x$

$$|u_{B_0}(y,3r_x)| \ge |u_{B_0}(z,r_x)| - |u_{B_0}(y,3r_x) - u_{B_0}(z_x,r_x)|$$

(5) $$\ge \gamma - C_{n,\beta}\, d(y,z_x)\, (\delta\, r_x)^{-1} \text{ (by Lemma 7) } \ge \gamma - C'_{n,\beta}.$$

Therefore, the proof is ended if we take $c_* = \max\{3C_{n,\beta}/\delta, C'_{n,\beta}\}$.

By Lemma 9, taking $\tilde\sigma(y) = \max\{\varepsilon, \sigma(y), \sigma_0(y), \sigma_1(y), \ldots, \sigma_m(y)\}$ where $\varepsilon := \min\{3r_1, \ldots, 3r_m\}$, and

$$\sigma_0(y) = \begin{cases} 0 & \text{if } d(x_0,y) \le r_0 \\ d(x_0,y) - r_0 & \text{if } d(x_0,y) \ge r_0 \end{cases}$$

$$\sigma_j(y) = \begin{cases} 3r_j & \text{if } d(x_j,y) \le r_j \\ 0 & \text{if } d(x_j,y) \ge 2r_j \\ \text{linear about } d(x_j,y) & \text{if } 2r_j \ge d(x_j,y) \ge r_j \end{cases} .$$

we can conclude that

$$|\{x \in B_0 : n_{B_0}(u)(x) > \gamma\}| \le C_{n,\beta}\, |\{x \in B_0 : \tilde\sigma(x) < r_0 \text{ and } |u_{B_0}(x,\tilde\sigma(x))| > \gamma - c_*\}|$$

because for $y \in B_0 \cap B_i$, $\tilde\sigma(y) = \sigma_i(y) = 3r_i$, so $|u_{B_0}(y,\tilde\sigma(y))| = |u_{B_0}(y,3r_i)| > \gamma - c_*$. Now, let $\mathcal{R} = \{(y,t) \in M^{\perp} : \tilde\sigma(y) < t < r_0\}$, $\partial^+ = \{(y,r_0) \in \partial\mathcal{R} : y \in M\}$, $\partial^- = \{(y,t) \in \partial\mathcal{R} : y \in M \text{ and } \tilde\sigma(y) < r_0\}$, then $\partial\ell = \partial^+ \cup \partial^-$, and $\ell \subset W \cap (B_{x_0}(2r_0) \times (0,r_0))$ since $\tilde{\sigma}(y) \geq \max(\sigma(y), \sigma_0(y))$, and, for $(y,r_0) \in \partial^+$, $|u_{B_0}(y,r_0)| \leq C_{n,\beta}/\delta$ by Lemma 7 and the fact that the geodesic line joining (y,r_0) and $(x_0,3r_0)$ is in W. Thus, applying Green's formula to ℓ (it is possible since $\tilde{\sigma}$ is in Lip_1), we get

$$\int_{\partial^-} |u_{B_0}|^2\, d\eta \leq \int_{\ell} |\nabla^{\perp}u|^2\, t dt dy + \int_{\partial^-} |u_{B_0}|\, |\nabla^{\perp}u|\, t d\eta + \int_{\partial^+}(|u_{B_0}|^2 + |u_{B_0}|\, |\nabla^{\perp}u|\, t)\, d\eta$$

$$\leq \int_{W\cap(B_0\times(0,r_0))} |\nabla^{\perp}u|^2\, t dt dy + \delta^{-2}\int_{\partial^+} d\eta + \int_{\partial^-} |u_{B_0}|\, d\eta\, \delta^{-1} \quad \text{(by Lemma 7)}$$

$$\lesssim \delta^{-2}|B_0| + (|B_0|\int_{\partial^-} |u_{B_0}|^2\, d\eta)^{1/2}\,\delta^{-1} \quad \text{(by Lemma 8).}$$

So, $\int_{\partial^-} |u_{B_0}|^2\, d\eta \leq C_{n,\beta}\,\delta^{-2}|B_0|$. And, finally, by (5), we get

$$|\{x\in B_0:\, n_{B_0}(u)(x) > \gamma\}| \lesssim (\gamma - c_*)^{-2}\int_{B_0} |u_{B_0}(x,\sigma(x))|^2\, dx$$

$$\lesssim (\gamma - c_*)^{-2}\int_{\partial^-} |u_{B_0}|^2\, d\eta \lesssim (\delta\gamma)^{-2}|B_0| \qquad (\text{for } \gamma \geq 2c)$$

because on $2B_0$, $|\nabla\tilde{\sigma}(y)| \leq 1$ which means $d\eta(x) = (1 + |\nabla\tilde{\sigma}(x)|^2)dx \sim dx$.

According to above discussions, we have set up following

Main Lemma 1. *For any ball $B_0 := B(x_0,r_0)$, $E := \{S_\beta(u) \leq 1/\delta\}$, $W := \{(x,t): d(x,E) < t\}$, u_{B_0} and $u_{B_0}(u)$ defined in Lemma 8, following inequality holds*

$$|\{x\in B_0:\, n_{B_0}(u)(x) > \gamma\}| \leq C_{n,\beta}\,(\delta\gamma)^{-2}\,|B_0|.$$

To prove Theorem 3, we need following Vitali type decomposition of a bounded open set.

Lemma 10. *For a bounded open set $G \subset M$, $G \neq M$, there is a family of balls $\{B_j\}_j$ such that*

1). $\{B_j\}_j$ are disjoint mutually

2). $\cup_j B_j \subset G \subset \cup_j 3B_j$

3). $\forall B_j$, $\exists x^*_j \in 2B_j \cap G^c$ s.t. $\Gamma(x^*_j) \supset 3B_j \times [5r_j,\infty)$ where x_j is the center of B_j and r_j is the radius of B_j.

Proof. For any $x \in G$, by boundedness of G, there is a ball $B_x(r) = B_x$ such that $B_x \subset G$ but $2B_x - G \neq \emptyset$. Then, $\{B_x\}_{x\in G}$ is a covering of G. By [15, Theorem 3.1], there is a subfamily, say $\{B_j\}_j$, such that $\{B_j\}_j$ are disjoint mutually and $G \subset \cup_j 3B_j$. Now, by definition of $\{B_x\}_{x\in G}$, there must be $x^*_j \in 3B_j \cap G^c$. For such x^*_j, $\Gamma(x^*_j)$ contains $3B_j \times [5r_j,\infty)$ because $x^*_j \in 2B_j$ means that for any $x \in 3B_j$, $d(x^*_j,x) < d(x^*_j,x_j) + d(x_j,x) \le 5r_j$. Q.E.D.

Proof of Theorem 3 (with $\tau = 1$, $\alpha = 1$ and $\beta > 1$). At first, we assume $\{N(u) > 1\} \neq M$. Set $G_R = \{N(u) > 1\} \cap B_{z_0}(R)$ where z_0 is fixed and R will tend to ∞. By Lemma 10, there is a family of balls $\{B_j\}_j$ such that $|G_R| \ge |\cup_j B_j| = \Sigma_j |B_j|$, and, there is $x^*_j \in 3B_j \cap G^c$ s.t. $\Gamma(x^*_j) \supset 3B_j \times [5r_j,\infty)$. The above second fact means that $\cup_j (3B_j)^\wedge \supset \{|u| > 1\}$ because for any $(x,t) \in \{|u| > 1\}$, $x \in 3B_{j'}$ for some j' and $t < d(x,G^c) \le d(x,x_{j'}) + d(x_{j'},x^*_{j'}) < 3r_{j'} + 2r_{j'} = 5r_{j'}$. It means that $(x,t) \in (3B_{j'})^\wedge$. Therefore, for $x \in 3B_j \cap \{S_\beta(u) \le 1/\delta\}$

$$
\begin{aligned}
N(u)(x) &= \sup\{\, |u(y,t)| : (y,t) \in \Gamma(x_j,5r_j) \,\} \\
&\le |u(x_j,15r_j)| + \sup\{|u(y,t)-u(x_j,15r_j)| : (y,t) \in \Gamma(x_j,5r_j) \cap W\} \qquad (6)\\
&\le N(u)(x^*_j) + n_{5B_j}(u)(x) \le 1 + n_{5B_j}(u)(x).
\end{aligned}
$$

Furthermore

$$|\{5B_j : N(u) > 2\gamma,\ S_\beta(u) \le \delta^{-1}\}| \le |\{5B_j : n_{5B_j}(u) > \gamma\}| \le C_{n,\beta}\,(\gamma\delta)^{-2}\,|B_j|$$

and

$$
\begin{aligned}
|\{G_R : N(u) > \gamma,\ S_\beta(u) \le \delta^{-1}\}| &\le \Sigma_j |\{3B_j : N(u) > \gamma,\ S_\beta(u) \le \delta^{-1}\}| \\
&\le C_{n,\beta}\,(\gamma\delta)^{-2}\,\Sigma_j |B_j| \le C_{n,\beta}\,(\delta\gamma)^{-2}\,|G_R|.
\end{aligned}
$$

Let $R \to \infty$, we get

$$|\{N(u) > \gamma,\ S_\beta(u) \le \delta^{-1}\}| \le C_{n,\beta}\,(\gamma\delta)^{-2}\,|\{N(u) > 1\}|.$$

Now, suppose $\{N(u) > 1\} = M$. If M is noncompact, everything is OK since $|M| = \infty$. If M is compact, we may assume $\{S_\beta(u) \le \delta^{-1}\} \neq \emptyset$ and let $2r_M$ = the diameter of M, we claim that there must be a $x^* \in M$ such that $u(x^*,3r_M) = 0$. Then, the above proof is OK if we take $\{B_j\}_j = \{M\}$ and $(x_j,15r_j) = (x^*,3r_M)$ in (6). Now, we prove our claim. It is enough to show that $f(t) := \int_M u(x,t)\,dx$ is constant. Now, identity

$$\frac{d^2}{dt^2} f(t) = \int_M u''_{t^2}(x,t)\,dx = -\int_M \Delta u(x,t)\,dx = 0$$

means that f(t) = at + b. But, inequalities

$$\int_{2r_M}^{\infty} t\,|\tfrac{\partial}{\partial t}f(t)|^2\,dt \le \int_{2r_M}^{\infty} t\,(\int_M |\tfrac{\partial}{\partial t}u(x,t)|\,dx)^2\,dt$$

$$\le \inf_{y\in M}|M|^2 \int_{2r_M}^{\infty} t\,(\int_M |\tfrac{\partial}{\partial t}u(x,t)|^2\,dx)V_y^{-1}(\beta t)\,dt$$

$$\le \inf_{y\in M}|M|^2 \int_{\Gamma_\beta(y)} t\,|\nabla^{\perp}u(x,t)|^2\,V_y^{-1}(\beta t)\,dxdt$$

$$= |M|^2 \inf_{y\in M} S_\beta(u)(y)^2 \le |M|^2/\delta < +\infty$$

means that a = 0. So, $f(t) = b = f(t_0) = 0$ for any $t > 0$. Q.E.D.

Note that in the case when M is compact, the condition $\int_M u(x,t)\,dx = 0$ for some $t = t_0$ is equivelent to $u(x,t) \to 0$ $(t \to \infty)$ for some $x \in M$.

If M is noncompact, then, we have better estimates.

Theorem 4. *Under the hypothesis of Theorem3, if M is noncompact, then*

$$|\{N_\alpha(u) > \tau\gamma,\ S_\beta(u) \le \tau\delta^{-1}\}| \le C_{n,\alpha,\beta}\, e^{-\gamma\delta/C_{n,\alpha,\beta}}\,|\{N_\alpha(u) > \tau\}|.$$

To prove it, we need following Lemmas.

Lemma 11. *Suppose* τ *and* $\delta > 1$, $\|f\|_{BMO} \le \delta^{-1}$, *then*

$$|\{x \in M : |f(x)| > \tau\}| \le C_n\, e^{-\gamma\delta/C_n}\,|\{x \in M : |f(x)| > 1\}|.$$

See [8, Lemma 5].

Lemma 12. *Under the notations of Main Lemma 1, suppose* $n(u)(x) := \sup\{|u(y,t)| : (y,t) \in W \cap \Gamma(x)\} \ne \infty$, *then* $\|n\|_{BMO} \le C_{n,\alpha,\beta}/(\delta\gamma)$.

Proof. For any ball $B_0 = B_{x_0}(r_0)$, by Lemma 7, we can take $r_1 = r_0$ or $3r_0$ such that

$$W \cap \Gamma(x,r_1) = \varnothing \qquad \text{for any } x \in B_0$$

in this case, $r_1 = r_0$; or

$$W \cap \Gamma(x,r_1) \ne \varnothing \qquad \text{for any } x \in B_0$$

in this case, $r_1 = 3r_0$. Now, let $n'_{r_1}(u)(x) = \sup\{|u(y,t)|: (y,t) \in \Gamma(x,r_1) \cap W\}$ and $n_{r_1}(u)(x) = \sup\{|u(y,t)|: (y,t) \in \Gamma(x,r_1) \cap W\}$, then $n(u)(x) = \max(n'_{r_1}(u)(x), n_{r_1}(u)(x))$ and for $x \in B_0$

$$|n'_{r_1}(u)(x) - n'_{r_1}(u)(x_0)| \overset{\text{say}}{=\!=} n'_{r_1}(u)(x) - n'_{r_1}(u)(x_0)$$

$$\leq 2(u(y,t) - n'_{r_1}(u)(x_0)) \qquad (\text{for some } (y,t) \in W \cap \Gamma(x,r_1))$$

$$\leq 2(u(y,t) - u(y,t+d(x,x_0))) \qquad (\text{since } (y,t+d(x,x_0)) \in W \cap \Gamma(x_0,r_1))$$

$$\leq 2\,|\tfrac{\partial}{\partial s}u(y,s)|\,|_{s=t+\varepsilon d(x,x_0)} \qquad (\text{for some } \varepsilon \in (0,1))$$

$$\lesssim d(x,x_0)/(\delta t) \leq r_0/(\delta t) \leq 1/\delta \qquad (\text{by Lemma 7}).$$

So

$$\inf_{a\in R^1} \int_{B_0} |n(u)(x) - a|\,dx = \inf_{a\in R^1} \int_{B_0} |\max(n'_{r_1}(u)(x), n_{r_1}(u)(x)) - a|\,dx$$

$$\lesssim \inf_{a\in R^1} \int_{B_0} |\max(n'_{r_1}(u)(x_0), n_{r_1}(u)(x)) - a|\,dx + |B_0|/\delta$$

$$\lesssim \inf_{a\in R^1} \int_{B_0} |n_{r_1}(u)(x) - a|\,dx + |B_0|/\delta \overset{\text{say}}{=\!=} J.$$

In the case when $r_1 = r_0$, $J = |B_0|$. In the case when $r_1 = 3r_0$, we have

$$J \leq \int_{B_0} |\, n_{r_1}(u)(x) - |u(x_0,3r_1)|\,|\,dx + |B_0|/\delta$$

$$\leq \int_{B_0} \sup\{|\, u(y,t) - |u(x_0,3r_1)|\,|: (y,t) \in W \cap \Gamma(x,r_1)\}\,dx + |B_0|/\delta$$

$$\leq \int_{B_0} n_{B_{x_0}}(r_1)(u)(x)\,dx + |B_0|/\delta \lesssim |B_0|/\delta$$

by Main Lemma 1. Q.E.D.

Proof of Theorem 4 (with $\tau = 1$, $\alpha = 1$ and $\beta > 1$).

$|\{N(u) > \gamma, S_\beta(u) \leq 1/\delta\}| = |\{x \in E: N(u) > \gamma\}|$

$$\leq |\{n(u) > \gamma\}| \qquad (\text{since } N(u)(x) = n(u)(x) \text{ for } x \in E)$$

$$\le C_{n,\beta}\, e^{-\delta\gamma/C_{n,\beta}}\, |\{n(u) > 1\}|$$

$$\le C_{n,\beta}\, e^{-\delta\gamma/C_{n,\beta}}\, |\{N(u) > 1\}| \qquad \text{(since } N(u) \ge n(u))$$

Q.E.D.

§4. Radial Maximal Function Controls Nontangential Maximal Function in Distribution Sense

For $M = T^1$, the unit circle, Wheeden proved (see [28])

Theorem C. *If u is harmonic on D, the unit disc,* $0 < \alpha < \beta < 1$, $\gamma > 1$, $\tau > 0$, *then*

$$|\{N_\alpha(u) > \tau,\ N_\beta(u) < \gamma N_\alpha(u)\}| \le C_{\alpha,\beta,\gamma}\, |\{N_+(u) > \tau/4\}|$$

and

$$|\{N_\beta(u) > \tau\}| \le C_{\alpha,\beta}\ |\{N_\alpha(u) > \tau\}$$

where N_α *is defined by a different approach.*

Here, we shall set up some corresponding inequalities on manifold M. We have (note that our $N_\alpha(u)$ is very different from Wheeden's $N_\alpha(u)$ for $M = T^1$ because, here, u is defined on $T^1 \times R^1_+$ not on D).

Theorem 5. *If u is harmonic on* $M^\perp$, $0 < \alpha < \beta < \infty$, $\gamma > 1$, $\tau > 0$, *then*

1). $|\{N_\alpha(u) > \tau,\ N_\beta(u) < \gamma N_\alpha(u)\}| \le C_{n,\alpha,\beta}\, \gamma^n\, |\{N_+(u) > \tau/4\}|$

2). $|\{N_\alpha(u) > t,\ S_\beta(u) < \tau\gamma\}| \le C_{n,\alpha,\beta}\, \gamma^n\, |\{N_+(u) > \tau/2\}|$

3). $|\{N_\beta(u) > \tau\}| \le C_{n,\alpha,\beta}\, |\{N_\alpha(u) > \tau\}|$.

Proof of 1). Let $E = \{N_\alpha(u) > \tau,\ N_\beta(u) < \gamma\tau\}$ and $A = N_\alpha(u)(x)$ where $x \in E$. By definition, there is $(y,t) \in \Gamma_\alpha(x)$ such that $|u(y,t)| > A/2$. On the other hand, for $\eta := (\beta-\alpha)/(1+\beta)$, $B((y,t),\eta t) \subset \Gamma_\beta(x)$ (where $B(w,s)$, for $w \in M^\perp$ and $s > 0$, denotes the geodesic ball with center w and radius r in $M^\perp$ with product metric) because for any $(z,s) \in B((y,t),\eta t)$, $s > t - \eta t = t(1+\alpha)/(1+\beta)$ and $d(z,x) \le d(z,y) + d(z,x) \le \beta t(1+\alpha)/(1+\beta) < \beta s$. So

$$\sup_{\text{on } B((y,t),\eta t)} |u(z,s)| \le N_\beta(u)(x) < A\gamma,$$

and, by Lemma 1

$$\sup_{\text{on } B((y,t),\eta t/2)} |\nabla^\perp u(z,s)| \le C_n\, (\eta t)^{-1} \sup_{\text{on } B((y,t),\eta t)} |u(z,s)| \le C^*_n\, A\gamma / (\eta t). \tag{7}$$

So, for $(z,s) \in B((y,t),\eta_0 t)$ where $\eta_0 := \min(\eta/2,\ \eta/(4C^*_n\gamma))$

$$|u(z,s)| \geq |u(y,t)| - \max\{|\nabla^{\perp}u(z',s')|\, d((y,t),(z',s')) : (z',s') \in \overline{(z,s),(y,t)}\}$$

$$\geq A/2 - \eta_0 t C^*_n A\gamma/(\eta t) = A(1/2 - \eta_0 C^*_n \gamma/\eta) \geq A/4 > \tau/4$$

where $\overline{(z,s),(y,t)}$ denotes the geodesic line, in $M^{\perp}$, joining (z,s) and (y,t); and, d((y,t),(z,s)) denotes the geodesic distance, in $M^{\perp}$, between (z,s) and (y,t). Consequently

$$\inf\{N_+(u)(z) : z \in B_y(\eta_0 t)\} \geq \tau/4.$$

Now, take $B^x = B_y((\alpha+\eta_0)t)$, then $x \in B^x \supset B_y(\eta_0 t) \subset E_0 := \{N_+(u) > \tau/4\}$. So

$$\mathcal{M}(\chi_{E_0})(x) > V_y(\eta_0 t) / V_y((\eta_0+\alpha)t) \geq C_n(\eta_0/(\eta_0+\alpha))^n$$

i.e., $x \in \{\mathcal{M}(\chi_{E_0}) > (C_n(\eta_0/(\eta_0+\alpha))^n\}$ where $\mathcal{M}$ is the Hardy-Littlewood maximal operator. Therefore $|E| \leq C_{n,\alpha,\beta}\gamma^n |E_0|$ by weak type (1,1)-boundedness of $\mathcal{M}$.

Proof of 2). It is similar to the proof of 1). The only difference is that we must replace "A" by τ, and the inequality (7) should be set up in following way

$$\sup_{\text{on } B((y,t),\eta t/2)} |\nabla^{\perp}u(z,s)|^2 \leq C_n V_{(y,t)}(\eta t/2)^{-1}$$

$$\leq C_n |B((y,t),\eta t/2)|^{-1} \int_{B((y,t),\eta t/2)} |\nabla^{\perp}u(z,s)|^2\, dzdt$$

$$\leq C_{n,\alpha,\beta}(\eta t)^{-2} \int_{\Gamma_\beta(x)} |\nabla^{\perp}u(z,s)|^2 V_x^{-1}(\alpha s)\, dzdt \leq C^*_{n,\alpha,\beta}\tau\gamma/(\eta t).$$

Proof of 3). Set $F = \{N_\alpha(u) > \tau\}$, $F^* = \{\mathcal{M}(\chi_F) > (\alpha/(\alpha+2\beta))^n\}$, then, $G := \{N_\beta(u) > \tau\} \subset F^*$. Actually, $\forall x \in F, \exists (y,t) \in \Gamma_\beta(x)$ s.t. $|u(y,t)| > \tau$. But, $B_y(\alpha t)$ must be contained in F since for any $z \in B_y(\alpha t)$, $\Gamma_\alpha(z) \ni (y,t)$, so, $N_\alpha(u)(z) > \tau$, i.e. $z \in F$. Therefore, by Lemma 5

$$\mathcal{M}(\chi_F)(x) \geq V_y(\alpha t) / V_x(\beta t+\alpha t) > V_y(\alpha t) / V_y(\alpha t+2\beta t) > \alpha^n / (\alpha+2\beta)^n,$$

i.e. $x \in F^*$, and thus $G \subset F^*$. So

$$|G| \leq |F^*| \leq C_{n,\alpha,\beta}\|\chi_F\|_1 = C_{n,\alpha,\beta}|E|$$

by weak type (1,1) boundedness of $\mathcal{M}$. Q.E.D.

§5. Nontangential Maximal Function Controls Square Function in Distribution Sense

Following distribution inequality is well-known. It was set up in [2] for $M = T^1$ and in [16] for $M = R^n$.

Theorem D. *For harmonic function u on R_+^{n+1} and $0 < \alpha < \beta < \infty, \tau > 0$*

$$|\{S_\alpha(u) > \tau\}| \leq C_{n,\alpha,\beta} \, (\tau^{-2} \int_{N_\beta(u) \leq \tau} N_\beta(u)^2 \, dx + |\{N_\beta(u) > \tau\}|).$$

And, its variants and improvements were also studied extensively. For example, see [4, 14, 17, 19, 23, 27, etc.]. In [13], we get

Theorem E. *For harmonic function u on $M^\perp$, and $0 < \alpha < \beta < \infty$ (without restriction $\alpha < \beta$), $\tau > 0$, following inequality holds*

$$|\{S_\alpha(u) > \tau\}| \leq C_{n,\alpha,\beta} \, (\tau^{-2} \int_{N_\beta(u) \leq \tau} N_\beta(u)^2 \, dx + |\{N_\beta(u) > \tau\}|).$$

Here, we shall make some further investigations on this kind of distribution inequalities. Our main results are Theorems 6 and 7. In the rest of this section, u always denotes a harmonic function on $M^\perp$,and v always denotes a subharmonic function on $M^\perp$. And

$$\bar{S}_\alpha(v)(x) := (\int_{\Gamma_\alpha(x)} t \, \Delta^\perp v(y,t) \, V_+^{-1}(\alpha t) \, dydt)^{1/2}$$

$$\bar{S}_\alpha^*(v)(x) := (\int_{M^\perp} \varphi(d(x,y)/(\alpha t)) \, t \, \Delta^\perp v(y,t) \, V_+^{-1}(\alpha t) \, dydt)^{1/2}$$

$$\bar{N}_\alpha(v)(x) := \sup_{(y,t) \in \Gamma_\alpha(x)} v(y,t)^{1/2}$$

where $\varphi \in C_c^\infty(R^1)$, $supp(\varphi) \subset [0,2]$, $\varphi \mid_{[0,2]} = 1$, and φ is fixed. Note that $\bar{S}_\alpha(v)(x) = 2^{1/2} S_\beta(u)$, and $\bar{N}_\alpha(v)(x) = N_\alpha(u)$ if $v = u^2$ since $\Delta^\perp u^2 = 2|\nabla^\perp u|^2$ for harmonic function u. Now, we first set up following

Mian Lemma 2. *For nonnegative subharmonic function v on $M^\perp$, $0 < \alpha < \beta < \infty$, let $E = \{\bar{N}_\beta(v) \leq 1/\delta\}$, $W = \{(x,t): d(x,E) < \alpha t\}$, and $d\mu_v = t \, \chi_W(x,t) \, \Delta^\perp v(x,t) \, dxdt$, then*

$$||d\mu_v||_{CM} \leq C_{n,\alpha,\beta} \, \delta^{-2}$$

when $\beta - \alpha$ is sufficiently large.

To prove it, we need some lemmas. Now, for simplicity, we only consider $\alpha = 1$ and β sufficiently large, say, $\beta = \beta_0 := 130$.

Lemma 13. *For a ball $B \subset M^\perp$ and a subharmonic function v on $M^\perp$, $0 \leq v \leq 1$, following inequalities hold*

$$\int_{(1/2)B} |\nabla^{\perp} v|^2 \, dxdt \leq 16\, r_B^{-2} |B| \quad \text{and} \quad \int_{(1/2)B} |\nabla^{\perp} v| \, dxdt \leq 4\, r_B^{-1} |B|$$

where r_B is the radius of B.

Proof. For any $\varphi \in C_c^1(B)$

$$0 \leq \int_B v(w)\, \varphi^2(w)\, \Delta^{\perp} v(w)\, dw = -\int_B \nabla^{\perp}(v(w)\, \varphi^2(w))\, \nabla^{\perp} v(w)\, dw$$

$$= -\int_B \varphi^2(w)\, |\nabla^{\perp} v(w)|^2\, dw - \int_B 2\varphi(w) v(w)\, \nabla^{\perp}\varphi(w)\, \nabla^{\perp} v(w)\, dw,$$

so

$$\int_B \varphi^2(w)\, |\nabla^{\perp} v(w)|^2\, dw \leq 2\Big(\int_B \varphi^2(w)\, |\nabla^{\perp} v(w)|^2\, dw\Big)^{1/2} . \Big(\int_B v^2(w)\, |\nabla^{\perp}\varphi(w)|^2\, dw\Big)^{1/2}.$$

Take φ such that $0 \leq \varphi \leq 1$, $\varphi(w) = 1$ for $w \in (1/2)B$, $= 0$ for $w \notin B$, = linear and Lip_1 elsewhere, and $|\nabla^{\perp}\varphi| \leq 2 / r$, then

$$\int_{(1/2)B} |\nabla^{\perp} v(w)|^2\, dw \leq 4 \int_B (2/r)^2\, v^2(w)\, dw \leq 16\, r_B^{-2}\, |B|,$$

$$\int_{(1/2)B} |\nabla^{\perp} v(w)|\, dw \leq |(1/2)B|^{1/2} \Big(\int_{(1/2)B} |\nabla^{\perp} v(w)|^2\, dw \leq 4\, r_B^{-1}\, |B|.$$

Lemma 14 (a kind of dyadic decomposition of $M^{\perp}$). *There is a sequence of balls $\{B_k\}_k$ such that*

$$1).\ M^{\perp} = \cup_k Q_k \qquad \text{and} \qquad 2).\ \|\Sigma_k \chi_{Q_k}\|_{\infty} \leq 18^n$$

where $Q_k = B_k \times (r_k, 2r_k]$, and r_k = the radius of B_k.

Proof. For any $d_m := 2^m$, $m \in \{0, \pm 1, \pm 2, \ \}$, take a family of balls $\{B_{m,j}\}_j$ in M such that

a). $B_{m,j} \not\ni$ the center of $B_{m,i}$ for $i \neq j$

b). every $B_{m,i}$ has radius 2^m

c). $\cup_j B_{m,j} = M$.

It is possible since M has dense numerable subsets. Now, we claim that $\|\Sigma_j \chi_{B_{m,j}}\|_{\infty} \leq 18^n$. Actually, if $B_{m,j} \cap B_{m,j_0} \neq \varnothing$, then

$$B_{m,j} \subset 3B_{m,j_0} \quad \text{and} \quad B_{m,j_0} \subset 3B_{m,j}$$

$$(1/2)B_{m,j} \cap (1/2)B_{m,i} = \emptyset \qquad (i \neq j)$$

$$|(1/2)B_{m,j}| \geq 6^{-n} |3B_{m,j}| \geq 6^{-n} |B_{m,j_0}| \geq 18^{-n} |3B_{m,j_0}|$$

thus

$$\Sigma_{B_{m,j} \cap B_{m,j_0} \neq \emptyset} |(1/2)B_{m,j}| \geq 18^{-n} |3B_{m,j_0}| \, \|\Sigma_j \chi_{B_{m,j}}\|_{L^\infty(B_{m,j_0})}$$

which means $\|\Sigma_j \chi_{B_{m,j}}\|_{L^\infty(B_{m,j_0})} \leq 18^n$. Now, set $Q_{m,j} = B_{m,j} \times (d_m, 2d_m]$, then, for $m' \neq m''$, $\{Q_{m',j}\}_j$ and $\{Q_{m'',j}\}_j$ have no members which have common points. So, 2) holds. And, 1) is obvious.

For a ball $B_0 = B_{x_0}(r_0) \in M$, set $\hat{B}_0 = B_0 \times [0, 2r_0] \subset M^\perp$. And, for $\eta > 0$, let $\psi_\eta(x) = (1/32) \cdot \max(d(x,E), d(x,B_0), \eta)$ where $d(x,E) := \inf_{y \in E} d(x,y)$ and $E := \{x \in M: \tilde{N}_\alpha(v)(x) \leq 1\}$, and

$$\varphi_\eta(x,t) = \begin{cases} 1 & \text{if } t \geq \psi_\eta(x) \\ 0 & \text{if } 2t \leq \psi_\eta(x) \\ (2t - \psi_\eta(x))/\psi_\eta(x) & \text{if } t \leq \psi_\eta(x) \leq 2t \end{cases}$$

$$h(t) = \begin{cases} 0 & \text{if } t \geq 4r_0 \\ 1 & \text{if } t \leq 2r_0 \\ (4r_0 - t)/(2r_0) & \text{if } 2r_0 \leq t \leq 4r_0 \end{cases}$$

$$\tilde{V}_\eta(x,t) = \varphi_\eta(x,t)\, h(t).$$

Then, we have

Lemma 15. *Under the above assumptions, we have*

1). $|\nabla^\perp \tilde{V}_\eta(x,t)| \leq C_n t^{-1} \chi_{\psi_\eta(x)/2 \leq t \leq 4r_0}(t)$

2). $\int_{\hat{B}_0} t\,\chi_W(x,t)\,\Delta^\perp v(x,t)\,dxdt \leq \lim_{t \to 0} \int_{B_0} t\,\tilde{V}_\eta(x,t)\,\Delta^\perp v(x,t)\,dxdt$

3). $\int_{supp(\nabla^\perp \tilde{V}_\eta)} t^{-1}\,dxdt \leq C_n |B_0|$

where $W := \{(x,t) \in M^\perp: d(x,E) < t\}$.

Proof. 1) is obvious since $\nabla^\perp h = O(1/r_0)\,\chi_{supp(h)}$ and $\nabla^\perp \varphi_\eta = O(1/t)\,\chi_{supp(\varphi_\eta)}$. 2) is an easy corollary of following fact.

$$\{(x,t): \tilde{V}_\eta(x,t) = 1\} = \{(x,t): \psi_\eta(x) \leq t \leq 2r_0\} \supset W \cap \hat{B}_0 \cap \{(x,t): t > \eta,\ x \in M\}$$

For 3), we have

$$\operatorname{supp}(\nabla^{\perp}V_{\eta}) \subset (\operatorname{supp}(\nabla^{\perp}\varphi_{\eta}) \cup \operatorname{supp}(\nabla^{\perp}h)) \cap \operatorname{supp}(\tilde{V}_{\eta})$$

$$\subset \{(x,t): 2r_0 \le t \le 4r_0 \text{ or } \psi_{\eta}(x)/2 \le t \le \psi_{\eta}(x)\} \cap \{(x,t): \psi_{\eta}(x)/2 \le t \le 4r_0\}$$

$$\subset \{(x,t) \in 4^6B_0 \times [0,4r_0]: 2r_0 \le t \le 4r_0 \text{ or } \psi_{\eta}(x)/2 \le t \le \psi_{\eta}(x)\}.$$

So

$$\int_{\operatorname{supp}(\nabla^{\perp}\tilde{V}_{\eta})} t^{-1}\,dxdt \le \int_{4^6B_0} \Big(\int_{2r_0}^{4r_0} + \int_{\psi_{\eta}(x)/2}^{\psi_{\eta}(x)} \Big) t^{-1}\,dtdx \le C_n\,|B_0|.$$

Q.E.D.

Now, set

$$\{Q^j\} = \{Q_k: \operatorname{supp}(\nabla^{\perp}\tilde{V}_{\eta}) \cap Q_k \ne \varnothing\}$$

Then

Lemma 16. a). *$\{Q^j\}$ is a finite family*

b). $\operatorname{supp}(\nabla^{\perp}\tilde{V}_{\eta})) \subset \cup_j Q^j \subset \cup_{x\in E} \Gamma_{130}(x)$

c). $\Sigma_j |B^j| \le C_n |B_0|$

where B^j is the base of Q^j, i.e., $Q^j = B^j \times (r_j, 2r_j]$ and r_j is the radius of B^j.

Proof. a) is obvious since $\operatorname{supp}(\nabla^{\perp}\tilde{V}_{\eta})$ is a compact subset of $M^{\perp}$. The first part of b) is an easy corollary of the fact that $\cup_j Q^j = M^{\perp}$. To prove c), we first note that $Q^j \cap \operatorname{supp}(\nabla^{\perp}\tilde{V}_{\eta}) \ne \varnothing$ implies $4r_0 \ge r_j$, thus $B^j \subset 2(4^6B_0)$, so

$$Q^j \subset 4^7B_0 \times [0,8r_0].$$

On the other hand, $\operatorname{supp}(\nabla^{\perp}\tilde{V}_{\eta}) \cap Q^j \ne \varnothing$ implies $\operatorname{supp}(\nabla^{\perp}h) \cap Q^j \ne \varnothing$ or $\operatorname{supp}(\nabla^{\perp}\varphi_{\eta}) \cap Q^j \ne \varnothing$. In the first case, $r_0 \le r_j \le 4r_0$, so

$$Q^j \subset \{(x,t): r_0 \le t \le 8r_0\}.$$

In the second case, take $(x_0,t_0) \in Q^j \cap \operatorname{supp}(\nabla^{\perp}\tilde{V}_{\eta})$, then

$$\psi_{\eta}(x_0)/2 \le t_0 \le \psi_{\eta}(x_0) \quad \text{and} \quad \psi_{\eta}(x_0)/4 \le r_j \le \psi_{\eta}(x_0).$$

So, for $(x,t) \in Q^j$

$$\psi_{\eta}(x_0)/4 \le r_j \le t \le 2r_j \le 2\psi_{\eta}(x_0) \quad \text{and} \quad d(x,x_0) < 2r_j\,.$$

Now

$$\psi_{\eta}(x_0) \le \psi_{\eta}(x) + |\psi_{\eta}(x_0) - \psi_{\eta}(x)| \le \psi_{\eta}(x) + 2r_j \le \psi_{\eta}(x) + t/16$$

$$\psi_\eta(x_0) \geq \psi_\eta(x) - |\psi_\eta(x_0) - \psi_\eta(x)| \geq \psi_\eta(x) - t/16$$

so

$$t \leq 2(\psi_\eta(x) + t/16 = 2\,\psi_\eta(x) + t/8$$

$$t \geq (\psi_\eta(x) - t/16)/4 = \psi_\eta(x)/4 - t/64$$

which means that

$$t \leq \psi_\eta(x)\,(16/7) \leq 3\,\psi_\eta(x)$$

$$t \geq \psi_\eta(x)\,(16/65) \geq \psi_\eta(x)/5.$$

Therefore

$$Q^j \subset \{(x,t): \psi_\eta(x)/5 \leq t \leq 3\,\psi_\eta(x)\}\,.$$

From all the above, we conclude that

$$\cup_j Q^j \subset \{(x,t) \in 4^7 B_0 \times [0,8r_0]: r_0 \leq t \leq 8r_0 \text{ or } \psi_\eta(x)/5 \leq t \leq 3\,\psi_\eta(x)\}.$$

Finally, it is easy to see that the number of m satisfying $r_0 \leq 2^m \leq 16r_0$ or $\psi_\eta(x)/5 \leq 2^m \leq 3\,\psi_\eta(x)$ doesn't exceed 6 + 6 = 12 for any fixed x. so

$$|\Sigma_{\text{above } Q^j} \chi_{B^j}(x)| \leq \Sigma_m \,(\Sigma_{r_j = 2^m} \chi_{B^j}(x)) \leq \Sigma_m\, 18^n = 12 \times 18^n < 18^{n+1}\,.$$

Therefore

$$\Sigma_j |B^j| \leq 18^{n+1} \int_M \chi_{\cup_j B^j}(y)\, dy \leq 18^{n+1} \int_{4^7 B_0} dy \leq 18^{n+1}\, 4^{7n}\, |B_0|.$$

Now, we prove the second part of b). We have

$$\text{supp}(\nabla^\perp \nabla_\eta)) \subset \text{supp}(\nabla_\eta)) \subset \{(x,t): \psi_\eta(x) \leq 2t\}$$

$$\subset \{(x,t): d(x,E) < 64\,t\} = \cup_{x\in E} \Gamma_{64}(x)\;.$$

So, for Q^j, $\exists\, x_0 \in E$ s.t. $Q^j \cap \Gamma_{64}(x_0) \neq \varnothing$, then

$$Q^j \subset \Gamma_{130}(x_0)$$

since for any $(x,t) \in Q^j$, $d(x,x_0) \leq d(x,x_*) + d(x_*,x_0) \leq 2r_j + 64 \times 2r_j = 130\, r_j \leq 130\, t$ where $(x_*,t_*) \in Q^j \cap \Gamma_{64}(x_0)$.

Proof of Main Lemma 2. For ball B_0, we have

$$\int_{\hat{B}_0} d\mu_v \leq \lim_{h\to 0} \int_{M^\perp} t\, \nabla_\eta(x,t)\, \Delta^\perp v(x,t)\, dxdt \quad \text{(by Lemma 15 2))}$$

$$\leq \lim_{h\to 0} \Big(\int_{M^\perp} t\, |\nabla\tilde{\nabla}_\eta(x,t)\, \nabla v(x,t)|\, dxdt \Big)$$

$$+ \int_{M^\perp} |\tfrac{\partial}{\partial t}\tilde{\nabla}_\eta(x,t)\, v(x,t) - t\tfrac{\partial}{\partial t}\tilde{\nabla}_\eta(x,t)\, \tfrac{\partial}{\partial t}v(x,t)|\, dxdt$$

$$\lesssim \lim_{h\to 0} \int_{supp(\nabla^\perp\tilde{\nabla}_\eta)} (|\nabla^\perp v(x,t)| + v(x,t)/t)\, dxdt \qquad \text{(by Lemma 15 1))}$$

$$\leq \lim_{h\to 0} \Big(\Sigma_j \int_{Q^j} |\nabla^\perp v(x,t)|\, dxdt + \int_{supp(\nabla^\perp\tilde{\nabla}_\eta)} t^{-1}\, dxdt \quad \text{(By Lemma 16 b))}$$

$$\leq C_n\, (|B_0| + \lim_{h\to 0} \Sigma_j\, |B^j|) \leq C_n\, |B_0|$$

by Lemma 13, Lemma 15 3) and Lemma 16. Q.E.D.

Now, we are in the position of seting up our desired distribution inequality.

Theorem 6. *For harmonic function u on $M^\perp$, $0 < \alpha < \beta < \infty$, γ and $\delta > 2$ (if M is compact, we assume γ and $\delta > C^*_M$, where C^*_M depending only on M)*

$$|\{S^*_\alpha(u) > \gamma\tau,\ N_\beta(u) \leq \tau/\delta\}| \leq C_{n,\alpha,\beta}\, (\gamma\delta)^{-2}\, |\{S^*_\alpha(u) > \tau\}|$$

when $\alpha - \beta$ is sufficiently large, where $\tau > 0$ and

$$S^*_\alpha(u)(x) := \Big(\int_{M^\perp} \varphi(d(x,y)/\alpha t)\, t\, |\nabla^\perp u(y,t)|^2\, V_x^{-1}(t)\, dydt \Big)^{1/2}$$

where $\varphi \in C_c^\infty(R^1)$, $supp(\varphi) \subset [0,2]$, $\varphi_{/[0,1]} = 1$, $0 \leq j \leq 1$ and φ is fixed.

Without loss of generalities, we may assume $\tau = 1$, $\alpha = 1$, $\beta = \beta_0 := 130$. For the ball $B_0 := B_{x_0}(r_0)$, set

$$\mathcal{S}(u)(x) = \Big(\int_{W\cap M^\perp} \varphi(d(x,y)/\alpha t)\, t\, |\nabla^\perp u(y,t)|^2\, V_x^{-1}(t)\, dydt \Big)^{1/2}$$

$$\mathcal{S}_{B_0}(u)(x) = \Big(\int_{W\cap(M\times(0,r_0))} \varphi(d(x,y)/\alpha t)\, t\, |\nabla^\perp u(y,t)|^2\, V_x^{-1}(t)\, dydt \Big)^{1/2}$$

$$\mathcal{S}'_{B_0}(u)(x) = \Big(\int_{W\cap(M\times(r_0,\infty))} \varphi(d(x,y)/\alpha t)\, t\, |\nabla^\perp u(y,t)|^2\, V_x^{-1}(t)\, dydt \Big)^{1/2}$$

where W and following E are defined in Main Lemma 2 with $v = u^2$ and $\beta = \beta_0$, $\alpha = 1$.

Lemma 17. *Under the above assumptions, if* $\exists\, x^* \in B_0$ *s.t.* $\mathcal{A}(u)(x^*) < 1$, *then*

$$|\{B_0\colon \mathcal{A}(u) > \gamma\}| \le C_n (\gamma\delta)^{-2} |B_0| .$$

Proof. At first, we have (for $x \in B_0$)

$$\mathcal{A}(u)^2(x) = \mathcal{A}^2_{B_0}(u)(x) + \mathcal{A}^2_{B_0}(u)(x^*) + \mathcal{A}'^2_{B_0}(u)(x) - \mathcal{A}'^2_{B_0}(u)(x^*)$$

$$\le 1 + \mathcal{A}^2_{B_0}(u)(x) + I$$

where

$$|I| = \Big| \int_{W \cap (M\times(r_0,\infty))} (\varphi(d(x,y)/t) - \varphi(d(x^*,y)/t))\, t\, |\nabla^\perp u(y,t)|^2\, V_x^{-1}(t)\, dydt \Big|$$

$$\le C_n \int_{W \cap (M\times(r_0,\infty))} (r_0 / t)\, t\, |\nabla^\perp u(y,t)|^2\, V_x^{-1}(t)\, \chi_{B_x(6t)}(y)\, dydt$$

(because $|\varphi(d(x,y)/t) - \varphi(d(x^*,y)/t)| \le C_n\, d(x,x^*)/t\, \chi_{B_x(4t)}(y) \le C_n\, (r_0 / t)\, \chi_{B_x(6t)}(y)$ for $t > r_0$)

$$\le C_n\, \delta^{-2} \int_{r_0}^{\infty} (r_0 / t)\, t\, t^{-2}\, V_x^{-1}(t)\, V_x(6t)\, dydt \le C_n\, \delta^{-2}$$

because

$$|\varphi(d(x,y)/t) - \varphi(d(x^*,y)/t)| \le C_n\, d(x,x^*)/t\, \chi_{B_x(4t)}(y) \le C_n\, (r_0 / t)\, \chi_{B_x(6t)}(y)$$

for $t > r_0$, and, for $(y,t) \in W$, say, $(y,t) \in \Gamma(x')$ for some $x' \in E$, we have (where $\beta = \beta_0$)

$$|\nabla^\perp u(y,t)| \le \sup\, \{C_{n,\beta}\, t^{-1}\, |u(z,s)|\colon (z,s) \in B((y,t), t(\beta-1)/(\beta+1))\}$$
$$\le C_{n,\beta}\, t^{-1} \sup\, \{|u(z,s)|\colon (z,s) \in \Gamma_\beta(x')\} \le C_{n,\beta}\, t^{-1}\, N_{\beta_0}(u)(x') \le C_n\, t^{-1}\, \delta^{-1} .$$

So, for $x \in B_0$

$$\mathcal{A}(u)^2(x) \le 1 + \mathcal{A}^2_{B_0}(u)(x) + C_n\, \delta^{-2}$$

and

$$|\{x \in B_0\colon \mathcal{A}(u) > \gamma\}| \le \int_{\{x \in B_0\colon \mathcal{A}(u) > \gamma\}} \mathcal{A}(u)^2(x)\, dx$$

$$\le \gamma^{-2}\, |\{x \in B_0\colon \mathcal{A}(u) > \gamma\}| + C_n\, (\gamma\delta)^{-2}\, |B_0| + \gamma^{-2} \int_{B_0} \mathcal{A}^2_{B_0}(u)(x)\, dx .$$

Thus

$$|\{x \in B_0\colon \mathcal{A}(u) > \gamma\}| \le C_n\, (\gamma\delta)^{-2}\, |B_0| + C_n\, \gamma^{-2} \int_{W \cap (2B_0)^\wedge} t\, |\nabla^\perp u(y,t)|^2\, dydt$$

$\leq C_n (\gamma\delta)^{-2} |B_0|$ (by Main Lemma 2).

Q.E.D.

Proof of Theorem 6 (with $\tau = 1$, $\alpha = 1$, $\beta = \beta_0 := 130$). Note that for $x \in \{N_\beta(u)(x) < \delta^{-1}\}$, $\mathcal{A}(u)(x) = S_1^*(u)(x)$. So

$$|\{S_1^*(u) > \gamma, N_\beta(u) \leq 1/\delta\}| \leq |\{\mathcal{A}(u) > \gamma\}|.$$

To prove Theorem 6, we first assume $\{\mathcal{A}(u) > 1\} \neq M$. In this case, applyiny Lemma 10 to $G := \{\mathcal{A}(u) > 1\}$, we get a family of balls $\{B_j\}_j$ such that $2B_j \cap G^c \neq \emptyset$, $B_j \subset G$, $\{B_j\}_j$ disjoint mutually and $G \subset \cup_j 3B_j$. Therefore

$$|\{\mathcal{A}(u) > \gamma\}| \leq \Sigma_j |\{3B_j: \mathcal{A}(u) > \gamma\}| \leq C_n (\gamma\delta)^{-2} \Sigma_j |3B_j| \qquad \text{(by Lemma 1)}$$
$$\leq C_n (\gamma\delta)^{-2} \Sigma_j |B_j| \leq C_n (\gamma\delta)^{-2} |\{\mathcal{A}(u) > 1\}| \leq C_n (\gamma\delta)^{-2} |\{S_1^*(u) > 1\}|$$

since $\mathcal{A}(u) \leq S_1^*(u)$.

Now, we consider the case when $\{\mathcal{A}(u) > 1\} = M$. If M is noncompact, everything is OK because $|M| = \infty$. If M is compact, then we may assume $\{N_\beta(u) < \delta^{-1}\} \neq \emptyset$ and take $\{B_j\}_j = \{M\}$ in the above proof. So, it is enough to show that Lemma 17 is true for $B_0 = M$ without $\mathcal{A}(u)(x^\bullet) < 1$. To do so, we only need to show

$$\mathcal{A}'_M(u)(x) \leq 1 \qquad \text{for every } x \in M.$$

Actually

$$\mathcal{A}'_M(u)(x) \leq \int_{2r_M}^{\infty} t\left(\int_{B_x(2t)} |\nabla^\perp u(y,t)|^2\, dy\right) V_x^{-1}(t)\, dt = |M|^{-1} \int_{2r_M}^{\infty} t f(x,t)\, dt$$

where

$$f(x,t) := \int_M |\nabla^\perp u(y,t)|^2\, dy = (1/2) \int_M \Delta^\perp u^2(y,t)\, dy$$

$$= (1/2) \int_M \left(\frac{\partial}{\partial t}\right)^2 u^2(y,t)\, dy \qquad \text{(by Green's formula)}$$

$$= \int_M \left(\frac{\partial}{\partial t} u^2(y,t) + u(y,t) \left(\frac{\partial}{\partial t}\right)^2 u(y,t)\right) dy.$$

Now, let $h(x) = u(x,r_M)$, then $\|h\|_\infty \leq \delta^{-1}$ since $\{N_\beta(u) < \delta^{-1}\} \neq \emptyset$ .By boundedness of $u(x,r_M+t)$ on $M^\perp$ and Lemma 2, $u(y,r_M+t) = P_t(h)$. So, by spectral theorem

$$\int_M u(y,r_M+t)\left(\frac{\partial}{\partial t}\right)^2 u(y,r_M+t)\,dy = \int_M P_t(h)\left(\frac{\partial}{\partial t}\right)^2 P_t(h)\,dy$$

$$= \int_0^\infty (\tau e^{-t\tau^{1/2}})\, e^{-t\tau^{1/2}}\, d\|E_\tau(h)\|_2 = O(1)\int_0^\infty d\|E_\tau(h)\|_2 \sup_{\tau>\tau_1(M)} \tau e^{-2t\tau^{1/2}}$$

$$= O(1)\,\|h\|_2\, e^{-\tau_1(M)^{1/2}t} \qquad (t \geq r_M)$$

where $\tau_1(M)$ is the first non-zero eigenvalue of M. Similarly

$$\int_M \left(\frac{\partial}{\partial t}u(y,t)\right)^2 dy = O(1)\,\|h\|_2\, e^{-\tau_1(M)^{1/2}t} \qquad (t \geq r_M).$$

So

$$|f(x,t)| \leq C_M\, \delta^{-1}\, e^{-\tau_1(M)^{1/2}t}$$

and thus, for any $x \in M$

$$\dot{s}_M(u)(x) \leq C_M\, \delta^{-1} \int_{r_M}^\infty t\, e^{-\tau_1(M)^{1/2}t}\, dt \leq C^*_M / \delta < 1$$

if $\delta > 1/C^*_M$. Q.E.D.

If M is non-compact, we have

Theorem 7. *Under the hypothesis of Theorem 6, if M is non-compact, then*

$$|\{S^*_\alpha(u) > \tau\gamma,\ N_\beta(u) \leq \tau/\delta\}| \leq C_{n,\alpha,\beta}\, e^{-(\gamma\delta)^2/C_{n,\alpha,\beta}}\, |\{S^*_\alpha > \tau\}|.$$

As in the proof of Theorem 6, we may assume $\tau = 1$, $\beta = \beta_0$ and $\alpha = 1$. Then, we also have notations $s(u)$, $s_{B_0}(u)$ and $\dot{s}_{B_0}(u)$, etc. But now, instead of Lemma 17, we have

Lemma 18. *Under the conditions of Lemma 17 (except that $\exists\, x^* \in B_0$ such that $s(u)(x^*) < 1$). If $s(u)(x) \neq \infty$, then $\|s^2(u)\|_{BMO} \leq C_{n,\beta}/\delta^2$.*

Proof. Takeing any ball $B_0 = B_{x_0}(r_0)$ as in the proof of Lemma 17, we have

$$|s^2_{B_0}(x_0) - s^2_{B_0}(x)| \leq C_n\, \delta^{-2} \qquad (\text{for } x \in B_0).$$

So

$$\inf_{a\in R^1} \int_{B_0} |s^2(u)(x) - a|\, dx \leq C_n \int_{B_0} (s^2_{B_0}(u)(x) + \delta^{-2})\, dx$$

$$\leq C_n\, |B_0|\, \delta^{-2} \qquad (\text{as in the proof of Lemma 17}).$$

Thus $\|S(u)\|_{BMO} \le C_n\, \delta^{-2}$. Q.E.D.

Proof of Theorem 7 (with $\tau = 1$, $\alpha = 1$ and $\beta = \beta_0$). Note that $\mathcal{S}(u)(x) = S_1^*(u)(x)$ for any $x \in \{N_\beta(u) \le 1/\delta\}$, so

$$|\{S_1^*(u) > \gamma, N_\beta(u) \le 1/\delta\}| \le |\{\mathcal{S}^2(u) \ge \gamma^2\}|$$

$$\le C_n\, e^{-(\gamma\delta)^2/C_n} |\{\mathcal{S}(u) \ge 1\}| \qquad \text{(by Lemma 10)}$$

$$\le C_n\, e^{-(\gamma\delta)^2/C_n} |\{S_1^*(u) \ge 1\}| \qquad \text{(since } S_1^*(u) \ge \mathcal{S}(u)\text{)}.$$

Q.E.D.

Finally, we point out that for subharmonic function, we have

Theorem 8. *Suppose M is noncompact, v is nonnegative subharmonic on $M^\perp$, $\alpha > 0$ and $\beta - \alpha$ is sufficiently large, γ and $\delta > 2$, $\tau > 0$, then*

$$|\{S_\alpha^*(v) > \gamma\tau, N_\beta(v) \le \tau/\delta\}| \le C_{n,\alpha,\beta}\, e^{-(\gamma\delta)^2/C_{n,\alpha,\beta}} |\{S_\alpha^*(v) > \tau\}|.$$

Proof. Without loss of generality, we assume $\tau = 1$. By Main Lemma 2, $d\mu_v := t\,\chi_W(x,t)\,\Delta^\perp v(x,t)\,dxdt$ is a Carleson measure, so, if we set

$$\mathcal{S}_\alpha^*(v)(x) = \int_{M^\perp} \varphi(\delta(x,y)/\alpha t)\, V_x^{-1}(\alpha t)\, d\mu_v$$

then, for any $g \in H^1(M)$

$$\left|\int_M \mathcal{S}_\alpha^{*2}(v)(x)\, g(x)\, dx\right| = \left|\int_{M^\perp} \left(\int_M g(x)\, \varphi(\delta(x,y)/\alpha t)\, V_x^{-1}(\alpha t)\, dx\right) d\mu_v\right|$$

$$\le C_{n,\alpha,\beta} \|g\|_{H^1} . \|d\mu_v\|_{CM} \le C_{n,\alpha,\beta} \|g\|_{H^1}\, \delta^{-2}$$

by duality of $H^1(M)$ and $BMO(M)$, and the characterization of BMO in terms of Carleson measure (see [10-12]). So

$$\|\mathcal{S}_\alpha^{*2}(v)\|_{BMO} \le C_{n,\alpha,\beta}\, \delta^{-2}.$$

Then, by Lemma 10 and fact that $\mathcal{S}_\alpha^*(v)(x) = S_\alpha^*(v)(x)$ for x satisfying $N_\beta(v)(x) \le \delta^{-1}$

$$|\{S_\alpha^*(v) > \gamma, N_\beta(v)(x) \le \delta^{-1}\}| \le |\{\mathcal{S}_\alpha^{*2}(v) > 4\gamma^2\}|$$

$$\leq C_{n,\alpha,\beta}\, e^{-(\gamma\delta)^2/C_{n,\alpha,\beta}}\, |\{\mathcal{I}_\alpha^{*2}(v) > 1\}| \leq C_{n,\alpha,\beta}\, e^{-(\gamma\delta)^2/C_{n,\alpha,\beta}}\, |\{S_\alpha^*(v) > 1\}|$$

since $S_\alpha^*(v) \geq \mathcal{I}_\alpha^*(v)$. Q.E.D.

§6. Characterizations of $H_\Phi(M)$

In this section, we shall give some equivalent norms of $H_\Phi(M)$-function, or, in other words, we shall prove that the $L_\Phi(M)$-norms of radial maximal function, nontangential maximal function and square function of functions on M are equivalent mutually, under suitable conditions on Φ.

Theorem 9. *Suppose Φ is H-increasing. For $f \in L_c^2(M)$ (if M is compact, we assume $\int_M f(x)\,dx = 0$), $0 < \alpha, \beta < \infty$*

$$C_{n,\alpha,\beta,\Phi}^{-1}\, \|f\|_{H_\Phi} \leq \|N_\alpha(u)\|_\Phi,\ \|S_\beta(u)\|_\Phi \leq C_{n,\alpha,\beta,\Phi}\, \|f\|_{H_\Phi}$$

where $u := P(f)$.

Proof. Following fact is basic that

$$\int_M \Phi(|g(x)|)\,dx = \int_0^\infty \mu_\tau(g)\,d\Phi(\tau)$$

where $\mu_\tau(g) := |\{x \in M: |g(x)| > \tau\}|$. Now, we first prove

$$\|N_\alpha(u)\|_\Phi \leq C_{n,\alpha,\beta,\Phi}\, \|f\|_{H_\Phi}.$$

Actually, by Theorem 5, we have

$$\mu_\tau(N_\alpha(u)) \leq C_{n,\alpha}\, \gamma^n\, \mu_{\tau/4}(N_+(u)) + \mu_{\tau\gamma}(N_{2\alpha}(u)) \leq C_{n,\alpha}\, (\gamma^n\, \mu_{\tau/4}(N_+(u)) + \mu_{\tau\gamma}(N_\alpha(u))).$$

So

$$\int_M \Phi(N_\alpha(u))(x)\,dx \leq C_{n,\alpha}\, (\gamma^n \int_M \Phi(4N_+(u))(x)\,dx + \int_M \Phi(N_\alpha(u)/\gamma)(x)\,dx)$$

$$\leq C'_{n,\alpha}\, (\gamma^n \int_M \Phi(4N_+(u))(x)\,dx + \phi(\gamma) \int_M \Phi(N_\alpha(u))(x)\,dx)$$

where $\phi(\gamma) := \sup_{t>0} (\Phi(t/\gamma) / \Phi(t))$. By (2), $\exists\ \gamma_\Phi$ sufficiently large such that $\phi(\gamma_\Phi) < 1/(2C'_{n,\alpha})$, so

$$\int_M \Phi(N_\alpha(u))(x)\,dx \le C_{n,\alpha,\Phi} \int_M \Phi(4N_+(u))(x)\,dx\,.$$

Replacing u by $u/(\tau\delta)$, we get

$$\int_M \Phi(N_\alpha(u)/(\tau\delta))(x)\,dx \le C_{n,\alpha,\Phi} \int_M \Phi(4N_+(u)/(\delta\tau))(x)\,dx$$

$$\le C_{n,\alpha,\Phi}\,\phi(\delta/4) \int_M \Phi(N_+(u)/\tau)(x)\,dx \ \le C_{n,\alpha,\Phi}\,\phi(\delta/4)\,\Phi(1)$$

for any $\tau > \|N_+(u)\|_\Phi$. Therefore, by (2), there is $\delta_0 = \delta_{n,\alpha,\Phi}$ sufficiently large such that $\phi(\delta_{n,\alpha,\Phi}/4) \le C^{-1}_{n,\alpha,\Phi}$. For such a δ_0, we have

$$\int_M \Phi(N_\alpha(u)/(\tau\delta_0))(x)\,dx \le \Phi(1)$$

i.e. $\|N_\alpha(u)\|_\Phi \le \delta_0\tau$. Let $\tau \to \|N_+(u)\|_\Phi$, we finally get $\|N_\alpha(u)\|_\Phi \le \delta_0\,\|N_+(u)\|_\Phi = \delta_0\,\|f\|_{H^\Phi}$. By same approach, we can prove

$$\|S_\alpha(u)\|_\Phi \le C_{n,\alpha,\beta,\Phi}\,\|N_\beta(u)\|_\Phi$$

and

$$\|N_+(u)\|_\Phi \le C_{n,\alpha,\beta,\Phi}\,\|N_{\alpha/2}(u)\|_\Phi \le C_{n,\alpha,\beta,\Phi}\,\|S_\alpha(u)\|_\Phi \quad ,$$

we omit details here.

Finally, we give two kinds of H-increasing functions. For an increasing function Φ on $[0,+\infty)$, $\Phi(+0) = 0$, $\Phi(+\infty) = +\infty$, we call Φ a quasi-convex function if

$(8)_a$ $$\Phi(\alpha t + \beta s) \le A_\Phi(\alpha\Phi(t) + \beta\Phi(s)) \qquad (\text{for } \alpha, \beta, t, s \ge 0 \text{ and } \alpha + \beta = 1)$$

$(8)_b$ $$\bar{B}_\Phi := \inf\{\gamma\colon \sup_{t>0} \Phi(t/\gamma)/\Phi(t) \le A_\Phi^{-1}\} < +\infty\,;$$

we call Φ a power-convex function if

(9) $$\Phi^\eta(x) \text{ is convex for some } \eta = \eta_\Phi > 0.$$

Of course, any convex function is both quasi- and power-convex, and x^p is power-convex for all $p > 0$ but quasi-convex only for $p \le 1$. It is not hard to check that following Φ is quasi-convex but not power-convex.

$$\Phi(t) := \begin{cases} x & \text{for } 0 \le x \le 1 \\ (x-1)^{-(x-1)^{-1/2}} & \text{for } 1 \le x \le 1 + e^{-2} \\ x + e^{2/e} - e^{-2} - 1 & \text{for } x \ge 1 + e^{-2} \end{cases} .$$

Lemma 19. *(a). Any quasi-convex function is H-increasing, and the best B_Φ in (3) doesn't exceed $\bar{B}_\Phi$ in $(8)_b$. (b). Any power-convex function is H-increasing, and the best B_Φ in (3) doesn't exceed*

$$\bar{\bar{B}}_\Phi := \inf\{\gamma : \sup_{t>0} \Phi(t/\gamma)/\Phi(t) \le 2^{1/\eta - 1}\}$$

where $\eta = \eta_\Phi$ is the one in (9).

Proof of (a). We first check (2), i.e. $\phi(\gamma) = \sup_{t>0} (\Phi(t/\gamma)/\Phi(t)) \to 0$ $(\gamma \to +\infty)$. In the case $A_\Phi > 1$, B_Φ must be larger that 1. Take $\beta \in (1, B_\Phi)$, then

$$\phi(\beta^n) = \sup_{t>0} \Pi_{k=1}^{n} \Phi(t/\beta^n)/\Phi(t/\beta^{n-1}) \le (A_\Phi^{-1})^n \to 0 \qquad (n \to \infty).$$

So, $\phi(\gamma) \to 0$ when $\gamma \to +\infty$ since $\phi(\gamma)$ is monotonous. Now, if $A_\Phi = 1$ ($A_\Phi < 1$ is impossible), we know, Φ is convex and $\Phi'(s)$ is increasing. So, for $\gamma > 1$

$$\Phi(t/\gamma) = \int_0^{t/\gamma} \Phi'(s)\,ds \le \Phi'(t/\gamma)(t/\gamma) \le (t/\gamma)(t - t/\gamma)^{-1} \int_{t/\gamma}^{t} \Phi'(s)\,ds$$

$$\le (\gamma - 1)^{-1} \int_0^t \Phi'(s)\,ds = (\gamma - 1)^{-1}\,\Phi(t).$$

It means that $\phi(\gamma) \le (\gamma - 1)^{-1} \to 0$ when $\gamma \to +\infty$.

Now, we estimate B_Φ. Let $\alpha > \|f\|_\Phi$, $\beta > \|g\|_\Phi$, $A > \bar{B}_\Phi$. We have

$$\Phi(|f+g|/(A(\alpha+\beta))) \le \Phi((|f|/(A\alpha))\,(\alpha/(\alpha+\beta)) + (|g|/(A\beta))\,(\beta/(\alpha+\beta)))$$

$$\le A_\Phi\,((\alpha/(\alpha+\beta))\,\Phi(|f|/(A\alpha)) + (\beta/(\alpha+\beta))\,\Phi(|g|/(A\beta)))$$

$$\le (\alpha/(\alpha+\beta))\,\Phi(|f|/\alpha) + (\beta/(\alpha+\beta))\,\Phi(|g|/\beta)\ .$$

By integrating over M, we get

$$\int_M \Phi(|f+g|/(A(\alpha+\beta)))\,dx \le \Phi(1).$$

So, $\|f+g\|_\Phi \le A(\alpha+\beta)$. Let $A \to \bar{B}_\Phi$, $\alpha \to \|f\|_\Phi$, $\beta \to \|g\|_\Phi$, we get the desired inequality.

Proof of (b). It is obvious that Φ satisfies (2) since Φ^η is convex. To consider B_Φ, we may assume $\eta > 1$ since $\Phi(t)$ must be convex for $\eta \le 1$. Now, let $\alpha > \|f\|_\Phi$, $\beta > \|g\|_\Phi$, $A > \bar{\bar{B}}_\Phi$, then

$$\Phi(|f+g|/(A(\alpha+\beta))) \le (\Phi^\eta((|f|/(A\alpha))\,(\alpha/(\alpha+\beta))) + \Phi^\eta((|g|/(A\beta))\,(\beta/(\alpha+\beta))))^{1/\eta}$$

$$\le 2^{1/\eta - 1}\,((\alpha/(\alpha+\beta))^{1/\eta}\,\Phi(|f|/\alpha) + (\beta/(\alpha+\beta))^{1/\eta}\,\Phi(|g|/\beta))\,.$$

By integrating over M, we get

$$\int_M \Phi(|f+g|)/(A(\alpha+\beta))dx \le 2^{\frac{1}{\eta-1}} \left(\left(\frac{\alpha}{\alpha+\beta}\right)\right)^{\frac{1}{\eta}} + \left(\frac{\beta}{(\alpha+\beta)}^{\frac{1}{\eta}}\right) \le \Phi(1).$$

So, $\|f+g\|_\Phi \le \bar{\bar{B}}_\Phi \left(\|f\|_\Phi + \|g\|_\Phi\right)$ if we let $\alpha \to \|f\|_\Phi, \beta \to \|g\|_\Phi$ and $A \to \bar{\bar{B}}_\Phi$. Q.E.D.

Note that for $\Phi(t) = t^p(0 < p \le)$, $\bar{\bar{B}}_\Phi$ is just the best $B_\Phi = 2^{1/p-1}$. and , for convex function $\Phi, \bar{B}_\Phi$ and $\bar{\bar{B}}_\Phi$ are just the best $B_\Phi = 1$.

ACKNOWLEDGEMENT. Some parts of this work were finished when the first author stayed in Kiel University, Germany, in 1991. Here, the first author would like to express his many thanks to Prof. M. -D. Cheng (i.e., M. T. Cheng), in Peking University, for his enthusiastic support. The first author would be also grateful to Prof. A. Irle, Prof. H. Konig, Prof. V. Wrobel for their hospitality, and to Kiel University for its financial support. In addition, we would like to have this opportunity to thank Prof. R. S. Strichartz, Prof. N. Th. Varapoulos, Prof. N. Louhoue, Dr. Bakry and Dr. J. -Ph.Anker for their helpful materials and friendship.

REFERENCE

1. M. Anderson and R. Schoen, *Positive harmonic functions on complete manifolds of negative curvature*, Ann. of Math. **121** (1985), 429-461.
2. D. L. Burkolder and R. F. Gundy, *Distribution function inequalities for the area integral*, Studia Math **44** (1972), 527-544.
3. D. L. Burkolder, R. F. Gundy and M. L. Silverstein, *A maximal function characterization of the class* H^p, Trans. Amer. Math. Soc. **157** (1971), 137-153.
4. A. P. Calderon and A. Torchinsky, *Parabolic maximal functions associated with a distribution*, Adv. in Math. **16** (1975), 1-64.
5. S. -Y. A. Chang and R. Fefferman, *A continuous version of duality of* H^1 *and BMO on the bi-disc*, Ann of Math. **112** (1980), 179-201.
6. J. Chen, Ph.D. Thesis (1987), Hangzhou University.
7. P. Yip, Ph.D. *Weak type (1, 1)-boundedness of Riesz transform on positively curved manifolds*, report given in the International Symposium on Number Theory and Analysis Dedicated to the Meory of Hua Loo-keng; to appear in Chinese Ann. of Math.

8. ---, L_{Φ}-boundedness of Riesz transform, **preprint.**

9. ---, A note on Riesz potentials and the first eigenvalue, to appear in **Proc. Amer. Math. Soc.**

10. ---, A note on BMO-functions and Carleson measures on manifolds, **preprint.**

11. --- and Ch. Luo, Duality of H^1 and BMO on positively curved manifolds and their characterizations, report given in the Academic Year 1987 - 1988 on Harmonic Analysis, Nankai Mathematical Institute; to appear in **Lecture Notes in Math., Springer-Verlag.**

12. J. - Ch. Chen and S. - L. Wang, Decomposition of BMO-functions on normal Lie groups, **Acta Math. Sinica**, 32 (1989), 345 - 357.

13. --- and ---,Comparisons between maximal function and square function on positively curved manifolds, **Science in China** (Formally, Scientia Sinica) (Series A), 33 (1990), 385 - 396.

14. R. R. Coifman, Distribution function inequalities for singular integrals, **Proc. Nat. Acad. Sci. U.S.A.**, 69 (1972), 2883 - 2839.

15. --- and G. Weiss, Extensions of Hardy spaces and their use in Analysis, **Bull. Amer. Math. Soc.**, 83 (1977), 569 - 645.

16. C. Fefferman and E. M. Stein, H^p spaces of several variables, **Acta Math.**, 129 (1972), 137 - 139.

17. R. Fefferman, R. F. Gundy, M. Silverstein and E. M. Stein, Inequalities for ratios of functionals of harmonic functions, **Proc. Nat. Acad. Sci. U.S.A.**, 79 (1982), 7985 - 7960.

18. G. B. Folland and E. M. Stein, << **Hardy spaces on homogeneous groups** >>, Princeton Univ. Press, 1982.

1 9. R. F. Gundy and E. M. Stein, H^p-theory for the poly-disc, **Proc. Nat. Acad. Sci. U.S.A.**, 76 (1979), 1026 - 1029.

20. C. Herz, H^p spaces of martigales, $0 < p \leq 1$, **Zeit. Warschein.**, 28 (1984), 280 - 301.

21. P. Li and R. Schoen, L^p and mean value properties of subharmonic functions on Riemannian manifolds, **Acta Math.**, 153 (1984), 280 - 301.

22. P. Li and S. T. Yau, On the parabolic kernel of the Schördinger operator, **Acta Math.**, 156 (1986), 153 - 201.

23. T. Murai and A. Uchiyama, good-λ inequalities for the area integral and the nontangential

maximal function, **Studia Math.**, 83 (1986), 251 - 262.

24. E. M. Stein, << **Singular integrals and differentiability properties of functions** >>, Princeton Univ. Press, 1970.

25. R. S. Strichartz, Analysis of the Laplacian on the complete Riemannian manifolds, **J. Funct. Anal.**, 52 (1983), 48 - 79.

26. J. -O. Strömberg, Bounded mean oscillation with Orlicz norms and duality of Hardy spaces, **Indiana Univ. Math. J.**, 28 (1979), 511 -544.

27. A. Uchiyama, On McConnell's inequality for functionals of subharmonic functions, **Pacific J. of Math.**, 128 (1987), 367 - 377.

28. R. L. Wheeden, On the radial and nontangential maximal functions for the disc, **Proc. Amer. Math. Soc.**, 42 (1974), 418 - 422.

29. Z. Yu, **Ph. D. Thesis**, Hangzhou University, 1989.

Calderon-Zygmund Operator Theory and Function Spaces

Deng Donggao*
(Zhongshan University)
Y.-S. Han
(Auburn University, U.S.A.)

Introduction Many fields in analysis require the study of specific function spaces. In harmonic analysis the Lebesgue spaces L^p, the Hardy spaces H^p, various forms of Lipschitz spaces and the space BMO are important. Similarly, the Sobolev spaces L^p_k are basic in the study of partial differential equations. From the original definitions of these spaces, it may not appear that they are closely related. There are, however, various unified approaches to their study. The Littlewood-Paley theory provides one of the most successful unifying perspectives on these and other function spaces.

Let ϕ be a function with the properties: $\phi \in \mathcal{S}$, supp $\hat{\phi} \subseteq \{ \xi \in R^n: \frac{1}{2} \leq |\xi| \leq 2 \}$, and $|\hat{\phi}(\xi)| \geq c > 0$ if $\frac{3}{5} \leq |\xi| \leq \frac{5}{3}$. The classical homogeneous Besov spaces $\dot{B}_p^{\alpha,q}$, for $\alpha \in R$ and $0 < p, q \leq \infty$, are the collection of $f \in \mathcal{S}'/\mathcal{P}$ such that

$$(0.1) \qquad \| f \|_{\dot{B}_p^{\alpha,q}} = \{ \sum_{k=-\infty}^{\infty} (2^{\alpha k} \| \phi_k * f \|_p)^q \}^{\frac{1}{q}} < \infty.$$

The Triebel-Lizorkin spaces $\dot{F}_p^{\alpha,q}$, for $\alpha \in R$ and $0 < p < \infty$, $1 \leq q \leq \infty$, are the collection of $f \in \mathcal{S}'/\mathcal{P}$ such that

$$(0.2) \qquad \| f \|_{\dot{F}_p^{\alpha,q}} = \| \{ \sum_{k=-\infty}^{\infty} (2^{\alpha k} | \phi_k * f |)^q \}^{\frac{1}{q}} \|_p < \infty,$$

where $\phi_k(x) = 2^{kn}\phi(2^k x)$.

It was well known from Littlewood-Paley theory that we have the following identifications:

(i) $L^p \approx \dot{F}_p^{0,2}$ when $1 < p < \infty$;
(ii) $H^p \approx \dot{F}_p^{0,2}$ when $0 < p \leq 1$;
(iii) BMO $\approx \dot{F}_\infty^{0,2}$;
(iv) $\dot{L}_\alpha^p \approx \dot{F}_p^{\alpha,2}$ when $\alpha > 0$ and $1 < p < \infty$;
(v) $\Lambda_\alpha \approx \dot{F}_\infty^{\alpha,\infty}$ when $\alpha > 0$.

---

*) The author is supported in part by the Foundation of Zhongshan University Advanced Research Center.

M. Cheng et al. (eds.), Harmonic Analysis in China, 55–79.

See [L1],[L2], and [Lu] for related results.

The key point of the study of these spaces is the following Calderon reproducing formula:

The Calderon reproducing formula: Suppose that ϕ satisfies the same conditions above. Then there exists a function ψ satisfying the same conditions as ϕ such that

$$f(x) = \sum_{k=-\infty}^{\infty} \psi_k * \phi_k * f(x) \tag{0.3}$$

where the series converges in L^2, $\mathcal{S}/\mathcal{P}$(test function modulo polynomials) and $(\mathcal{S}/\mathcal{P})'$(the dual of $\mathcal{S}/\mathcal{P}$).

For instance, to see that the definitions of the Besov spaces $\dot{B}_p^{\alpha,q}$, for $-1 < \alpha < 1$ and $1 \leq p, q \leq \infty$, are independent of the choice of functions ϕ, suppose that $\tilde{\phi}$ satisfies the same conditions as ϕ, by Calderon reproducing formula, then

$$\tilde{\phi}_j * f(x) = \tilde{\phi}_j * \sum_{k=-\infty}^{\infty} \psi_k * \phi_k * f(x) = \sum_{k=-\infty}^{\infty} \tilde{\phi}_j * \psi_k * \phi_k * f(x)$$

which implies

$$\{\sum_{j=-\infty}^{\infty} (2^{\alpha j}\| \tilde{\phi}_j * f \|_p)^q\}^{\frac{1}{q}} = \{\sum_{j=-\infty}^{\infty} (2^{\alpha j}\|\sum_{k=-\infty}^{\infty} \tilde{\phi}_j * \psi_k * \phi_k * f \ \|_p)^q\}^{\frac{1}{q}}$$

$$\leq \{\sum_{j=-\infty}^{\infty} (2^{\alpha j}\sum_{k=-\infty}^{\infty} \|\tilde{\phi}_j * \psi_k\|_{p,p} \|\phi_k * f \ \|_p)^q\}^{\frac{1}{q}}$$

where $\|\tilde{\phi}_j * \psi_k\|_{p,p}$ denotes the norm of the convolution operator $\tilde{\phi}_j * \psi_k$.

By a simple calculation we have the following estimate (see [DH1] and [H1]):

$$\|\tilde{\phi}_j * \psi_k\|_{p,p} \leq c\, 2^{-| k - j |}$$

which together with Minkowski's inequality shows

$$\{\sum_{j=-\infty}^{\infty} (2^{\alpha j}\| \tilde{\phi}_j * f \|_p)^q\}^{\frac{1}{q}} \leq c\{\sum_{k=-\infty}^{\infty} (2^{\alpha k}\| \phi_k * f \|_p)^q\}^{\frac{1}{q}} = c\| f \|_{B_p^{\alpha,q}}.$$

It has been known for some time that the Calderon reproducing formula on R^n is a powerful tool to study the Littlewood-Paley theory, Calderon-Zygmund operator theory and as well as Waveletes analysis. Further applications of Calderon's reproducing formula on R^n can be found in [C1], [C2], [CF], [DJ], [FJ1], [FJ2], [FJW], [GM], [P], [R1], [R2], [T] and [U].

Our goal is to develop the theory of the Besov and Triebel-Lizorkin spaces on spaces of homogeneous type in the sense of Coifman and Weiss ([CW]). The key fact to do this is that we need the Calderon type reproducing formula on spaces of homogeneous type. The classical Calderon reproducing formula was established by the Fourier transform. Since there are no translations and dilations on spaces of homogeneous type, and no analogue of the Fourier transform and convolution operation, the new idea to establish the Calderon type reproducing formula on spaces of homogeneous type is to use the Calderon-Zygmund operator theory. To show how this idea works, in section 1 we give a presentation of basic notions of Calderon-Zygmund operator theory on R^n and show that by using the Calderon-Zygmund theory, namely the T1 theorems, the classical Besov and Triebel-Lizorkin spaces can be characterized by the family of operators which are not convolution operators. In section 2 we use the Cauchy kernel and its high order deravitives on Lipschitz curves to provide a reproducing formula on Lipschitz curves and use this reproducing formula to introduce the Besov and Triebel-Lizorkin spaces on Lipschitz curves. In the last section we first introduce a new test function space on spaces of homogeneous type and a new class of the Calderon-Zygmund operators, and then, prove a new boundedness of this new class of Calderon-Zygmund operators on these test functions. Using this boundedness we prove a Calderon reproducing formula on spaces of homogeneous type. As in the case of R^n this reproducing formula allows us to develop the Besov and Triebel-Lizorkin spaces on spaces of homogeneous type which include atomic decomposition, interpolation, dualities, and the boundedness of the Calderon-Zygmund operators for these spaces.

1 The T1 theorems and Besov and Triebel-Lizorkin spaces on R^n

Definition(1.1)([M]) A continuous complex-valued function K(x,y) defined on $\Omega = \{(x,y) \in R^n \times R^n: x \neq y\}$ is called a standard kernel if there exist $0 < \epsilon \leq 1$, and $C < \infty$ such that for all x, $y \in R^n$ with $x \neq y$,

(i) $| K(x,y) | \leq C | x - y |^{-n}$,

(ii) $| K(x,y) - K(x,y') | \leq C | y - y' |^{\epsilon} | x - y |^{-(n+\epsilon)}$ for $| y - y' | \leq \frac{1}{2} | x - y |$,

(iii) $| K(x,y) - K(x',y) | \leq C | x - x' |^{\epsilon} | x - y |^{-(n+\epsilon)}$ for $| x - x' | \leq \frac{1}{2} | x - y |$.

Definition(1.2)([M]) A continuous linear operator $T: \mathfrak{D} \to (\mathfrak{D})'$ is a singular integral operator if there is a standard kernel K such that

$$< Tf, g > = \int \int K(x,y)\, f(y)\, g(x)\, dy\, dx$$

for all f, g $\in \mathfrak{D}$ with supp f $\cap$ supp g $= \phi$. We then write T $\in$ CZK(ϵ).

Definition(1.3)([M]) A singular integral operator T is a Calderon-Zygmund operator if T can be extended to be a bounded operator on L^2. For such operators T we write T $\in$ CZO.

It was well known from Calderon-Zygmund theory that any Calderon-Zygmund operator is bounded on L^p for $1 < p < \infty$ and is of weak type (1, 1), and also bounded from L^∞ to BMO. A basic, important problem is how to establish the L^2 boundedness of a singular integral operator. This problem was solved by the celebrated T1 theorem of David and Journé. To state their result we need the following definition:

Definition(1.4) A operator T is said to have the weak boundedness property if there exist $0 < \eta \leq 1$ and $C < \infty$ such that

$$(1.5) \qquad | < Tf, g > | \leq C\, r^n$$

for all f and g $\in \mathfrak{D}$ with supp f, g $\subseteq B(x_1, r)$, $x_1 \in R^n$, $\| f \|_\infty \leq 1$, $\| g \|_\infty \leq 1$, $\| f \|_\eta \leq r^{-\eta}$, and $\| g \|_\eta \leq r^{-\eta}$. We write T $\in$ WBP.

Theorem(1.6)(The T1 theorem)([DJ]) Suppose that T is a singular integral operator. Then T is bounded on L^2 if and only if (a) T1 $\in$ BMO; (b) $T^*1 \in$ BMO; (c) T has the weak boundedness property.

This T1 theorem was first generalized to the Besov spaces by Lemarie([L]).

Theorem(1.7)(The T1 theorem for Besov spaces) Suppose that T is a singular integral operator whose kernel satisfies the conditions (i) and (iii) in the definition (1.1) and T1 = 0, and T has the weak boundedness property. Then T is bounded on the Besov spaces $\dot{B}_p^{\alpha,q}$ for $0 < \alpha < \epsilon$ and $1 \leq p, q \leq \infty$.

The T1 theorem for the Triebel-Lizorkin spaces was proved in [FHJW], [HS] and [HJTW].

Theorem(1.8)(The T1 theorem for Triebel-Lizorkin spaces) Suppose that T is a singular integral operator whose kernel satisfies the conditions (i) and (iii) in the definition (1.1) and T1 = 0, and T has the weak boundedness property. Then T is bounded on the Triebel-Lizorkin spaces $\dot{F}_p^{\alpha,q}$ for $0 < \alpha < \epsilon$ and $1 \leq p, q \leq \infty$.

Using these T1 theorems for the Besov and Triebel-Lizorkin spaces, the classical Besov and Triebel-Lizorkin spaces can be characterized by the family of operators which are not convolution operators. To state these results we begin with a Coifman's idea.

Let $S_k(x, y)$ be functions on R^n satisfying the following conditions:

(i) $S_k(x,y) = 0$ if $| x - y | \geq C\, 2^{-k}$ and $| S_k(x,y) | \leq C\, 2^{nk}$;

(ii) $| S_k(x,y) - S_k(x',y) | \leq C\, | x - x' |^{\epsilon} 2^{k(n+\epsilon)}$;

(iii) $| S_k(x,y) - S_k(x,y') | \leq C\, | x - x' |^{\epsilon} 2^{k(n+\epsilon)}$;

(iv) $\int S_k(x,y)\, dy = 1$,

(v) $\int S_k(x,y)\, dx = 1$.

See [DJS] for a similar family of operators on homogeneous type. Let $D_k = S_k - S_{k-1}$. Then Coifman's idea is to write the identity operator I as

$$I = \sum_k D_k = \sum_k D_k \sum_j D_j = \sum_k \sum_j D_k D_j$$

$$= \sum_{| k - j | \leq N} D_k D_j + \sum_{| k - j | > N} D_k D_j = T_N + R_N.$$

It was proved in [DJ] that T_N tends to I as N tends to ∞. In [DH1] and [HJTW] it was proved that R_N is a Calderon-Zygmund operator satisgying theorems (1.7) and (1.8) with the operator norm at most $c2^{-N\delta}$ for some $\delta > 0$. This shows that T_N is bounded and invertible on the Besov and Triebel-Lizorkin spaces by the T1 theorems above. By the Calderon reproducing formula on R^n it is not hard to show

$$\{ \sum_k (2^{\alpha k} \| D_k(f) \|_p)^q \}^{\frac{1}{q}} \leq c \| f \|_{\dot{B}_p^{\alpha,q}} \qquad \text{for } 1 \leq p, q \leq \infty,$$

and

$$\| \{ \sum_k (2^{\alpha k} | D_k(f) |)^q \}^{\frac{1}{q}} \|_p \leq c \| f \|_{\dot{F}_p^{\alpha,q}} \qquad \text{for } 1 \leq p, q \leq \infty.$$

Conversely,

$$\| f \|_{\dot{B}_p^{\alpha,q}} \leq c \, \| T_N f \|_{\dot{B}_p^{\alpha,q}}.$$

By the duality,

$$\| T_N f \|_{\dot{B}_p^{\alpha,q}} = \sup_{\| g \|_{\dot{B}_{p'}^{-\alpha,q'}} \leq 1} | < T_N f, g > |$$

$$\leq \sup_{\| g \|_{\dot{B}_{p'}^{-\alpha,q'}} \leq 1} \{ \sum_k (2^{\alpha k} \| D_k(f) \|_p)^q \}^{\frac{1}{q}} \{ \sum_k (2^{-\alpha k} \| D_k^N(g) \|_{p'})^{q'} \}^{\frac{1}{q'}}$$

where $D_k^N = \sum_{| j | \leq N} D_{k+j}$

$$\leq c \{ \sum_k (2^{\alpha k} \| D_k(f) \|_p)^q \}^{\frac{1}{q}} .$$

The proof for the Triebel-Lizorkin spaces is similar.

Now we can state the main results in this section.

Theorem(1.9)([DH1] and [HJTW]) Suppose that D_k satisfy the above conditions. Then

(1.10) $\| f \|_{\dot{B}_p^{\alpha,q}} \approx \{ \sum_{k=0}^{\infty} (2^{\alpha k} \| D_k(f) \|_p)^q \}^{\frac{1}{q}}$ for $-\epsilon < \alpha < \epsilon$ and $1 \leq p, q \leq \infty$,

(1.11) $\| f \|_{\dot{F}_p^{\alpha,q}} \approx \| \{ \sum_{k=0}^{\infty} (2^{\alpha k} | D_k(f) |)^q \}^{\frac{1}{q}} \|_p$ for $-\epsilon < \alpha < \epsilon$ and $1 \leq p < \infty$ and $1 \leq q \leq \infty$.

If only considering the " half conditions " of S_k, namely $S_k(x, y)$ satisfy the conditions (i), (ii) and (iv) above, then we have

Theorem(1.12)([HS2]) Suppose that $S_k(x, y)$ satisfy the " half conditions " and $D_k = S_k - S_{k-1}$. Then

(1.13) $\| f \|_{\dot{B}_p^{\alpha,q}} \approx \{ \sum_{k=0}^{\infty} (2^{\alpha k} \| D_k(f) \|_p)^q \}^{\frac{1}{q}}$ for $0 < \alpha < \epsilon$ and $1 \leq p, q \leq \infty$,

(1.14) $\| f \|_{\dot{F}_p^{\alpha,q}} \approx \| \{ \sum_{k=0}^{\infty} (2^{\alpha k} | D_k(f) |)^q \}^{\frac{1}{q}} \|_p$ for $0 < \alpha < \epsilon$ and $1 \leq p < \infty$ and $1 \leq q \leq \infty$.

2 The Besov and Triebel-Lizorkin spaces on Lipschitz curves

In this section we consider the special spaces of homogeneous type, namely Lipschitz curves on complex plane $\mathbb{C}$. Suppose that A: R→R is a real-valued Lipschitz function, that is, there exists a constant M such that for all x and $y \in R$, $|A(x) - A(y)| \leq M |x - y|$, and Γ is the graph of function x +iA(x) in complex plane $\mathbb{C}$. If we consider the Euclidean distance d and the Lebesgue arclength measure ds on Γ, then (Γ, d, ds) is a space of homogeneous type in the sense of Coifman and Weiss (see Definition (3.2) below). By using the Cauchy kernel and its high order derivatives on Γ, we can establish a Calderon type reproducing formula on Γ, and then develop the theory of the Besov and Triebel-Lizorkin spaces on Γ. We first introduce the following definition.

Definition(2.1)([DH2] and [DH4]) Suppose that $0 < \beta \leq 1$, $\gamma > 0$ and $K \geq 0$ is an integer. A function f: $\mathbb{C} \to \mathbb{C}$ is said to be a (w_0, r, β, γ, K) type test function centered at $w_0 \in \Gamma$ with the width $r > 0$ if there exists a constant c such that

$$\text{(i)} \quad \left| \frac{d^\nu f}{dw^\nu}(w) \right| \leq c \frac{r^\gamma}{(r + |w - w_0|)^{\nu+\gamma+1}} \text{ for all } w \in \Gamma \text{ and } 0 \leq \nu \leq K,$$

$$\text{(2.2)} \qquad \text{(ii)} \quad \left| \frac{d^K f}{dw^K}(w) - \frac{d^K f}{dw^K}(w') \right| \leq c \left(\frac{|w - w_0|}{r + |w - w_0|} \right)^\beta \frac{r^\gamma}{(r + |w - w_0|)^{K+\gamma+1}}$$

for all w and $w' \in \Gamma$ with $|w - w'| \leq \frac{1}{2}(r + |w - w_0|)$,

(iii) $\int_\Gamma f(w) w^\theta \, dw = 0$ for all $0 \leq \theta \leq J$ where $j = [\gamma - 1]$, the integer less than or equal to $\gamma - 1$.

Denote $\mathbb{M}^{(\beta,\gamma,K)}(w_0, r)$ the collection of all (β,γ,K) type test functions centered at w_0 with the width $r > 0$. If $f \in \mathbb{M}^{(\beta,\gamma,K)}(w_0, r)$, the norm of f in $\mathbb{M}^{(\beta,\gamma,K)}(w_0, r)$ is defined to be

$$\| f \|_{\mathbb{M}^{(\beta,\gamma,K)}(w_0, r)} = \inf\{c \geq 0, \text{ c satisfies (i) and (ii) of (2.2)}\}.$$

We denote $\mathbb{M}^{(\beta,\gamma,K)} = \mathbb{M}^{(\beta,\gamma,K)}(iA(0), 1)$. It is easy to see that $\mathbb{M}^{(\beta,\gamma,K)}$ is a Banach space under the norm $\| f \|_{\mathbb{M}^{(\beta,\gamma,K)}} < \infty$. Just as the space of distributions $\mathcal{S}'$ is defined on R^n, the dual space $(\mathbb{M}^{(\beta,\gamma,K)})'$ consists of all linear functionnals $\mathcal{L}$ from $\mathbb{M}^{(\beta,\gamma,K)}$ to $\mathbb{C}$ with the property that there exists a constant c such that for all $f \in \mathbb{M}^{(\beta,\gamma,K)}$, $|\mathcal{L}(f)| \leq c \| f \|_{\mathbb{M}^{(\beta,\gamma,K)}}$. We denote $< h, f >$ the natural pairing of elements $h \in (\mathbb{M}^{(\beta,\gamma,K)})'$ and $f \in \mathbb{M}^{(\beta,\gamma,K)}$. It is easy to see that for $w_1 \in \Gamma$ and $r_1 > 0$ $\mathbb{M}^{(\beta,\gamma,K)}(w_1, r_1) = \mathbb{M}^{(\beta,\gamma,K)}$ with equivalent norms. Thus, for all $h \in (\mathbb{M}^{(\beta,\gamma,K)})'$ and all $f \in \mathbb{M}^{(\beta,\gamma,K)}(w_1, r_1)$ with $w_1 \in \Gamma$ and $r_1 > 0$ $< h, f >$ is well defined.

Now we establish a Calderon type reproducing formula on Γ. For $f \in \mathbb{M}^{(\beta,\gamma,K)}$, let

$$F(z) = \frac{1}{2\pi i} \int \frac{f(w)}{w - z}\, dw, \; z \in \mathbb{C}\backslash\Gamma,$$

be the Cauchy integral of f on Γ. Integrating by parts gives

$$\int_{\epsilon}^{M} t^{(\ell-1)} F^{(\ell-1)}(w+ it) = \tfrac{1}{i} t^{(\ell-1)} F^{(\ell-1)}(w+ it)|_{\epsilon}^{M} + \sum_{j=1}^{\ell-1} (-1)^{j} \frac{(\ell-1)(\ell-2)...(\ell-j)}{i^{j+1}} t^{(\ell-j-1)} F^{(\ell-j-1)}(w+it)|_{\epsilon}^{M}.$$

Since $t^{(\ell-j-1)} F^{(\ell-j-1)}(w+it) = \frac{(\ell-j-1)!}{2\pi i} \int \frac{t^{(\ell-j-1)} f(z)}{(z - w - it)^{(\ell-j)}}\, dz$ and $\frac{t^{(\ell-j-1)}}{(z - w - it)^{(\ell-j)}}$ on Γ decays like the Poisson kernel for $0 \le j \le \ell - 2$, and $\int \frac{t^{(\ell-j-1)}}{(z - w - it)^{(\ell-j)}}\, dz = 0$ for all $t \neq 0$ and $w \in \Gamma$, thus, for $0 \le j \le \ell - 2$, $t^{(\ell-j-1)} F^{(\ell-j-1)}(w + it) \to 0$ a. e. as $t \to 0$ and $t^{(\ell-j-1)} F^{(\ell-j-1)}(w + it) \to 0$ a. e. as $|t| \to \infty$ for $f \in \mathbb{M}^{(\beta,\gamma,K)}$ by direct computation. Therefore

$$\lim_{\epsilon\to 0, M\to\infty} \Big[\Big(\int_{\epsilon}^{M} + \int_{-M}^{\epsilon}\Big) t^{(\ell-1)} F^{(\ell)}(w + it)\, dt\Big]$$

$$= (-1)^{(\ell-1)} \frac{(\ell-1)!}{i^{\ell}} \lim_{\epsilon\to 0, M\to\infty} [\, F(w + iM) - F(w - iM) - F(w + i\epsilon) + F(w - i\epsilon)\,].$$

Because $\frac{1}{z - w - it} - \frac{1}{z - w + it} = \frac{2it}{(z - w - it)(z - w + it)}$, and $\| \sup_{t>0} [\, F(w+it) - F(w-it)] \|_p \le c\| f \|_p$ and $\lim_{M\to\infty} [\, F(w+it) - F(w-it)] = 0$. By the residue theorem, $\frac{1}{2\pi i} \int \frac{2it}{(z - w - it)(z - w + it)} dz = 1$ for all $t > 0$ and $w \in \Gamma$, and hence $\frac{1}{2\pi i} \frac{2it}{(z - w - it)(z - w + it)}$ is the kernel of an approximation of the identity, so that $F(w + i\epsilon) - F(w - i\epsilon) \to -f(w)$ a. e. as $\epsilon \to 0$. This shows

$$\lim_{\epsilon\to 0, M\to\infty} \Big[\Big(\int_{\epsilon}^{M} + \int_{-M}^{\epsilon}\Big) t^{(\ell-1)} F^{(\ell)}(w + it)\, dt\Big] = (-1)^{(\ell-1)} \frac{(\ell-1)!}{i^{\ell}} f(w)$$

for a. e. $w \in \Gamma$ and $f \in \mathbb{M}^{(\beta,\gamma,K)}$.

For $f \in \mathbb{M}^{(\beta,\gamma,K)}$ define $J_t(f)(z) = t^{\ell} F^{(\ell)}(z + it)$,, then $J_t^2(f)(z) = t^{2\ell} F^{(2\ell)}(z + it)$. Thus, the preliminary facts tell us that

$$\text{p. v.} \int_{-\infty}^{\infty} J_t^2(f) \frac{dt}{t} = (-1)^{\ell} 2^{-2\ell} (2\ell - 1)!\, f$$

where the principal value integral converges pointwise a. e., and in L^p norm for $1 < p < \infty$ (see [DJS] or [M] for the case where $\ell = 1$). The key point is that the principal value integral converges in the norm of $\mathbb{M}^{(\beta,\gamma,K)}$. More precisely, we have

Theorem(2.3)([DH2] and [DH4]) Suppose that $\{ J_t \}_{t \neq 0}$ is a family of operators whose kernels are given by

(2.4) $$J_t(z, w) = \frac{\ell!}{2\pi i} \frac{t^\ell}{(w - z - it)^{(\ell+1)}}, \text{ where } \ell \geq 1 \text{ is an integer.}$$

Then for $f \in M^{(\beta,\gamma,K)}$ with $0 < \beta \leq 1$ and $0 < \gamma < \ell$, and $K < \ell$,

(2.5) $$\text{p. v.} \int_{-\infty}^{\infty} J_t^2 (f) \frac{dt}{t} = (-1)^\ell 2^{-2\ell}(2\ell - 1)!\, f$$

where the principal value integral converges in the norm of $M^{(\beta',\gamma',K')}$ with $0 < \beta' < \beta$ and $0 < \gamma' < \gamma$, and $K' \leq K$.

By the duality argument we obtain

Theorem(2.6)([DH4]) Suppose that $\{ J_t \}_{t \neq 0}$ is a family of operators whose kernels are given by

(2.7) $$J_t(z, w) = \frac{\ell!}{2\pi i} \frac{t^\ell}{(w - z - it)^{(\ell+1)}}, \text{ where } \ell \geq 1 \text{ is an integer.}$$

Then for $f \in (M^{(\beta,\gamma,K)})'$ with $0 < \beta \leq 1$ and $0 < \gamma < \ell$, and $K < \ell$,

(2.8) $$\text{p. v.} \int_{-\infty}^{\infty} J_t^2 (f) \frac{dt}{t} = (-1)^\ell 2^{-2\ell}(2\ell - 1)!\, f$$

where the principal value integral converges in $(M^{(\beta',\gamma',K')})'$ with $\beta' > \beta$ and $\gamma' > \gamma$, and $K' \geq K$.

Definition(2.9)([DH2,3,4]) Suppose that $\{ J_t \}_{t \neq 0}$ is a family of operators whose kernels are given by

$$J_t(z, w) = \frac{\ell!}{2\pi i} \frac{t^\ell}{(w - z - it)^{(\ell+1)}}$$

where $\ell \geq 1$ is an integer and $| \alpha | < \ell$, $1 \leq p, q \leq \infty$. For $f \in (M^{(\beta,\gamma,K)})'$, we define the following norms

(2.10) $$\| f \|_{\dot{B}_p^{\alpha,q}} = \left\{ \int_{-\infty}^{\infty} (| t |^{-\alpha} \| J_t(f) \|_p)^q \frac{dt}{| t |} \right\}^{\frac{1}{q}} \quad \text{for } 1 \leq p, q \leq \infty,$$

(2.11) $$\| f \|_{\dot{F}_p^{\alpha,q}} = \left\| \left\{ \int_{-\infty}^{\infty} (| t |^{-\alpha} | J_t(f) |)^q \frac{dt}{| t |} \right\}^{\frac{1}{q}} \right\|_p \quad \text{for } 1 \leq p < \infty \text{ and } 1 \leq q \leq \infty.$$

Theorem(2.12)([DH2,3,4]) The norms defined in definition (2.9) are independent of β, γ and K with $\max(0, \alpha) < \beta + K < \ell$ and $\max(0, -\alpha) < \gamma < \ell$.

Now we can introduce the Besov and Triebel-Lizorkin spaces on Lipschitz curves.

Definition(2.13)([DH2,3,4]) Suppose that $\{ J_t \}_{t \neq 0}$ is a family of operators whose kernels are given by

$$J_t(z, w) = \frac{\ell!}{2\pi i} \frac{t^\ell}{(w - z - it)^{(\ell+1)}}$$

where $\ell \geq 1$ is an integer and $| \alpha | < \ell$. The Besov space $\dot{B}_p^{\alpha,q}$ is a collection of all $f \in (M^{(\beta,\gamma,K)})'$ with $\max(0, \alpha) < \beta + K < \ell$ and $\max(0, -\alpha) < \gamma < \ell$ such that

$$(2.14) \qquad \| f \|_{\dot{B}_p^{\alpha,q}} = \{ \int_{-\infty}^{\infty} (| t |^{-\alpha} \| J_t(f) \|_p)^q \frac{dt}{|t|} \}^{\frac{1}{q}} < \infty \quad \text{for } 1 \leq p, q \leq \infty,$$

The Triebel-Lizorkin space $\dot{F}_p^{\alpha,q}$ is a collection of all $f \in (M^{(\beta,\gamma,K)})'$ with $\max(0, \alpha) < \beta + K < \ell$ and $\max(0, -\alpha) < \gamma < \ell$ such that

$$(2.15) \qquad \| f \|_{\dot{F}_p^{\alpha,q}} = \| \{ \int_{-\infty}^{\infty} (| t |^{-\alpha} | J_t(f) |)^q \frac{dt}{|t|} \}^{\frac{1}{q}} \|_p < \infty \quad \text{for } 1 \leq p < \infty \text{ and } 1 \leq q \leq \infty.$$

Using the Calderon type reproducing formula in theorem(2.6), one can prove the T1 theorems, dual spaces and interpolation for the Besov and Triebel-Lizorkin spaces on Lipschitz curves. See [DH2,3,4] for the details.

Finally, using the Calderon type reproducing formula in theorem (2.6), we can give new characterizations of the Besov and Triebel-Lizorkin spaces on Lipschitz curves.

Theorem(2.16)([DH4]) Suppose that $1 < p < \infty$ and k is a positive integer. Then $L_k^p(\Gamma) = \dot{F}_p^{k,2}(\Gamma) \cap L^p(\Gamma)$ where $L_k^p(\Gamma)$ is the Sobolev space on Γ.

Theorem(2.17)([DH4]) Suppose that $0 < \alpha < 1$ and $1 < p < \infty$. Then for all $f \in L^p(\Gamma)$,

$$\| f \|_{\dot{B}_p^{\alpha,p}} \approx \int_\Gamma \int_\Gamma \frac{| f(z) - f(w) |^p}{| z - w |^{1 + \alpha p}} | dz | | dw |.$$

3 The Besov and Triebel-Lizorkin spaces on spaces of homogeneous type

The key fact to develop the Besov and Triebel-Lizorkin spaces on Lipschitz curves is the Calderon type reproducing formula which follows from the Cauchy kernel on Lipschitz curves. In this section we will use the Calderon-Zygmund operator theory to provide a Calderon type reproducing formula on spaces of homogeneous type introduced by Coifman and Weiss([CW]).

We begin by recalling the definitions on spaces of homogeneous type. A quasi-metric d on a set X is a function d: $X \times X \rightarrow R$ satisfying

(3.1)

(i) $d(x, y) = 0$ if and only if $x = y$,

(ii) $d(x, y) = d(y, x)$ for all x and $y \in X$,

(iii) There exists a constant $A < \infty$ such that for all x, y and $z \in X$,

$$d(x, y) \leq A\, [d(x, z) + d(z, y)].$$

Any quasi-metric d defines a topology, for which the balls $B(x, r) = \{ y \in X: d(x, y) < r \}$ form a base. However, the balls themselves need not to be open when $A > 1$.

Definition(3.2)([CW]) A space of homogeneous type (X, d, μ) is a set X together with a quasi-metric d and a nonnegative measure μ on X such that $\mu(B(x, r)) < \infty$ for all $x \in X$ and all $r > 0$, and such that there exists a constant $A' < \infty$ such that for all $x \in X$ and all $r > 0$,

$$\mu(B(x, 2r)) \leq A'\mu(B(x, r)). \tag{3.3}$$

In [MS] Macias and Segovia have shown that one can replace d by another quasi-metric ρ such that there exist $c > 0$ and some θ, $0 < \theta < 1$, satisfying

$$\rho(x, y) \sim \inf\{ \mu(B): \text{B is a ball containing x and y}\}, \tag{3.4}$$

$$| \rho(x, y) - \rho(x', y) | \leq c\, \rho(x, x')^{\theta} [\rho(x, y) + \rho(x', y)]^{1-\theta} \tag{3.5}$$

for all x, x' and $y \in X$.

There are many important examples of spaces of homogeneous type. Any C^{∞} compact Riemannian manifold with the Riemannian metric and volume; non-isotropic Euclidian space; some Lie groups; Heisenberg group and any bounded Lipschitz domains in R^n are spaces of homogeneous type. See [Ch] for more examples of spaces of homogeneous type.

We first consider $\mu(X) = \infty$ and $\mu(\{x\}) = 0$ for all $x \in X$. Let C_0^η, $\eta > 0$, be the space of all continuous functions on X with bounded support such that

$$\|f\|_\eta = \sup_{x \neq y} \frac{|f(x)-f(y)|}{\rho(x,y)} < \infty. \tag{3.6}$$

Endow C_0^η with the natural topology and let $(C_0^\eta)'$ be its dual space.

Definition(3.7)([CW]) A continuous complex-valued function K(x,y) defined on $\Omega = \{(x,y) \in X \times X: x \neq y\}$ is called a standard kernel if there exist $0 < \epsilon \leq \theta$, and $C < \infty$ such that for all x, $y \in X$ with $x \neq y$,

(i) $|K(x,y)| \leq C |x - y|^{-1}$,

(ii) $|K(x,y) - K(x,y')| \leq C |y - y'|^\epsilon |x - y|^{-(1+\epsilon)}$ for $|y - y'| \leq \frac{1}{2A} |x - y|$,

(iii) $|K(x,y) - K(x',y)| \leq C |x - x'|^\epsilon |x - y|^{-(1+\epsilon)}$ for $|x - x'| \leq \frac{1}{2A} |x - y|$.

Definition(3.8)([CW]) A continuous linear operator T: $C_0^\eta \to (C_0^\eta)'$ is a singular integral operator if there is a standard kernel K such that

$$< Tf, g > = \int\int K(x,y)\, f(y)\, g(x)\, dy\, dx$$

for all f, $g \in C_0^\eta$ with supp $f \cap$ supp $g = \phi$. We then write $T \in CZK(\epsilon)$.

Definition(3.9)([CW]) A singular integral operator T is a Calderon-Zygmund operator if T can be extended to be a bounded operator on L^2. For such operators T we write $T \in CZO$.

It was well known from Calderon-Zygmund theory that any Calderon-Zygmund operator is bounded on L^p for $1 < p < \infty$ and is of weak type (1, 1), and also bounded from L^∞ to BMO.

Definition(3.10)([HS2]) A operator T with the kernel K is said to have the strong weak boundedness property if there exist $0 < \eta \leq \theta$ and $C < \infty$ such that

$$|< K, f >| \leq C\, r \tag{3.11}$$

for all $f \in C_0^\eta$ with supp $f \subseteq B(x_1, r)$, $x_1 \in X$, $\| f \|_\infty \le 1$, $\| f(.,y) \|_\eta \le r^{-\eta}$, and $\| f(x, y) \|_\eta \le r^{-\eta}$. We write $T \in SWBP$.

Now we introduce a class of test functions on X.

Definition(3.12)([HS2]) Fix $\gamma > 0$, and $0 < \beta \le \theta$. A function f defined on X is said to be a test function of type (x_0, d, β, γ), $x_0 \in X$ and $d > 0$, if f satisfies the following conditions:

$$\text{(i)} \quad | f(x) | \le C \frac{d^\gamma}{(d+\rho(x,x_0))^{1+\gamma}};$$

$$\text{(ii)} \quad | f(x) - f(y) | \le C \left(\frac{\rho(x,y)}{d+\rho(x,x_0)} \right)^\beta \frac{d^\gamma}{(d+\rho(x,x_0))^{1+\gamma}} \qquad \text{for } \rho(x,y) \le \frac{1}{2A} [d+\rho(x,x_0)],$$

$$\text{(iii)} \quad \int f(x)\, d\mu(x) = 0.$$

If f is a test function of type (x_0, d, β, γ) we write $f \in \mathfrak{M}(x_0, d, \beta, \gamma)$, and the norm of f in $\mathfrak{M}(x_0, d, \beta, \gamma)$ is defined by

$$(3.13) \qquad \| f \|_{\mathfrak{M}(x_0, d, \beta, \gamma)} = \inf \{ C: \text{(i) and (ii) hold} \}.$$

Now fix $x_0 \in X$ and denote $\mathfrak{M}(\beta, \gamma) = \mathfrak{M}(x_0, 1, \beta, \gamma)$. It is easy to see that $\mathfrak{M}(x_1, d, \beta, \gamma) = \mathfrak{M}(\beta, \gamma)$ with the equivalent norms for all $x_1 \in X$ and $d > 0$. Furthermore, it is also easy to check that $\mathfrak{M}(\beta, \gamma)$ is a Banach space with respect to the norm in $\mathfrak{M}(\beta, \gamma)$. Just as the space of distributions $\mathcal{S}'$ is defined on R^n, the dual space $(\mathfrak{M}(\beta, \gamma))'$ consists of all linear functionnals $\mathcal{L}$ from $\mathfrak{M}(\beta, \gamma)$ to C with the property that there exists a constant c such that for all $f \in \mathfrak{M}(\beta, \gamma)$, $| \mathcal{L}(f) | \le c \| f \|_{\mathfrak{M}(\beta, \gamma)}$. We denote $< h, f >$ the natural pairing of elements $h \in (\mathfrak{M}(\beta, \gamma))'$ and $f \in \mathfrak{M}(\beta, \gamma)$. It is easy to see that for $x_1 \in X$ and $d > 0$ $\mathfrak{M}(x_1, d, \beta, \gamma) = \mathfrak{M}(\beta, \gamma)$ with equivalent norms. Thus, for all $h \in (\mathfrak{M}(\beta, \gamma))'$ and all $f \in \mathfrak{M}(x_1, d, \beta, \gamma)$ with $x_1 \in X$ and $d > 0$ $< h, f >$ is well defined.

The one main result in this section is the following theorem.

Theorem(3.14)([DH5] and [H2]) Suppose that $T \in CZK(\epsilon) \cap SWBP$, and $T(1) = T^*(1) = 0$. Suppose further that K(x, y), the kernel of T, satisfies the following condition:

$$(3.15) \quad | [K(x,y) - K(x',y)] - [K(x,y') - K(x',y')] | \le C\, \rho(x,x')^\epsilon \rho(y,y')^\epsilon \rho(x,y)^{-(1+2\epsilon)}$$
$$\text{for } \rho(x,x'),\ \rho(y,y') \le \frac{1}{3A^2}\rho(x,y).$$

Then $T(f) \in \mathfrak{M}(x_1, d, \beta, \gamma)$ if $f \in \mathfrak{M}(x_1, d, \beta, \gamma)$ for $x_1 \in X$, $d > 0$, and $0 < \beta, \gamma < \epsilon$. Moreover, there exists a constant C such that

$$\| T(f) \|_{\mathfrak{M}(x_1, d, \beta, \gamma)} \leq C \| T \| \| f \|_{\mathfrak{M}(x_1, d, \beta, \gamma)} \tag{3.16}$$

where $\| T \|$ denotes the smallest constant in the estimates of the kernel of T.

Notice that the condition of (3.15) is also necessary for Calderon-Zygmund operators which map test function to test function(see [R2] for the case of R^n and [H2] for spaces of homogeneous type).

To establish a Calderon type reproducing formula on spaces of homogeneous type, we first introduce the following definition.

Definition(3.17)([HS2]) A sequence of operators $\{ S_k \}_{k \in \mathbb{Z}}$ is said to be an approximation to the identity if $S_k(x, y)$, the kernels of S_k, are functions defined on $X \times X$ satisfying the following conditions:

(3.18)

(i) $S_k(x, y) = 0$ if $\rho(x, y) \geq c2^{-k}$ and $\| S_k \|_\infty \leq c2^k$;

(ii) $| S_k(x, y) - S_k(x', y) | \leq c\rho(x, x')^\epsilon 2^{(1 + \epsilon)}$;

(iii) $| S_k(x, y) - S_k(x, y') | \leq c\rho(y, y')^\epsilon 2^{(1 + \epsilon)}$;

(iv) $| [S_k(x, y) - S_k(x', y)] - [| S_k(x, y') - S_k(x', y') | \leq c\rho(x, x')^\epsilon \rho(y, y')^\epsilon 2^{(1 + 2\epsilon)}$;

(v) $\int S_k(x, y)\, d\mu(y) = 1$;

(vi) $\int S_k(x, y)\, d\mu(y) = 1$.

See [DJS] for a construction of this sequence of operators on spaces of homogeneous type. The condition (iv) is new(see [H1] for the case of R^n). It is easy to check that construction of operators in [DJS] satisfies the condition (iv) above. Let $D_k = S_k - S_{k-1}$. Again, using Coifman's idea, we write the identity operator I as follows:

$$I = \sum_k D_k = \sum_k D_k \sum_j D_j = \sum_k \sum_j D_k D_j$$

$$= \sum_{| k - j | \leq N} D_k D_j + \sum_{| k - j | > N} D_k D_j = T_N + R_N.$$

Theorem(3.19)([DH5], [H1] and [HS2]) Suppose that $\{ S_k \}_{k \in \mathbb{Z}}$ is an approximation to the identity and $D_k = S_k - S_{k-1}$. Let $R_N = \sum_{|k-j|>N} D_k D_j$ where N is a positive integer. Then R_N is a Calderon-Zygmund operator. More precisely, $R_N(x, y)$, the kernel of R_N, satisfies the following estimates: for $0 < \epsilon' < \epsilon$ there exist a constant c and some $\delta > 0$ such that

(3.20) $$| R_N(x, y) | \leq c2^{-N\delta} \rho(x, y)^{-1};$$

(3.21) $$| R_N(x, y) - R_N(x, y') | \leq c2^{-N\delta} \rho(y, y')^{\epsilon'} \rho(x, y)^{-(1+\epsilon')} \text{ for } \rho(y, y') \leq \frac{1}{2A}\rho(x, y);$$

(3.22) $$| R_N(x, y) - R_N(x', y) | \leq c2^{-N\delta} \rho(x, x')^{\epsilon'} \rho(x, y)^{-(1+\epsilon')} \text{ for } \rho(y, y') \leq \frac{1}{2A}\rho(x, y);$$

(3.23) $$| [R_N(x, y) - R_N(x, y')] - [R_N(x, y) - R_N(x, y') | \leq c2^{-N\delta} \rho(x, x')^{\epsilon'} \rho(y, y')^{\epsilon'} \rho(x, y)^{-(1+\epsilon')}$$

for $\rho(x, x')$ and $\rho(y, y') \leq \frac{1}{3A^2}\rho(x, y)$;

(3.24) $$| < R_N, f > | \leq c2^{-N\delta} r$$

for all $f \in C_0^\eta (X \times X)$ with supp $f \subseteq B(x_1, r)$, $x_1 \in X$, $\| f \|_\infty \leq 1$, $\| f(.,y) \|_\eta \leq r^{-\eta}$, and $\| f(x, .) \|_\eta \leq r^{-\eta}$.

Using theorem (3.14) together with the fact that $T_N^{-1} = \sum_{m=0}^{\infty} (R_N)^m$, we can prove that T_N^{-1} is a Calderon-Zygmund operator and maps test function to test function. To be precise, we have

Theorem(3.25)([DH5], [H1] and [HS2]) Suppose that $\{ S_k \}_{k \in \mathbb{Z}}$ is an approximation to the identity and $D_k = S_k - S_{k-1}$. Let $T_N = \sum_{|k-j| \leq N} D_k D_j$ where N is a positive integer. Then T_N^{-1} exists and maps test function to test function. More precisely, there exists a constant c such that for $f \in \mathfrak{M}(x_1, d, \beta, \gamma)$ with $x_1 \in X$, $d > 0$ and $0 < \beta, \gamma < \epsilon$,

(3.26) $$\| T_N^{-1}(f) \|_{\mathfrak{M}(x_1, d, \beta, \gamma)} \leq c \| f \|_{\mathfrak{M}(x_1, d, \beta, \gamma)}$$

whenever N is sufficiently large.

Now, if denote $D_k^N = \sum_{|j| \leq N} D_{k+j}$, formally, we have

$$I = T_N^{-1} T_N = T_N^{-1} \sum_k D_k^N D_k = \sum_k (T_N^{-1} D_k^N) D_k.$$

Since $D_k^N \in \mathfrak{M}(x, 2^{-k}, \epsilon, \epsilon)$ and hence, by theorem (3.25), $T_N^{-1} D_k^N$ is a test function in $\mathfrak{M}(x, 2^{-k}, \epsilon, \epsilon)$. Denote $\tilde{D}_k = T_N^{-1} D_k^N$, then

$$I = \sum_k \tilde{D}_k D_k.$$

More precisely, we have

Theorem(3.27)(Calderon type reproducing formula, [DH5], [H1] and [HS2]) Suppose that $\{ S_k \}_{k \in \mathbb{Z}}$ is an approximation to the identity and $D_k = S_k - S_{k-1}$. Then there exists a sequence of operators $\{ \tilde{D}_k \}$(or $\{ \tilde{\tilde{D}}_k \}$) such that for $f \in \mathfrak{M}(\beta, \gamma)$,

$$f = \sum_k \tilde{D}_k D_k(f) = \sum_k D_k \tilde{\tilde{D}}_k(f) \tag{3.28}$$

where the series converges in the norm of L^p for $1 < p < \infty$ and $\mathfrak{M}(\beta', \gamma')$ for $0 < \beta' < \beta$ and $0 < \gamma' < \gamma$. Moreover, the kernel of $\tilde{D}_k$ satisfy the conditions (i) and (ii) of (3.18), and $\int \tilde{D}_k(x, y)\, d\mu(y) = \int \tilde{D}_k(x, y)\, d\mu(x) = 0$. (the kernel of $\tilde{\tilde{D}}_k$ satisfy the conditions (i) and (iii) of (3.18), and $\int \tilde{\tilde{D}}_k(x, y)\, d\mu(y) = \int \tilde{\tilde{D}}_k(x, y)\, d\mu(x) = 0$.)

By the duality argument we also have

Theorem(3.29)(Calderon type reproducing formula, [DH5], [H1] and [HS2]) Suppose that $\{ S_k \}_{k \in \mathbb{Z}}$ is an approximation to the identity and $D_k = S_k - S_{k-1}$. Then there exists a sequence of operators $\{ \tilde{D}_k \}$(or $\{ \tilde{\tilde{D}}_k \}$) such that for $f \in (\mathfrak{M}(\beta, \gamma))'$,

$$f = \sum_k \tilde{D}_k D_k(f) = \sum_k D_k \tilde{\tilde{D}}_k(f) \tag{3.30}$$

where the series converges in $(\mathfrak{M}(\beta', \gamma'))'$ for $\beta' > \beta$ and $\gamma' > \gamma$. Moreover, the kernel of $\tilde{D}_k$ satisfy the conditions (i) and (ii) of (3.18), and $\int \tilde{D}_k(x, y)\, d\mu(y) = \int \tilde{D}_k(x, y)\, d\mu(x) = 0$. (the kernel of $\tilde{\tilde{D}}_k$ satisfy the conditions (i) and (iii) of (3.18), and $\int \tilde{\tilde{D}}_k(x, y)\, d\mu(y) = \int \tilde{\tilde{D}}_k(x, y)\, d\mu(x) = 0$.)

See [DH5] for more general conditions on the kernel of S_k, the approximation to the identity, and the new continuous version of the Calderon type reproducing formula on spaces of homogeneous type. As in the case of R^n, using these Calderon type reproducing formulas one can develop the theory of the

Besov and Triebel-Lizorkin spaces on spaces of homogeneous type. To be precise, we can introduce the following norms:

Definition(3.31)([HS2]) Suppose that $\{ S_k \}_{k \in \mathbb{Z}}$ is an approximation to the identity and $D_k = S_k - S_{k-1}$. For for $f \in (\mathfrak{M}(\beta, \gamma))'$ with $0 < \beta, \gamma < \epsilon$, we define

(3.32) $$\| f \|_{\dot{B}_p^{\alpha,q}} \approx \{ \sum_{k=0}^{\infty} (2^{\alpha k} \| D_k(f) \|_p)^q \}^{\frac{1}{q}} \quad \text{for } -\epsilon < \alpha < \epsilon \text{ and } 1 \leq p, q \leq \infty,$$

(3.33) $$\| f \|_{\dot{F}_p^{\alpha,q}} \approx \| \{ \sum_{k=0}^{\infty} (2^{\alpha k} | D_k(f) |)^q \}^{\frac{1}{q}} \|_p \quad \text{for } -\epsilon < \alpha < \epsilon \text{ and } 1 \leq p < \infty \text{ and } 1 \leq q \leq \infty.$$

Using the Calderon type reproducing formula in theorem (3.29), we have

Theorem(3.34)([DH5] and [HS2]) The norms defined in definition (3.31) are independent of the choice of the approximations, and β and γ with $\max(0, \alpha) < \beta < \epsilon$ and $\max(0, -\alpha) < \gamma < \epsilon$.

Now we can introduce the Besov and Triebel-Lizorkin spaces on spaces of homogeneous type.

Definition(3.35)([DH5] and [HS2]) Suppose that $\{ S_k \}_{k \in \mathbb{Z}}$ is an approximation to the identity and $D_k = S_k - S_{k-1}$, and $-\epsilon < \alpha < \epsilon$, where ϵ is the exponent of the regularity of the kernels of S_k. The Besov spaces $\dot{B}_p^{\alpha,q}$, for $1 \leq p, q \leq \infty$, are the collection of $f \in (\mathfrak{M}(\beta, \gamma))'$ with $\max(0, \alpha) < \beta < \epsilon$ and $\max(0, -\alpha) < \gamma < \epsilon$, such that

(3.36) $$\| f \|_{\dot{B}_p^{\alpha,q}} = \{ \sum_{k=-\infty}^{\infty} (2^{\alpha k} \| D_k(f) \|_p)^q \}^{\frac{1}{q}} < \infty.$$

The Triebel-Lizorkin spaces $\dot{F}_p^{\alpha,q}$, for $1 \leq p < \infty$, $1 \leq q \leq \infty$, are the collection of $f \in (\mathfrak{M}(\beta, \gamma))'$ with $\max(0, \alpha) < \beta < \epsilon$ and $\max(0, -\alpha) < \gamma < \epsilon$, such that

(3.37) $$\| f \|_{\dot{F}_p^{\alpha,q}} = \| \{ \sum_{k=-\infty}^{\infty} (2^{\alpha k} | D_k(f) |)^q \}^{\frac{1}{q}} \|_p < \infty.$$

See [DH5], [H1] and [HS2] for other characterizations(includes atomic decomposition), the T1 type theorems, embedding theorems, dual spaces and interpolations of these spaces. The above Triebel-Lizorkin spaces on spaces of homogeneous type were generalized to the case where $p_0 < p \leq 1 \leq q < \infty$ by using Littlewood-Paley S function and a new atomic decomposition was also obtained. See [H7] for the details.

If $\{ S_k \}$ is an approximation to the identity and $D_k = S_k - S_{k-1}$ for $k \geq 1$ and $D_0 = S_0$, and $D_k = 0$

for $k < 0$, as the above we can write the identity operator as follows:

$$I = \sum_k D_k = \sum_k D_k \sum_j D_j = \sum_k \sum_j D_k D_j$$

$$= \sum_{|k-j| \leq N} D_k D_j + \sum_{|k-j| > N} D_k D_j = T_N + R_N.$$

Notice that the above equalities still hold even for $\mu(X) < \infty$. It is easy to see that R_N still satisfies theorem (3.19). Formally, we have

$$I = T_N^{-1} T_N = T_N^{-1} \sum_k D_k^N D_k = \sum_k (T_N^{-1} D_k^N) D_k.$$

Since $D_k^N \in \mathfrak{M}(x, 2^{-k}, \epsilon, \epsilon)$ for $k > N$ and hence, by theorem (3.25), for $k > N$ $T_N^{-1} D_k^N$ is a test function in $\mathfrak{M}(x, 2^{-k}, \epsilon, \epsilon)$. Denote $\tilde{D}_k = T_N^{-1} D_k^N$ for $k > N$ and $\tilde{S}_k = T_N^{-1} D_k^N$ for $0 \leq k \leq N$. By a careful estimate $\tilde{S}_k$ satisfy the similar conditions. We obtain the inhomogeneous Calderon type reproducing formula:

$$I = \sum_{0 \leq k \leq N} \tilde{S}_k D_k + \sum_{k > N} \tilde{D}_k D_k$$

More precisely, we first need the following definition.

Definition(3.38)([H2]) Fix $\gamma > 0$, and $0 < \beta \leq \theta$. A function f defined on X is said to be a test function of type (x_0, d, β, γ), $x_0 \in X$ and $d > 0$, if f satisfies the following conditions:

(i) $| f(x) | \leq C \dfrac{d^\gamma}{(d+\rho(x,x_0))^{1+\gamma}}$;

(ii) $| f(x) - f(y) | \leq C \left(\dfrac{\rho(x,y)}{d+\rho(x,x_0)} \right)^\beta \dfrac{d^\gamma}{(d+\rho(x,x_0))^{1+\gamma}}$ for $\rho(x,y) \leq \frac{1}{2A} [d+\rho(x,x_0)]$,

If f is a test function of type (x_0, d, β, γ) we write $f \in \mathfrak{S}(x_0, d, \beta, \gamma)$, and the norm of f in $\mathfrak{S}(x_0, d, \beta, \gamma)$ is defined by

(3.39) $$\| f \|_{\mathfrak{S}(x_0, d, \beta, \gamma)} = \inf \{ C: \text{(i) and (ii) hold} \}.$$

Now fix $x_0 \in X$ and denote $\mathfrak{S}(\beta, \gamma) = \mathfrak{S}(x_0, 1, \beta, \gamma)$. It is easy to see that $\mathfrak{S}(x_1, d, \beta, \gamma) = \mathfrak{S}(\beta, \gamma)$ with the equivalent norms for all $x_1 \in X$ and $d > 0$. Furthermore, it is also easy to check that $\mathfrak{S}(\beta, \gamma)$ is a Banach space with respect to the norm in $\mathfrak{S}(\beta, \gamma)$). Just as the space of distributions $\mathcal{S}'$ is defined on R^n, the dual space $(\mathfrak{G}(\beta, \gamma))'$ consists of all linear functionnals $\mathcal{L}$ from $\mathfrak{G}(\beta, \gamma)$ to $\mathbb{C}$ with the property that there exists a constant c such that for all $f \in \mathfrak{G}(\beta, \gamma)$, $| \mathcal{L}(f) | \leq c \, \| f \|_{\mathfrak{G}(\beta, \gamma)}$. We denote $< h, f >$ the natural pairing of elements $h \in (\mathfrak{G}(\beta, \gamma))'$ and $f \in \mathfrak{G}(\beta, \gamma)$. It is easy to see that for $x_1 \in X$ and $d > 0$ $\mathfrak{G}(x_1, d, \beta, \gamma) = \mathfrak{G}(\beta, \gamma)$ with equivalent norms. Thus, for all $h \in (\mathfrak{G}(\beta, \gamma))'$ and all $f \in \mathfrak{G}(x_1, d, \beta, \gamma)$ with $x_1 \in X$ and $d > 0$ $< h, f >$ is well defined.

Theorem(3.40)(the inhomogeneous Calderon type reproducing formula, [H2]) Suppose that $\{ S_k \}_{k \geq 0}$ is an approximation to the identity and $D_k = S_k - S_{k-1}$ for $k \geq 1$ and $D_0 = S_0$. Then there exists a sequence of operators $\{ \tilde{S}_k \}_{0 \leq k \leq N}$ and $\{ \tilde{D}_k \}_{k > N}$ (or $\{ \tilde{\tilde{S}}_k \}_{0 \leq k \leq N}$ and $\{ \tilde{\tilde{D}}_k \}_{k > N}$) such that for $f \in \mathfrak{G}(\beta, \gamma)$ with $0 < \beta, \gamma < \epsilon$,

$$(3.41) \qquad f = \sum_{0 \leq k \leq N} \tilde{S}_k D_k(f) + \sum_{k > N} \tilde{D}_k D_k(f)$$

$$= \sum_{0 \leq k \leq N} D_k \tilde{\tilde{S}}_k(f) + \sum_{k > N} D_k \tilde{\tilde{D}}_k(f).$$

where the series converges in the norm of L^p for $1 < p < \infty$ and $\mathfrak{G}(\beta', \gamma')$ for $0 < \beta' < \beta$ and $0 < \gamma' < \gamma$. Moreover, the kernels of $\tilde{S}_k$ and $\tilde{D}_k$ satisfy the conditions (i) and (ii) of (3.18) with ϵ replaced by ϵ', $0 < \epsilon' < \epsilon$, and $\int \tilde{S}_k(x, y) \, d\mu(y) = \int \tilde{S}_k(x, y) \, d\mu(x) = 1$ and $\int \tilde{D}_k(x, y) \, d\mu(y) = \int \tilde{D}_k(x, y) \, d\mu(x) = 0$. (the kernels of $\tilde{\tilde{S}}_k$ and $\tilde{\tilde{D}}_k$ satisfy the conditions (i) and (iii) of (3.18) with ϵ replaced by ϵ', $0 < \epsilon' < \epsilon$, and $\int \tilde{\tilde{S}}_k(x, y) \, d\mu(y) = \int \tilde{\tilde{S}}_k(x, y) \, d\mu(x) = 1$ and $\int \tilde{\tilde{D}}_k(x, y) \, d\mu(y) = \int \tilde{\tilde{D}}_k(x, y) \, d\mu(x) = 0$.)

By the duality argument we also have

Theorem(3.42)(The inhomogeneous Calderon type reproducing formula, [H2]) Suppose that $\{ S_k \}_{k \geq 0}$ is an approximation to the identity and $D_k = S_k - S_{k-1}$ for $k \geq 1$ and $D_0 = S_0$. Then there exists a sequence of operators $\{ \tilde{S}_k \}_{0 \leq k \leq N}$ and $\{ \tilde{D}_k \}_{k > N}$ (or $\{ \tilde{\tilde{S}}_k \}_{0 \leq k \leq N}$ and $\{ \tilde{\tilde{D}}_k \}_{k > N}$) such that for $f \in (\mathfrak{G}(\beta, \gamma))'$ with $0 < \beta, \gamma < \epsilon$,

$$(3.43) \qquad f = \sum_{0 \leq k \leq N} \tilde{S}_k D_k(f) + \sum_{k > N} \tilde{D}_k D_k(f)$$

$$= \sum_{0 \leq k \leq N} D_k \tilde{\tilde{S}}_k(f) + \sum_{k > N} D_k \tilde{\tilde{D}}_k(f).$$

where the series converges in $\mathfrak{G}(\beta', \gamma')$ for $\beta' > \beta$ and $\gamma' > \gamma$. Moreover, the kernels of $\tilde{S}_k$ and $\tilde{D}_k$ satisfy the conditions (i) and (ii) of (3.18) with ϵ replaced by ϵ', $0 < \epsilon' < \epsilon$, and $\int \tilde{S}_k(x, y) \, d\mu(y) =$

$\int \tilde{S}_k(x, y)\, d\mu(x) = 1$ and $\int \tilde{D}_k(x, y)\, d\mu(y) = \int \tilde{D}_k(x, y)\, d\mu(x) = 0$. (the kernels of $\tilde{\tilde{S}}_k$ and $\tilde{\tilde{D}}_k$ satisfy the conditions (i) and (iii) of (3.18) with ϵ replaced by ϵ', $0 < \epsilon' < \epsilon$, and $\int \tilde{\tilde{S}}_k(x, y)\, d\mu(y) = \int \tilde{\tilde{S}}_k(x, y)\, d\mu(x) = 1$ and $\int \tilde{\tilde{D}}_k(x, y)\, d\mu(y) = \int \tilde{\tilde{D}}_k(x, y)\, d\mu(x) = 0$.)

Using these inhomogeneous Calderon type reproducing formulas one can develop the theory of the inhomogeneous Besov and Triebel-Lizorkin spaces on spaces of homogeneous type with finite or infinite measures. See [H8] for the details.

To deal with the case where p and q are less than 1, one needs the discrete Calderon type reproducing formula. This discretization step on R^n was done by using an argument similar to the Shannon sampling theorem, namely the Fourier transform(see [FJW]). To obtain the discrete Calderon type reproducing formula on spaces of homogeneous type, we introduce the Riemann sum operator and prove that the Riemann sum operator is a Calderon-Zygmund operator. The discretization step of the continuous version of the Calderon type reproducing formula then follows from the boundedness of the inverse Riemann sum operator. All these results are new even for R^n. To be precise, we say that a cube $Q \subseteq R^n$ is a dyadic cube if $Q = Q_{k,\ell} = \{x \in R^n: 2^{-k}\ell_i \le x_i < 2^{-k}(\ell_i+1)\}$, $i = 1, 2, \ldots, n$, and $k \in \mathbb{Z}$, and $\ell = (\ell_1, \ell_2, \ldots, \ell_n) \in \mathbb{Z}^n$. Denote by $Q^{\nu}_{k,\ell}$, $\nu = 1, 2, \ldots, 2^{jn}$ for $j \in \mathbb{Z}^+$, the all cubes $Q_{k+j,\ell} \subseteq Q_{k,\ell}$. Choose $y^{\nu}_{k,\ell} \in Q^{\nu}_{k,\ell}$. The Riemann sum operator then is defined by

(3.44) $$S(f)(x) = \sum_{k \in \mathbb{Z}} \sum_{\ell \in \mathbb{Z}^n} \sum_{\nu=1}^{2^{jn}} 2^{-jn}\, 2^{-kn}\, \tilde{D}_k(x, y^{\nu}_{k,\ell})\, D_k(f)(y^{\nu}_{k,\ell})$$

where D_k and $\tilde{D}_k$ are operators given in the above Calderon type reproducing formula.

To see that operator S is well defined, we have

Theorem(3.45)([H3]) The Riemann sum operator S is bounded on $L^2(R^n)$. Moreover, there exists a constant c such that

(3.46) $$\| S(f) \|_2 \le c \, \| f \|_2.$$

Theorem(3.47)([H3]) If j is a sufficiently large integer, then S^{-1}, the inverse of the Riemann sum operator of S, maps test function to test function. Moreover, there exists a constant c which is independent of f, such that

(3.48) $$\| S^{-1}(f) \|_{\mathfrak{M}(x_1, d, \beta, \gamma)} \le c \, \| f \|_{\mathfrak{M}(x_1, d, \beta, \gamma)}$$

Theorem(3.47) provides the following discrete Calderon type reproducing formula.

Theorem(3.49)(The discrete Calderon type reproducing formula, [H3]) Suppose that $\{ S_k \}_{k \in \mathbb{Z}}$ is an approximation to the identity and $D_k = S_k - S_{k-1}$. Then there exists a sequence of operators $\{ \tilde{\tilde{D}}_k \}$ such that for $f \in \mathfrak{M}(\beta, \gamma)$ with $0 < \beta, \gamma < \epsilon$,

$$(3.50) \qquad f(x) = \sum_{k \in \mathbb{Z}} \sum_{\ell \in \mathbb{Z}^n} \sum_{\nu=1}^{2^{jn}} 2^{-jn}\, 2^{-kn}\, \tilde{\tilde{D}}_k(x, y_{k,\ell}^{\nu})\, D_k(f)(y_{k,\ell}^{\nu})$$

where the series converges in the norm of L^p for $1 < p < \infty$ and $\mathfrak{M}(\beta', \gamma')$ for $0 < \beta' < \beta$ and $0 < \gamma' < \gamma$. Moreover, the kernels of $\tilde{\tilde{D}}_k$ satisfy the conditions (i) and (ii) of (3.18) with ϵ replaced by ϵ', $0 < \epsilon' < \epsilon$, and $\int \tilde{\tilde{D}}_k(x, y)\, d\mu(y) = \int \tilde{\tilde{D}}_k(x, y)\, d\mu(x) = 0$.

By the duality argument we also have

Theorem(3.51)(The discrete Calderon type reproducing formula, [H3]) Suppose that $\{ S_k \}_{k \in \mathbb{Z}}$ is an approximation to the identity and $D_k = S_k - S_{k-1}$. Then there exists a sequence of operators $\{ \tilde{\tilde{D}}_k \}$ such that for $f \in (\mathfrak{M}(\beta, \gamma))'$ with $0 < \beta, \gamma < \epsilon$,

$$(3.52) \qquad f = \sum_{k \in \mathbb{Z}} \sum_{\ell \in \mathbb{Z}^n} \sum_{\nu=1}^{2^{jn}} 2^{-jn}\, 2^{-kn}\, \tilde{\tilde{D}}_k(x, y_{k,\ell}^{\nu})\, D_k(f)(y_{k,\ell}^{\nu})$$

where the series converges in $(\mathfrak{M}(\beta', \gamma'))'$ for $\beta' > \beta$ and $\gamma' > \gamma$. Moreover, the kernels of $\tilde{\tilde{D}}_k$ satisfy the conditions (i) and (ii) of (3.18) with ϵ replaced by ϵ', $0 < \epsilon' < \epsilon$, and $\int \tilde{\tilde{D}}_k(x, y)\, d\mu(y) = \int \tilde{\tilde{D}}_k(x, y)\, d\mu(x) = 0$.

Similarly we have the following discrete inhomogeneous Calderon type reproducing formula.

Theorem(3.53)(The discrete inhomogeneous Calderon type reproducing formula, [H5]) Suppose that $\{ S_k \}_{k \geq 0}$ is an approximation to the identity and $D_k = S_k - S_{k-1}$ for $k \geq 1$ and $D_0 = S_0$. Then there exists a sequence of operators $\{ \tilde{\tilde{S}}_k \}_{0 \leq k \leq N}$ and $\{ \tilde{\tilde{D}}_k \}_{k > N}$ such that for $f \in \mathfrak{S}(\beta, \gamma)$ with $0 < \beta, \gamma < \epsilon$,

$$\begin{aligned}(3.54) \qquad f = & \sum_{0 \leq k \leq N} \sum_{\ell \in \mathbb{Z}^n} \sum_{\nu=1}^{2^{jn}} \tilde{\tilde{S}}_k(x, y_{k,\ell}^{\nu}) \int_{Q_{k,\ell}^{\nu}} D_k(f)(y)\, dy \, + \\ & \sum_{0 \leq k \leq N} \sum_{\ell \in \mathbb{Z}^n} \sum_{\nu=1}^{2^{jn}} D_k(f)(y_{k,\ell}^{\nu}) \int_{Q_{k,\ell}^{\nu}} \tilde{\tilde{S}}_k(f)(y)\, dy \, - \\ & \sum_{0 \leq k \leq N} \sum_{\ell \in \mathbb{Z}^n} \sum_{\nu=1}^{2^{jn}} 2^{-jn}\, 2^{-kn}\, \tilde{\tilde{S}}_k(x, y_{k,\ell}^{\nu})\, D_k(f)(y_{k,\ell}^{\nu}) \, + \end{aligned}$$

$$\sum_{k>N} \sum_{\ell \in \mathbb{Z}^n} \sum_{\nu=1}^{2^{jn}} 2^{-jn} 2^{-kn} \tilde{\tilde{D}}_k(x, y^\nu_{k,\ell}) D_k(f)(y^\nu_{k,\ell})$$

where the series converges in the norm of L^p for $1 < p < \infty$ and $\mathfrak{S}(\beta', \gamma')$ for $0 < \beta' < \beta$ and $0 < \gamma' < \gamma$. Moreover, the kernels of $\tilde{\tilde{S}}_k$ and $\tilde{\tilde{D}}_k$ satisfy the conditions (i) and (ii) of (3.18) with ϵ replaced by ϵ', $0 < \epsilon' < \epsilon$, and $\int \tilde{\tilde{S}}_k(x, y)\, d\mu(y) = \int \tilde{\tilde{S}}_k(x, y)\, d\mu(x) = 1$ and $\int \tilde{\tilde{D}}_k(x, y)\, d\mu(y) = \int \tilde{\tilde{D}}_k(x, y)\, d\mu(x) = 0$.

By the duality argument we also have

Theorem(3.55)(The discrete inhomogeneous Calderon type reproducing formula, [H5]) Suppose that $\{ S_k \}_{k \geq 0}$ is an approximation to the identity and $D_k = S_k - S_{k-1}$ for $k \geq 1$ and $D_0 = S_0$. Then there exists a sequence of operators $\{ \tilde{\tilde{S}}_k \}_{0 \leq k \leq N}$ and $\{ \tilde{\tilde{D}}_k \}_{k > N}$) such that for $f \in (\mathfrak{S}(\beta, \gamma))'$ with $0 < \beta, \gamma < \epsilon$,

$$f = \sum_{0 \leq k \leq N} \sum_{\ell \in \mathbb{Z}^n} \sum_{\nu=1}^{2^{jn}} \tilde{\tilde{S}}_k(x, y^\nu_{k,\ell}) \int_{Q^\nu_{k,\ell}} D_k(f)(y)\, dy + \tag{3.56}$$
$$\sum_{0 \leq k \leq N} \sum_{\ell \in \mathbb{Z}^n} \sum_{\nu=1}^{2^{jn}} D_k(f)(y^\nu_{k,\ell}) \int_{Q^\nu_{k,\ell}} \tilde{\tilde{S}}_k(f)(y)\, dy -$$
$$\sum_{0 \leq k \leq N} \sum_{\ell \in \mathbb{Z}^n} \sum_{\nu=1}^{2^{jn}} 2^{-jn} 2^{-kn} \tilde{\tilde{S}}_k(x, y^\nu_{k,\ell}) D_k(f)(y^\nu_{k,\ell}) +$$
$$\sum_{k>N} \sum_{\ell \in \mathbb{Z}^n} \sum_{\nu=1}^{2^{jn}} 2^{-jn} 2^{-kn} \tilde{\tilde{D}}_k(x, y^\nu_{k,\ell}) D_k(f)(y^\nu_{k,\ell})$$

where the series converges in $(\mathfrak{S}(\beta', \gamma'))'$ for $\beta' > \beta$ and $\gamma' > \gamma$. Moreover, the kernels of $\tilde{\tilde{S}}_k$ and $\tilde{\tilde{D}}_k$ satisfy the conditions (i) and (ii) of (3.18) with ϵ replaced by ϵ', $0 < \epsilon' < \epsilon$, and $\int \tilde{\tilde{S}}_k(x, y)\, d\mu(y) = \int \tilde{\tilde{S}}_k(x, y)\, d\mu(x) = 1$ and $\int \tilde{\tilde{D}}_k(x, y)\, d\mu(y) = \int \tilde{\tilde{D}}_k(x, y)\, d\mu(x) = 0$.

All these result still hold on spaces of homogeneous type. See [H3] and [H5] for the details.

As an application of the discrete Calderon type reproducing formulas one can develop the theory of the Besov $\dot{B}^{\alpha,q}_p$ and Triebel-Lizorkin spaces $\dot{F}^{\alpha,q}_p$ on spaces of homogeneous type for the case where p and q are less than 1. See [H4] and [H8] for the details.

Bibliography

[C1] A. P. Calderon, Intermediate spaces and interpolation, the complex method, Studia

Math. 24 (1964), 113-190.

[C2] A. P. Calderon, An atomic decomposition of distributions in parabolic H^p spaces, Adv. in Math. 25 (1977), 216-225.

[CF] A. S.-Y. Chang and R. Fefferman, The Calderon-Zygmund decomposition on product domains, Amer. J. Math. 104 (1982), 445-458.

[Ch] M. Christ, Singular integral operator, NSF-CBMS Regional Conf. at Missoula, Montana, Aug. 1989.

[CW] R. R. Coifman and G. Weiss, Analyse harmonique noncommutative sur certains espaces homogenes, Lecture Notes in Math. Vol. 242, Springer-Verlager Berlin, Heidelberg, New York, 1971.

[DH1] D. G. Deng and Y.- S. Han, The characterizations of the Besov and Triebel-Lizorkin spaces and ϵ family of operators,

[DH2] D. G. Deng and Y.- S. Han, The Besov and Triebel-Lizorkin spaces on Lipschitz curves (I), Acta Math. Sinica, 35 (1992), 608-619.

[DH3] D. G. Deng and Y.- S. Han, The Besov and Triebel-Lizorkin spaces on Lipschitz curves (II), Acta Math. Sinica, 36 (1993), 122-135.

[DH4] D. G. Deng and Y.- S. Han, The Besov and Triebel-Lizorkin spaces with high order on Lipschitz curves, Approx. Theory and its Appl. 9 (1993), 89-106.

[DH5] D. G. Deng and Y.- S. Han, The Calderon reproducing formula on spaces of homogeneous type, to appear.

[DJ] G. David and J.-L. Journe, A boundedness criterion for generalized Calderon-Zygmund operators, Ann. of Math. 120 (1984), 371-397.

[DJS] G. David, J.-L. Journe and S. Semmes, Operateurs de Calderon-Zygmund, fonctions para-accretives et interpolation, Revista Mat. Iberoamericana, 4 (1985), 1-56.

[FJW] M. Frazier, Y.-S. Han, B. Jawerth and G.Weiss, The T1 theorem for Triebel-Lizorkin spaces, Harmonic Analysis and Partial Differential Equations, Lecture Notes in Math. 1384, 1989, 168-181.

[FJ1] M. Frazier and B. Jawerth, Decomposition of Besov spaces, Indiana Univ. Math. J., 34 (1985), 777-799.

[FJ2] M. Frazier and B. Jawerth, A discrete transform and decompositions of distribution spaces, J. Fun. Anal. 93 (1990), 34-170.

[FJW] M. Frazier, B. Jawerth and G.Weiss, Littlewood-Paley theory and the study of function spaces, CBMS-AMS Regional Conf., at Auburn Univ., 1989.

[GM] A. Grossman and J. Morlet, Decomposition of Hardy functions into square integrable wavelets of constant shape, SIAM. J. Math. Anal. 15 (1984), 723-736.

[H1] Y.-S. Han, Calderon-type reproducing formula and the Tb theorem, Revista Mat. Iberoamericana, 10 (1994), 59-99.

[H2] Y.-S. Han, Inhomogeneous Calderon reproducing formula on spaces of homogeneous type, to appear.

[H3] Y.-S. Han, Discrete Calderon reproducing formula on spaces of homogeneous type, to appear.

[H4] Y.-S. Han, Besov and Triebel-Lizorkin spaces on spaces of homogeneous type, to appear.

[H5] Y.-S. Han, Discrete inhomogeneous Calderon reproducing formula on spaces of homogeneous type, to appear.

[H6] Y.-S. Han, The embedding theorem for Besov and Triebel-Lizorkin spaces on spaces of homogeneous type, to appear in Proc. of AMS.

[H7] Y.-S. Han, Triebel-Lizorkin spaces on spaces of homogeneous type, Studia Math., 108 (1994), 247-273.

[H8] Y.-S. Han, Inhomogeneous Besov and Triebel-Lizorkin spaces on spaces of homogeneous type, to appear.

[HJTW] Y.-S. Han, B. Jawerth, M. Taibleson and G. Weiss, Littlewood-Paley thory and ϵ-family of operators, Colloq. Math., 50/51 (1990), 1-39.

[HS1] Y.-S. Han and E. Sawyer, Para-accretive functions, the weak boundedness property and the Tb theorem, Revista Mat. Iberoamericana, 6 (1990), 17-41.

[HS2] Y.-S. Han and E. Sawyer, Littlewood-Paley theory on spaces of homogeneous type and classical function spaces, to appear in Memoirs of AMS.

[L] P. Lemarie, Continuite sur les espaces de Besov des operateurs definis par des integrales singulieres, Ann. Inst. Fourier, 35 (1985), 175-187.

[Lo1] R. L. Long, The spaces generated by block, Sinica, 7 (1983), 594-603.

[Lo2] R. L. Long, The theory of H^p on martingales, Peking Univ. Press, 1985.

[Lu] S. Z. Lu, On block decomposition of functions, Sinica, 12 (1983), 1089-1098.

[LY] R. L. Long and L. Yang, The BMO functions on spaces of homogeneous type, Sinica, 4 (1984), 301-312.

[M] Y. Meyer, Ondelettes et operateurs, I, II, III, Hermann ed., Paris 1990.

[MS] R. A. Macias and C. Segovia, Lipschitz functions on spaces of homogeneous type, Adv. in Math., 33 (1979), 257-270.

[P] J. Peetre, New thoughts on Besov spaces, Duke Univ. Math. Series, Durham NC., 1976.

[R1] R. Rochberg, Toeplitz and Hankel operators, Wavelets, NWO Sequences, and

Almost Diagonalization of Operators, Proc. Symp. Pure Math., 51 (1990), 425-444.

[R2] R. Rochberg, Size estimates for eigenvectors of singular integral operators andd Schrodinger operators and for derivatives of quasiconformal mappings, to appear.

[T] H. Triebel, Theory of function spaces, Monographs in Maht., 78, Birkhauser, Verlag, Basel, 1983.

[U] A. Uchiyma, A constructive proof of the Fefferman-Stein decomposition for $BMO(R^n)$, Acta Math., 148 (1982), 215-241.

H^p THEORY ON COMPACT LIE GROUPS

DASHAN FAN

Auhui University

1. INTRODUCTION

The purpose of this article is to summarize some recent developments of H^p spaces and their related theorems on a compact Lie group. I would like to thank Professor Sheng Gong for encouraging me to write this paper. Many results in this paper originally come from my Ph. D. thesis in Washington University, I would like to thank my thesis advisor, Professor Brian Blank for introducing me to this interesting topic. I also owe a great deal to Dr. Z. Xu, who provided me with a lot of materials for writing this paper. Since the article is going to present the theorems rather give the proofs, only part of the theorems will be sketched their proofs. Interested readers may see the listed references for further details and proofs. In some sections of the article, I will post some open questions to be solved. Atomic decomposition of Hardy spaces of real functions on Euclidean spaces first arose in the work of R. Coifman [*Coi*] and R. Laufm ter [*L*]. An abstract theory of atomic Hardy spaces was later developed by R. Coifman and G. Weiss [*CW*] in the context of spaces of homogeneous type. These spaces include Euclidean spaces and compact Lie groups but do not in general have the structure on which to base a theory of Hardy space defined by maximal functions. It was noted by Coifman and Weiss in [*CW*] and by Uchiyama in [*U*] that when a space of homogeneous type admits a certain family of kernels, a maximal function based Hardy space can be defined and shown equivalent to atomic Hardy space. Although the kernels in question are well-suited to an argument of L. Carleson [*Ca*], they are not necessarily intrinsic to any additional geometry (such as Riemannian structure) that a space of homogeneous may posses. For example, compact Riemannian manifolds have Laplace-Beltrami operators which give rise via Poisson kernels to maximal function based Hardy spaces. In such cases it is of interest to obtain the atomic decomposition of Hardy spaces defined by maximal functions as were done for spheres by L. Colzani [*Co*] and for the homogeneous groups by Folland and Stein [*FoS*]. Moreover, the atomic Hardy spaces in [*CW*] and [*U*] are, of necessity because of the more general structure, defined by duality. Where polynomials are available, such as is the case with compact Lie groups, it is desirable to have a direct definition of atoms in analogy with those in [*Coi*] and [*L*].

In this paper, we will list the theorems which establish the equivalence of the Hardy spaces on compact semisimple Lie groups that arise from their homogeneous

M. Cheng et al. (eds.), Harmonic Analysis in China, 80–102.

type structure with those that arise from their Riemannian structure. After this equivalence is established, we will consider further theorems on H^p spaces, such as Calderón-Zygmund operators theory, Hölmander multiplier theorem, Lipschitz spaces theory, and so on.

To establish the above mentioned theories, we must point out that because two main pillars of H^p theory on $\mathbf{R}^n$, dilation and Fourier transform, are largely unavailable in compact Lie groups, some basic ideas on $\mathbf{R}^n$ are adaptable but are technically more difficult to execute; other ideas do not transfer at all and must be replaced with new ones. Another difficulty arises in the handling of polynomials which are necessarily more cumbersome than in the Euclidean seufm ting such as $\mathbf{R}^n$ and the unit sphere $\Sigma(n)$.

2. Maximal Functions

Let G be a connected, simply connected, compact semisimple Lie group of dimension n. Let $\mathfrak{g}$ be the Lie algebra of G and $\mathfrak{t}$ the Lie algebra of a fixed maximal torus $\mathbf{T}$ in G of dimension ℓ . Let A be a system of positive roots for $(\mathfrak{g}, \mathfrak{t})$, so that $\mathrm{Card}(A) = \frac{n-\ell}{2}$ and let $\beta = \frac{1}{2}\sum_{\alpha\in A}\alpha$. Let $|\cdot|$ be the norm on $\mathfrak{g}$ induced by the negative of the Killing form B on $\mathfrak{g}^{\mathbf{C}}$, the complexification of $\mathfrak{g}$, then $|\cdot|$ induces a bi-invariant metric d on G. Furthermore, since $B|_{\mathfrak{t}^{\mathbf{C}}\times\mathfrak{t}^{\mathbf{C}}}$ is nondegenerate, given $\lambda \in \hom_{\mathbf{C}}(\mathfrak{t}^{\mathbf{C}}, \mathbf{C})$ there is a unique H_λ in $\mathfrak{t}^{\mathbf{C}}$ such that $\lambda(H) = B(H, H_\lambda)$ for each $H \in \mathfrak{t}^{\mathbf{C}}$. We let $< \cdot, \cdot >$ and $\|\ \|$ denote the inner product and norm transferred from $\mathfrak{t}$ to $\mathfrak{t}^*$(the dual of $\mathfrak{t}$) by means of this canonical isomorphism and let ξ be this natural map of $\mathfrak{t}$ onto $\mathfrak{t}^*$. Let $\mathbf{N} = \{H \in \mathfrak{t}, \exp H = e\}$, e being the identity in G . The weight laufm tice P is defined by $P = \{\lambda \in \mathfrak{t}, < \lambda, x > \in 2\pi\mathbf{Z}\ , \text{any } x \in \mathbf{N}\}$ with dominant weights defined by $\Lambda = \{\lambda \in P, < \lambda, \alpha > \geq 0 \text{ any } \alpha \in A\}$. We can identify $\widehat{G}$ with this Λ because Λ provides a full set of parameters for the equivalence classes of unitary irreducible representations of G. For $\lambda \in \Lambda$, the representation U_λ has dimension

$$d_\lambda = \prod_{\alpha\in A} \frac{< \lambda + \beta, \alpha >}{< \beta, \alpha >} \tag{2.1}$$

and its associated character is

$$\chi_\lambda(x) = \frac{\sum_{w\in W} \epsilon(w) e^{i<w(\lambda+\beta), x)}}{\sum_{w\in W} \epsilon(w) e^{i<w\beta, x>}} \tag{2.2}$$

where $x \in \mathfrak{t}$, W is the Weyl group and $\epsilon(w)$ is the signature of w $\in$W. Let $X_1, X_2, \cdots, X_n$ be an orthonormal basis of $\mathfrak{g}$. Form the Casimir operator

$$\Delta = \sum_{i=1}^{n} X_i^2. \tag{2.3}$$

This is an elliptic bi-invariant operator on G which is independent of the choice of the orthonormal basis of $\mathfrak{g}$. The solution of the heat equation on $G \times \mathbf{R}^+ = G \times (0, \infty)$

$$\Delta\Phi(x,t) = \frac{d}{dt}\Phi(x,t), \Phi(x,0) = f(x)$$

for $f \in L^1(G)$ is given by

$$\Phi(x,t) = (W_t*f)(x),$$

where

$$W_t(x) = \sum_{\lambda\in\Lambda} e^{-t(\|\lambda+\beta\|^2-\|\beta\|^2)} d_\lambda \chi_\lambda(x) \tag{2.4}$$

is the Gauss-Weierstrass kernel (heat kernel).

Another well-known kernel on G is the Poisson kernel

$$P_t(x) = \sum_{\lambda\in\Lambda} e^{t(\|\lambda+\beta\|^2-\|\beta\|^2)^{1/2}} d_\lambda \chi_\lambda(x) \tag{2.5}$$

which is defined by E.M. Stein $[S1]$.

We also need the Bochner-Riesz kernel:

$$S_t^\delta(x) = \sum_{\lambda\in\Lambda} \{1 - t^{-1} \parallel \lambda+\beta \parallel^2\}_+^\delta d_\lambda\chi_\lambda(x), (\delta \geq 0). \tag{2.6}$$

All these kernels are central functions on G and are determined by their restrictions to $\mathbf{T}$ as given in (2.4), (2.5) and (2.6). The relationships among these three kernels are formulated by the following(see $[S1]$, $[BF1]$):

$$P_t(x) = \pi^{-1/2} \int_0^\infty u^{-1/2} e^{-u} W_{t^2/4u}(x)dx \tag{2.7}$$

$$W_t(x) = Ce^{t\|\beta\|^2} \int_0^\infty s^\beta e^{-s} S_{t/s}^\delta(x)ds, \tag{2.8}$$

where C is a constant independent of t, (see $[BF1]$).

For any distribution f in $\mathcal{S}'(G)$(see $[BF]$ for the detail of the distribution on G, P_t*f is a measurable function on $G^+ = G \times (0,\infty)$. With the interpretation of G as the boundary of G^+, we have maximal functions associated with three different boundary approaches. The radial maximal function P^+f, the nontangential maximal function P_γ^*f, and the tangential maximal function $P_M^{**}f$ are defined for x in G by

$$P^+f(x) = \sup_{t>0} |P_t*f(x)| \tag{2.9}$$

$$P_\gamma^*f(x) = \sup_{(y,t)\in\Gamma_\gamma(x)} |P_t*f(y)| \tag{2.10}$$

and

$$P_M^{**}f(x) = \sup_{(y,t)\in G+} |P_t*f(y)| \{\frac{t}{d(y,x)+t}\}^M \tag{2.11}$$

where

$$\Gamma(x) = \{(y,t) : d(y,x) < \gamma t\}.$$

When $\gamma = 1$ we write P^*f for P_1^*f.

We also define local maximal functions for $\epsilon_0 > 0$ as follows:

$$P^+_{\epsilon_0} f(x) = \sup_{o \le t \le \epsilon_0} |P_t * f(x)|$$

$$P^*_{\gamma,\epsilon_0} f(x) = \sup_{(y.t)\in\Gamma_\gamma(x), 0\le t\le\epsilon_0} |P_t * f(y)|, \quad P^*_{\epsilon_0} f = P^*_{1,\epsilon_0} f$$

and

$$P^{**}_{M,\epsilon_0} f(x) = \sup\{|P_t * f(y)|(\frac{t}{d(x,y)+t})^M : (y,t) \in G \times (0, \epsilon_0)\}$$

Similarly, we can use the heat kernel to define the associated maximal functions $W^+ f, W^* f$ and $W^{**}_M f$ by different boundary approaches.

As an analogy of the classical definition on $\mathbf{R}^n$, the Hardy space $H^p(G)$ on a compact Lie group G is the collection of all distribution $f \in S'(G)$ for which $P^* f \in L^p(G)$. The H^p "norm" of f is defined as

$$\| f \|_{H^p} = \| P^* f \|_p .$$

Although not a norm in general, $\| \cdot \|_{H^p(G)}$ provides a complete metrizable topology in $H^p(G)$. The spaces $H^p(G)(1 < p < \infty)$ are known to be Lebesgue spaces ($[B], [Sr]$) and so we assume that $0 < p \le 1$ throughout this paper. To prove the above definition is equivalent to the definitions defined by other maximal function, we need to define another maximal function associated with distributions on G which is more flexible in some ways than those discussed above. For $N \in \mathbf{N}^+$ and $x \in G$ we define the subclass $K_N(x)$ of $\mathcal{S}(G)$ as all those $\phi \in \mathcal{S}(G)$ satisfying i) supp $\phi \subseteq B(x,h)$, where $B(x,h)$ is the ball centered at x with radius h; ii) $\sup_{t,x} |\frac{d^N}{dt^N}(P_t * \phi)(x)| \le h^{-N-n}$; and iii) $\| \phi \|_\infty \le h^{-n}$ for some $h > 0$.

For distributions f in $\mathcal{S}'(G)$ and $\phi \in K_N(x)$ we use the pairing $< f, \phi >= \int_G f(g)\overline{\phi(g)}dg$. The grand maximal function f^* is defined by

$$f^*(x) = \sup\{| < f, \phi > | : \ \phi \in K_N(x)\}(f \in \mathcal{S}'(G)) \tag{2.12}$$

in analogy with the Fefferman-Stein grand maximal function for $\mathbf{R}^n([FS])$. Of course, the dependence on N is concealed by the notation.

For each $x \in G$ we choose τ small enough so that $\exp^{-1} \circ L_{x^{-1}} : \ B(x,\tau) \to \mathfrak{g}($ where L denotes left multiplication) gives a coordinate chart $(y_1, \cdots, y_n)$, where $x^{-1}y = \exp(y_1 X_1 + \cdots + y_n X_n)$ and $d(y,x) \cong y_1^2 + \cdots + y_n^2$. Since G is compact, τ may be uniformly chosen. We do not fix such a τ at this time but it is to be understood that any further reference to a parameter labeled τ entails that it is small enough for this condition to be satisfied.

Theorem 1 ($[BF1]$) Let f be a distribution in $\mathcal{S}'(G)$. Suppose that $M > n/p$ and $N > n/p$ with N even. The following are equivalent:

i) $f \in H^p(G)$; ii) $P^+ f \in L^p(G)$; iii) $P^+_\epsilon f \in L^p(G)$ for some $\epsilon > 0$;
iv) $P^{**}_{M,\epsilon} f \in L^p(G)$ $(0 < \epsilon \le \tau)$; v) $P^*_\epsilon f \in L^p(G)$ $(0 < \epsilon \le \tau)$; vi) $f^* \in L^p(G)$

Moreover, we have, for $0 < \epsilon \leq \tau$:

$$\| f \|_{H^p(G)} \cong \| P^+ f \|_{L^p(G)} \cong \| P_\epsilon^* f \|_{L^p(G)} \cong \| P_{M,\epsilon}^{**} f \|_{L^p(G)} \cong \| f^* \|_{L^p(G)}$$

As in the Euclidean space, it is also useful to have a heat kernel characterization of Hardy space. We define a modified heat kernel $\widetilde{W_t}$ by $\widetilde{W_t} = e^{-t\|\beta\|^2} W_t$. The associated maximal functions is $\widetilde{W}^+ f(x) = \sup_{t>0} |\widetilde{W_t} * f(x)|$. We define a nontangential maximal function by $\widetilde{W}^* f(x) = \sup_{d(x,y)^2<t} |\widetilde{W_t} * f(y)|$.

Theorem 2 ([$BF1$]) The following are equivalent:

$$\text{i) } f \in H^p; \quad \text{ii) } W_1^* f \in L^p(G); \quad \text{iii) } \widetilde{W}_1 f \in L^p(G);$$
$$\text{iv) } \widetilde{W}^* f \in L^p(G); \ \text{v) } \widetilde{W}^+ f \in L^p(G); \quad \text{vi) } W^+ f \in L^p(G)$$

Besides using Poisson kernel and heat kernel to characterize the Hardy spaces, one also can use other kernels to do so. Let $\phi \in \mathcal{S}(\mathfrak{g})$ be a radial function satisfying $\int_{\mathfrak{t}} \phi(H) dH = 1$, we define a central function

$$\phi_t(x) = \sum_{\lambda \in \Lambda} \phi^\wedge(t \| \lambda + \beta \|) d_\lambda \chi_\lambda(x) (t > 0). \tag{2.13}$$

Theorem 3 ([$BF1$]) If $0 < p < \infty$, then for all $f \in \mathcal{S}'(G)$ the following are equivalent:

i) $\sup_{t>0} |\phi_t * f| \in L^p(G)$ for some ϕ satisfying (2.13),

ii) $\sup_{d(x,y)<t} |\phi_t * f(y)| \in L^p(G)$ for some ϕ satisfying (2.13),

iii) $f^\#(x) = \sup_{\phi \in A_0} \sup_{d(x,y)<t} |\phi_t * f(y)| \in L^p(G)$, where

$$A_0 = \{\phi \in \mathcal{S}(\mathbf{R}^\ell) : \int_{\mathfrak{t}} (1 + |\theta|)^{N_0} \sum_{|\alpha| \leq N_0} |\frac{\partial^\alpha}{\partial \theta^\alpha} \phi(\theta)|^2 d\theta \leq 1\}$$

for some large number N_0 depending on p and n.

Heretofore our kernels have been infinitely differentiable. The following theorem shows that such smoothness is not a necessary condition for the characterization of $H^p(G)$.

Theorem 4 ([$BF1$]) For the maximal Bochner-Riesz operator, if $\delta > n/p - (n+1)/2$, then

$$S_*^\delta f(x) = \sup_{t>0} |S_t^\delta * f(x)|, \quad \| S_*^\delta f \|_{L^p(G)} \cong \| f \|_{H^p(G)}$$

We can further relax the regularity assumption, admiufm ting even non-smooth kernels, by generalizing a Euclidean result of Han [H] concerning the characterization of Hardy space by kernels satisfying a Dini condition. Readers can see [BF] for further details.

Note. The theorems on $\mathbf{R}^n$ for which the theorems in this section are in analogy with can be found in [BGS], [FS] and [T].

3. Atomic Decompositions of Hardy Spaces

As we mentioned in the introduction, another important space when $0 < p \leq 1$ is the atomic H^p space which was discovered by Coifman and Laufm ter on $\mathbf{R}^n$, and later introduced by Coifman and Weiss on more general spaces of homogeneous type [CW]. Clerc modified their definitions to compact Lie groups on which polynomials play an important role to define the "cancellation" conditions. The following definitions of the atomic H^p can be found in Clerc's paper [C].

An exceptional $(1, \infty)$ atom a(x) is an L^∞ function bounded by 1. In order to define a regular atom, we consider a faithful unitary representation Π of G. Then G can be identified as a submanifold in a real vector space E underlying End($\mathbf{C}^L$). A regular (p, q) atom for $0 < p \leq 1 \leq q \leq \infty$ is a function a(x) supported in some ball $B(y, \rho)$ such that

$$\| a \|_q \leq \rho^{-n(1/p-1/q)}, \tag{3.1}$$

$$\int_G a(x) P(\Pi(x)) dx = 0, \tag{3.2}$$

where P is any polynomial on E of degree less than or equal to $[n(1/p - 1)]$.

The atomic Hardy space $H_a^{p,q}, 0 < p \leq 1$, is the space of all $f \in \mathcal{S}'(G)$ having the form

$$f = \sum c_k a_k, \quad \text{with } \sum |c_k|^p < \infty \tag{3.3}$$

where each $a_k(x)$ is either a regular (p, q) atom or an exceptional atom. The "norm" $\| f \|_{H_a^{p,q}}$ is the infimum of all expressions $(\sum |c_k|^p)^{1/p}$ for which we have such a representation (3.3) of $f(x)$. In [CW], in the more general context of space of homogeneous type we know that $H_a^{p,q} = H_a^{p,\infty}$ and

$$\| f \|_{H_a^{p,q}} \cong \| f \|_{H_a^{p,\infty}}$$

for all $q > 1$.

Theorem 5 ([$BF1$]) For $0 < p \leq 1$, we have

$$H^p(G) = H_a^p(G) \quad \text{and} \quad \| f \|_{H^p(G)} \cong \| f \|_{H_a^p(G)}$$

Sketch of the Proof: We first use a geometric argument to study the classical group $U(n)$ of unitary isometries, which is the direct product of a semisimple compact Lie group and a one-dimensional center. A result from approximation theory is then used to prove that any distribution in $H^p(U(n))$ has an atomic decomposition. Next, unitary embedding, a well-known consequence of the Peter-Weyl theorem, allows us to transfer this result to G, yielding

$$H^p(G) = H_a^p(G)$$

Obviously, the above theorem is analogous to the Coifman-Laufm ter's theorem on $\mathbf{R}^n$. Some Related theorems on other Lie groups and spaces can be founded in [FoS] and [Co].

Recall that the exponential map exp is an analytic diffeomorphism on an open neighbourhood of the origin of $\mathfrak{g}$. Choose ϵ_0 and ϵ_0' to be the maximal positive number so that exp is such a diffeomorphism of $B(0,\epsilon_0')$ onto $B(e,\epsilon_0)$. Let $T_e(G)$ be the tangent plane of G at e, then $T_e(G)$ can be identified with the Lie algebra $\mathfrak{g}$. For a positive integer k we set

$$\begin{aligned} p_k =\{p:\ p(x) = q(\exp^{-1} x) \text{ for } x\in B(e,\epsilon_0),\ q \text{ is a polynomial on}\\ T_e(G) \text{ with degree less than or equal to } k\} \end{aligned} \tag{3.4}$$

We define

$$\begin{aligned} p_k(y) =\{p(x) : p(x) = q(\exp^{-1}(y^{-1}x),\ q \text{ is } a \text{ polynomial on}\\ T_e(G) \text{ with degree} \le k\} \end{aligned} \tag{3.5}$$

as the set of all polynomials with degree $\le k$ on $B(y,\epsilon_0)$. Then we can use this definition of polynomials to define the regular (p,q) atoms and the atomic Hardy spaces $H_a^{p,q}(G)$. We can further prove this new definition is equivalent to the forementioned definition. More naturally, we can use the representation polynomials to define atoms and Hardy spaces. It can be proved that on $SU(2)$, this definition coincides with the ones defined before. But for a general compact Lie group, we have the following open problem:

Problem. If we use the representation polynomials to define an atomic H^p space, does it coincide with the Maximal function based H^p space?

4. S-FUNCTIONS AND $\mathfrak{g}_\lambda$-FUNCTIONS

In this section we introduce the S-function characterization of the Hardy spaces. We define a C^∞ central function on G by

$$\Psi_t(x) = \sum_{\lambda\in\Lambda} 2t^2 \parallel \lambda+\beta \parallel^2 e^{-t^2\|\lambda+\beta\|^2} d_\lambda \chi_\lambda(x) \tag{4.1}$$

Then by checking the Fourier series, we can obtain a Calderón reproducing formula on G:

$$f = c\int_0^\infty \Psi_t{*}\Psi_t{*}ft^{-1}dt \tag{4.2}$$

where c is a constant independent of f and the equality (4.2) is in the distribution sense. The S-function of any function $f(x)\in S'(G)$ is defined by

$$S_\Psi f(x) = \{\int_{\Gamma(x)} |f{*}\Psi_t(y)|^2 t^{-n-1} dydt\}^{1/2} \tag{4.3}$$

In $[BF2]$, we obtained the following result:

Theorem 6 ($[BF2]$) For $0 < p \le 1$, the following three statements are equivalent:

(a) $P^*f \in L^p(G)$,

(b) $S_\Psi f \in L^p(G)$,

(c) $f \in H_a^{p,\infty}(G)$. Moreover

$$\| P^* f \|_p \cong \| S_\Psi f \|_p \cong \| f \|_{H_a^{p,\infty}}$$

Sketch of the Proof: If $f \in H^p(G)$, by Theorem 5, $f = \Sigma c_k a_k$ with all a_k's being atoms. We use a standard argument to obtain that

$$\| S_\Psi f \|_L^p p \leq C \sum |c_k|^p \| S_\Psi(a_k) \|_{L^p}^p \leq C \sum |c_k|^p \leq C \| f \|_{H_a^{p,\infty}}$$

To prove the converse part, we take a radial function $\Phi \in \mathcal{S}(\mathbf{R}^\ell)$ satisfying supp $\Phi \subseteq \{\theta : \ |\theta| \leq 1\}$, $\int_{\mathbf{R}^\ell} \theta^J \Phi(\theta) d\theta = 0$ for all multi-indices J with $|J| \leq 3n + 3 + 2n(1/p - 1/2)$, and $\int_0^\infty \Phi^\wedge(t)\Psi^\wedge(t)t^{-1}dt = 1$. Then we can prove that

$$\begin{aligned} f(x) &= \int_{G\times\mathbf{R}_+} f*\Psi_t(y)\Phi_t(xy^{-1})dyt^{-1}dt \\ &= \int_0^\epsilon \int_G + \int_\epsilon^\infty \int_G \} f*\Psi_t(y)\Phi_t(xy^{-1})dyt^{-1}dt \\ &= I_1(x) + I_2(x) \end{aligned}$$

where ϵ is a small positive number. By the compactness of G we can prove that $I_2(x) = C_{\epsilon,p}a(x)$, where $a(x)$ is an exceptional atom and $|C_{\epsilon,p}| \leq C \| S_\Psi(f) \|_p$. To estimate $I_1(x)$, we let xy^{-1} conjugate to an element $\exp\theta$ in $\mathbf{T}$ and let $D(xy^{-1}) = \prod_{\alpha\in A} \sin \frac{<\alpha,\theta>}{2}$. Then by Taylor's formula, we have

$$\begin{aligned} I_1(x) =& C \int_0^\epsilon \int_G f*\Phi_t(y)\{\prod_{\alpha\in A} <\alpha,\theta> + \sum_{\alpha_j\in A} C_j <\alpha_j,\theta>^3 \prod_{\alpha\in A} <\alpha,\theta> + \\ &+ \cdots + P_{n(p)}(\theta)\}D^{-1}(\theta)\Psi_t(xy^{-1})dyt^{-1}dt + \\ &+ C \int_0^\epsilon \int_G f*\Psi_t(y)R(\theta)D^{-1}(\theta)\Phi_t(xy^{-1})dyt^{-1}dt \end{aligned} \tag{4.4}$$

where $P_{n(p)}$ is a polynomial with degree $2n + 2 + n(1/p - 1/2)$, and

$$\begin{aligned} R(\theta) = R(xy^{-1}) =& D(xy^{-1}) - \{\prod_{\alpha\in A} <\alpha,\theta> + \\ &+ \sum_{\alpha_j\in A} C_j <\alpha,\theta>^3 \prod_{\alpha\in A} <\alpha,\theta> + + \cdots + P_{n(p)}(\theta)\} \end{aligned}$$

is a functionin $C^\infty(\mathbf{R}^\ell)$, $C_j's$ are constants. we can prove that the first term in (4.4)

$$I_{1,1}(x) = C \int_0^\epsilon \int_G f*\Psi_t(y)\Phi_t(xy^{-1}) \prod_{\alpha\in A} \frac{<\alpha,\theta>}{\sin\frac{<\alpha,\theta>}{2}} dyt^{-1}dt$$

has an atomic decomposition $I_{1,1}(x) = \sum \lambda_j a_j(x)$, where each a_j is a $(p,2)$ atom and $\sum |\lambda_j|^p \leq C \| S_\Psi(f) \|_p^p$. In a similar way, we can prove that each term in

(4.4), except the last term, has the similar atomic decomposition as $I_{1,1}(x)$. For the last term $\mathcal{L}(x)$ in (4.4), we can prove that it has an atomic decomposition $\mathcal{L}(x) = \sum \gamma_j a_j(x)$, where each $a_j(x)$ is an exceptional atom and $\sum |\gamma_j|^p \leq C \parallel S_\Psi(f) \parallel_p^p$. Since there are only finite number of terms in (4.4), the theorem follows easily from Theorem 5.

The g_λ^*-function of a distribution f is defined by

$$g_\lambda^*(f)(x) = (\int_0^\infty \int_G (\frac{t}{d(x,y)+t})^{\lambda n} |f * \Psi_t(y)|^2 t^{-n-1} dy dt)^{1/2}. \tag{4.5}$$

We can use the g_λ^*- function to characterize the Hardy spaces.

Theorem 7 ([$BF2$]) When $0 < p \leq 1$, and $\lambda > 2/p$, we have

$$\parallel g_\lambda^*(f) \parallel_p \cong \parallel S_\Psi(f) \parallel_p \cong \parallel f \parallel_{H^p(G)}$$

By Theorem 7 together with using interpolation (Theorem E of [CW]), we further have:

Theorem 8 ([$BF2$]) For $p > 1$ and $\lambda > 2/p$, we have $\parallel g_\lambda^*(f) \parallel_p \leq C \parallel f \parallel_p$.

Theorem 7 was first established on $\mathbf{R}^n$ by C. Fefferman and Stein [FS]. It was later generalized to the homogeneous group by Folland and Stein [FoS]. Theorem 8 was proved by N. Weiss [W] for the case $p \geq 2$. The Euclidean version of Theorem 8 can be found in [$S2$]. The g-functions on compact Lie groups were defined and studied first in [$S1$]. For more details about the background and history of the S-function and Liufm tlewood-Paley theory, readers can see [Fe], [$S2$] and [FJW].

The S-function characterization sometime is more flexible to use than the maximal function characterization. It is not difficult to see, by checking the proof of Theorem 7 (see [$BF2$]), that Theorem 7 holds on the n-dimension torus $\mathbf{T}^n$, though it is well-known that $\mathbf{T}^n$ is not semisimple. Using this fact, we further can obtain a deLeeuw's theorem [deL] which discusses the relation of multipliers on H^p between $\mathbf{R}^n$ and $\mathbf{T}^n$.

Theorem 9 ([$F2$]) Suppose that T is a multiplier associated with a continuous function on $\mathbf{R}^n$. Then T is bounded in $H^p(\mathbf{R}^n)$ if and only if its periodlization $\widetilde{T}_\epsilon$ is uniformly bounded in $H^p(\mathbf{T}^n)$ for $\epsilon > 0$.

Sketch of the Proof: Using the S-function characterizations of H^p spaces, we can prove that for any $f \in \mathcal{D}(\mathbf{R}^n) \cap H^p$, $\|Tf\|_{H^p(\mathbf{R}^n)} \leq \sup_{\epsilon>0} \|T_\epsilon\| \|f\|_{H^p(\mathbf{R}^n)}$. To prove the converse, we notice that $(Tf)^\wedge(x) = m(x) f^\wedge(x)$ and define an operator T_ϵ by $(T_\epsilon f)^\wedge(x) = m(\epsilon x) f^\wedge(x)$. Then a simple computation shows that $\parallel T_\epsilon \parallel = \parallel T \parallel$. So it suffices to prove the boundedness for $\widetilde{T}_1 = \widetilde{T}$. For $\tilde{f} \in C^\infty(\mathbf{T}^n) \cap H^p(\mathbf{T}^n)$, $\tilde{f}$ has an atomic decomposition $\tilde{f}(x) = \sum c_k a_k(x)$ with $\sum |c_k|^p < \infty$, where each a_k is either a regular (p, ∞, s) atom or an exceptional atom. We need to prove that $\parallel \widetilde{T} a \parallel_{H^p(\mathbf{T}^n)} \leq C$ with a constant C independent of the atom. Notice that T is bounded in $H^p(\mathbf{R}^n)$ implies that T is a bounded operator in $L^2(\mathbf{R}^n)$. Hence $m(x)$ is a bounded function and $\widetilde{T}$ is a bounded operator in the $L^2(\mathbf{T}^n)$ space.

If a(x) is an exceptional atom, then using Hölder's inequality, we easily see that $\| \widetilde{T}a \|_{H^p(\mathbf{T}^n)} \leq C \| a \|_{L^2(\mathbf{T}^n)} \leq C$ with a constant C independent of a(x). So it remains to show the above inequality for any regular atom.

Let a(x) be a regular atom on $\mathbf{T}^n$ with support in $B(0, \rho)$. We can consider this function as an atom $a'(x)$ on $\mathbf{R}^n$ with support in the fundamental cube Q. Then the function $b(x) = (Ta')(x)$ is in $H^p(\mathbf{R}^n)$ and

$$\| b \|_{H^p(\mathbf{R}^n)} = \| \sup_{t>0} |P_t * b| \|_{L^p} \leq \| T \| \| a' \|_{H^p(\mathbf{R}^n)} \cong \| T \| \| a \|_{H^p(\mathbf{T}^n)}$$

It is easy to observe that if we view $\widetilde{T}(a)$ as a periodical function on $\mathbf{R}^n$, then $\widetilde{T}a(x) = \tilde{b}(x) = \sum_{k \in \Lambda} b(x+k)$. So we define for $y \in Q$

$$\gamma(y) = \int_Q \tilde{b}(x) \tilde{P}_t(y-x) dx$$

where $\tilde{P}_t$ is the Poisson kernel on $\mathbf{T}^n$ which is considered to be a periodical function on $\mathbf{R}^n$ and all we have to show is that

$$\| \widetilde{T}a \|_{H^p(\mathbf{T}^n)} = \left(\int_Q |\gamma(y)|^p dy \right)^{1/p} \leq \| T \| \| a \|_{H^p(\mathbf{T}^n)}$$

But

$$\sup_{t>0} \left| \int_Q \tilde{b}(x) \tilde{P}_t(y-x) dx \right| = \sup_{t>0} \left| \int_{\mathbf{R}^n} \tilde{b}(x) P_t(y-x) dx \right|$$

$$\leq \sup_{t>0} \sum_{k \in \Lambda} |b * P_t(y+k)| \leq \sum_{k \in \Lambda} \sup_{t>0} |b * P_t(y+k)|$$

This means that

$$\| \gamma \|_{L^p(\mathbf{T}^n)} \leq \| \sup_{t>0} |b * P_t(y)| \|_{L^p(\mathbf{R}^n)}$$

So we prove the inequality for any atom centered at 0. But $\widetilde{T}$ commutes with shift operators, hence the inequality valids for all atoms.

5. Lipschitz Spaces and the Dual of $H^p(G)$

Suppose $\alpha \geq 0$, we introduce Lipschitz spaces on G.

Definition 5.1 Denote by $\dot{\Lambda}_\alpha$ the homogeneous Lipschitz spaces and Λ_α the (inhomogeneous) Lipschitz spaces.

i) For $0 < \alpha < 1$, $\dot{\Lambda}_\alpha = \{f : \; |f(x) - f(y)| \leq c d(x,y)^\alpha, \; x, y \in G\}$ and $\| f \|_{\dot{\Lambda}_\alpha}$ is the infimum of all c for which the above estimate holds. ii) For $\alpha = 1$, $\dot{\Lambda}_1 = \{f : \| f \|_{\dot{\Lambda}_1} < \infty\}$, where

$$\| f \|_{\dot{\Lambda}_1} = \sup\left\{ \frac{|f(xy) - 2f(x) + f(xy^{-1})|}{d(y,e)} : \; x, y \in G, y \neq e \right\}$$

iii) For $\alpha > 1$, let k be a positive integer such that $k < \alpha \leq k+1$. Then $\dot{\Lambda}_\alpha = \{f : \| f \|_{\dot{\Lambda}_\alpha} < \infty\}$, where

$$\| f \|_{\dot{\Lambda}_\alpha} = \sum_{|\nu|=k} \| X^\nu f \|_{\dot{\Lambda}_{\alpha-k}}$$

iv) For any $\alpha > 0$ we define

$$\| f \|_{\Lambda_\alpha} = \| f \|_1 + \| f \|_{\dot{\Lambda}_\alpha} \text{ and } \Lambda_\alpha = \{f \in L^1(G) : \| f \|_{\Lambda_\alpha} < \infty\}.$$

For an integer $k \geq 0$ we set

$$A_k = \{\phi \in \mathcal{S}(\mathfrak{t}) : \phi \text{ radial, supp } \phi \subseteq \{H \in \mathfrak{t} : \| H \| \leq 1\} \text{ and } \int_{\mathfrak{t}} \phi(H) H^\nu dH = 0 \text{ for all} |\nu| \leq k\}$$

and

$$\tilde{A}_k = \{\phi \in A_k : \int_0^\infty \phi^\wedge(tH)^2 t^{-1} dt = 1, H \neq 0\}$$

where $|\nu| = \nu_1 + \cdots + \nu_\ell$ for a positive ℓ-tuple $\nu = (\nu_1, \cdots, \nu_\ell)$. By [$FJW$], we know $\tilde{A}_K$, and therefore A_k are not empty. For $t > 0$ let

$$\phi_t(x) = \sum_{\lambda \in \Lambda} \phi^\wedge(t \| \lambda + \beta \|) d_\lambda \chi_\lambda(x) \tag{5.1}$$

Definition 5.2 For $0 < \alpha < 1$, and $\phi \in \tilde{A}_0$ we set

$$\| f \|_{\dot{B}_\alpha} = \sup\{t^{-\alpha} |\phi_t * f(x)| : x \in G, t > 0\}$$

$$\| f \|_{B_\alpha} = \| f \|_1 + \| f \|_{\dot{B}_\alpha}$$

and define

$$\dot{B}_\alpha = \{f : \| f \|_{\dot{B}_\alpha} < \infty\}, \ B_\alpha = \{f : \| f \|_{B_\alpha} < \infty\}$$

Then the following theorems are in analogy with some well-known theorems on $\mathbf{R}^n$ ([FJW]).

Theorem 10 ([$FX1$]) For $0 < \alpha < 1$ we have $\Lambda_\alpha = B_\alpha$ and $\dot{\Lambda}_\alpha = \dot{B}_\alpha$.

Theorem 11 ([$FX1$]) Let $\alpha > 0$ and $k \in \mathbf{Z}$, $k \geq [\alpha]$([α] is the greatest integer in α). The followings are equivalent: (i) $f \in \Lambda_\alpha$, (ii) for all $\phi \in A_{2k+n}$, $\sup\{t^{-\alpha} |\phi_t * f(x)| : x \in G, t > 0\} \leq C_{\phi,f}$ with $C_{\phi,f}$ being a constant independent of t, (iii) there exists $\phi \in \tilde{A}_{2k+n}$ such that

$$\sup\{t^{-\alpha} |\phi_t * f(x)| : x \in G, t > 0\} \leq C_{\phi,f}$$

Moreover, if we define

$$\| f \|_{B_\alpha} = \| f \|_1 + \sup_{x,t} t^{-\alpha} |\phi_t * f(x)|$$

for some $\phi \in A_{2k+n}$, then $\| f \|_{\Lambda_\alpha} \cong \| f \|_{B_\alpha}$.

Recalling the polynomial spaces $\mathcal{P}_k$ defined in (3.4), we define the Campanato spaces as following:

Definition 5.3 Let $1 \leq q < \infty, \lambda > 0$ and $k \in \mathbf{Z}^+$. The Campanato spaces are defined as

$$L_k^{q,\lambda}(G) = \{f : \| f \|_{L_k^{q,\lambda}} < \infty\}$$

where

$$\| f \|_{L_k^{q,\lambda}(G)} = \| f \|_1 +$$

$$+ \sup_{0<r<\epsilon_0,\ x\in G} \inf_{p\in\mathcal{P}_k} \{|B(x,r)|^{-\lambda/n} \int_{B(x,r)} |f(y) - p(x^{-1}y)|^q dy\}$$

and $|B(x,r)|$ is the Harr measure of the ball $B(x,r)$.

The following theorem gives the relationship between the Lipschitz spaces, the Hardy spaces $H^p(G)$ and the Campanato spaces.

Theorem 12 ($[FX1]$) Let $1 \leq q < \infty, \alpha > 0, k = [\alpha]$ and $p = n/(n+p)$. Then the following statements are equivalent:

(i) $f \in \Lambda_\alpha$,

(ii) there exists $c > 0$ such that for each $x \in G$ and $0 < r < \epsilon_0$, there is a polynomial $p_k = p_{x,r,k} \in \mathcal{P}_k$ satisfying

$$\sup_{y\in B(x,r)} |f(y) - p_k(x^{-1}y)| \leq cr^\alpha,$$

(iii) $f \in L_k^{q,n+q\alpha}(G)$,

(iv) $f \in (H^p)^*$, where $(H^p)^*$ denotes the dual of $H^p(G)$.

For the background and results of the Lipschitz spaces theory on $\mathbf{R}^n$ one can see $[FJW]$ for further reference.

6. Convergence Theorems Of Fourier Series

Changing the parameter, we write the Bochner-Riesz kernel as

$$S_R^\delta(x) = \sum_{\lambda\in\Lambda} (1- \| \lambda + \beta \|^2 / R^2)_+^\delta d_\lambda \chi_\lambda(x). \tag{6.1}$$

The associated maximal operator is

$$S_*^\delta f(x) = \sup_{R>0} |f * S_R^\delta(x)| \tag{6.2}$$

The pioneer work of H^p spaces on compact Lie groups is the following convergence theorem which was proved by J. L. Clerc in 1987.

Theorem 13 ([*C*]) Suppose that $0 < p < 1$ and $\delta = n/p - (n+1)/2$. Then the maximal operator $S_*^\delta f(x)$ is of weak-type (H^p, L^p).

As a standard result of the above theorem, for $0 < p < 1$ and $\delta \geq n/p - (n+1)/2, S_R^\delta f(x)$ converges to $f(x)$ almost everywhere as R tends to infinity for each $f \in H^p(G)$.

Motivated by the generalized Bochner-Riesz kernel on $\mathbf{R}^n$ ([*Lu*]), the generalized Bochner-Riesz kernel on G is defined by [*Wa*]

$$S_R^{\delta,\alpha}(x) = \sum_{\lambda \in \Lambda} (1 - \| \lambda + \beta \|^\alpha / R^\alpha)_+^\delta d_\lambda \chi_\lambda(x), \delta \geq 0, \alpha > 0. \tag{6.3}$$

So the associated maximal operator is

$$S_*^{\delta,\alpha} f(x) = \sup_{R>0} |S_R^{\delta,\alpha} f(x)| \tag{6.4}$$

Using a new classification of the positive roots system, Wang modified Clerc's method to generalize Theorem 13 to the more general seufm ting:

Theorem 14 ([*Wa*]) Suppose that $0 < p < 1, \alpha > 0$ and $\delta = n/p - (n+1)/2$. Then the maximal operator $S_*^{\delta,\alpha} f(x)$ is of weak type (H^p, L^p).

From the definition of the generalized Bochner-Riesz means, one easily defines the generalized Poisson kernels by

$$P_t^\alpha(x) = \sum_{\lambda \in \Lambda} e^{-(t\mu(\lambda)^{1/2})^\alpha} d_\lambda \chi_\lambda(x) \tag{6.5}$$

where $\alpha > 0$, $\mu(\lambda) = \| \lambda + \beta \|^2 - \| \beta \|^2$. The generalized Abel means is now defined by $P_t^\alpha * f$ and its associated maximal operator is $P_*^\alpha f(x) = \sup_{t>0} |P_t^\alpha * f(x)|$. For this generalized Able means, Xu obtain an almost everywhere convergence theorem for any $f \in H^p$.

Theorem 15 ([*X*]) Let $f \in H^p(G)$ and $0 < p \leq 1$, then

$$\| P_*^\alpha f \|_p \leq C_{p,\alpha} \| f \|_{H^p(G)}$$

with $C_{p,\alpha}$ is a constant depending only on p and α.

As usual, the theorem implies the almost everywhere convergence of $P_t^\alpha f(x)$ as t tends to zero for $f \in H^p(G)$, $0 < p \leq 1$.

Moreover, Xu obtained the following approximation theorem on H^1 spaces:

Theorem 16 ([*X*]) Let $f \in H^1(G)$. Then as t tends to zero

$$\| P_t^\alpha f - f \|_{H^1(G)} \leq \begin{cases} C\omega(f,t)_{H^1(G)}, & \alpha > 1, \\ C\omega(f, t \ln t^{-1})_{H^1(G)}, & \alpha = 1, \\ C\omega(f, t^\alpha)_{H^1(G)}, & \alpha < 1, \end{cases}$$

where $\omega(f,t) = \sup_{d(x,e)\leq t} \| L_x f - f \|_{H^1(G)}$ is the H^1-modulus of continuity.

The following theorems were recently proved by Bloom and Xu.

Theorem 17 ([BX]) Suppose that $0 < p \leq 1$ and $\delta > \delta_p = n/p - (n+1)/2$. Then $f \mapsto S_R^\delta(f)$ is of type (H^p, H^p) and $\| S_R^\delta(f) \|_{H^p(G)} \leq C \| f \|_{H^p(G)}$ with constant C independent of f and R. Moreover for $f \in H^p(G)$, $\| S_R^\delta(f) - f \|_{H^p(G)} \to 0$ as $R \to \infty$.

In the same paper, Bloom and Xu also obtained some weak type boundedness and weak type approximation theorems. All these results can be found in [BX].

7. Calderon-Zygmund Operators

We now introduce the Riesz potential and Riesz transforms on G. These operators were first defined in this seufm ting by Stein [$S1$].

The Riesz potential $(-\Delta)^{-z}$ is defined for $z \in \mathbf{C}, Re(z) > 0$ by

$$(-\Delta)^{-z} f = \Gamma(z)^{-1} \int_0^\infty t^{z+1}(W_t - 1)*f dt \tag{7.1}$$

and extended to the complex plane by $(-\Delta)^{-z} = -\Delta(-\Delta)^{-z-1}$. We are interested in particular in the case $z = 1/2$,

$$(-\Delta)^{-1/2} f = (\pi)^{-1/2} \int_0^\infty t^{-1/2}(W_t - 1)*f dt \tag{7.2}$$

The Riesz transforms $R_j (j = 1, 2, \cdots, n)$ are defined by

$$R_j f = X_j(-\Delta)^{-1/2} f = (\pi)^{-1/2} \int_0^\infty t^{-1/2} X_j W_t * f \text{ dt} \tag{7.3}$$

When in the case $z = i\gamma$, $\gamma \in \mathbf{R}\backslash\{0\}$,

$$(-\Delta)^{-i\gamma} f = -\Gamma(1+i\gamma)^{-1} \int_0^\infty t^{i\gamma} \Delta W_t * f \text{ dt} \tag{7.4}$$

In [CMR], both operators $(-\Delta)^{-i\gamma}$ and R_j are shown to be Calderón-Zygmund singular integrals. Combining the L^2 boundedness proved by Stein [$S1$], it is easy to see, by a standard Calderón-Zygmund decomposition method, that these two kinds of operators are of strong type (p,p) for $1 < p < \infty$. Thus it will be interesting to study the boundedness properties for these operators in $H^p(G)$ for $0 < p \leq 1$. Now we will introduce more general convolution operators which include the above mentioned Riesz potentials and Riesz transforms. First we need

Theorem 18 ([$BF1$]) $\mathcal{S}(G)$ is dense in $H^p(G)$.

Thus a linear operator T defined on $\mathcal{S}(G)$ can be extended to a bounded operator from the space H^p to the space H^q if

$$\| Tf \|_{H^q(G)} \leq C \| f \|_{H^p(G)} \tag{7.5}$$

for all $f \in H^p(G) \cap C^\infty(G)$, where C is a constant independent of $f(x)$.

The convolution kernels we shall be considering are the following. Suppose that $0 < \alpha < n$ and r is a positive number. A kernel of type (α, r) is a function K on G which is of class C^r on $G \backslash \{e\}$ and satisfies

$$|X^J K(x)| \leq C_J d(x,e)^{\alpha - n - |J|} \tag{7.6}$$

for all $J = (j_1, \cdots, j_n)$ with $|J| \leq r$, and $x \neq e$. Here, $X^J = X_1^{j_1} X_2^{j_2} \cdots X_n^{j_n}$ and C_J is a constant independent of $x \in$ G. ¿From this definition it is easy to see that K is an L^1 function , so we can define the Calderón-Zygmund operator with the kernel K by

$$Tf(x) = T_K f(x) = \int_G f(y) K(xy^{-1}) dy \tag{7.7}$$

for any $L^1(G)$ function $f(x)$.

More interesting and more subtle is the limiting case $\alpha = 0$. Here the size condition (7.6) of K does not imply the integrability of K(hence K defines a distribution), nor, does it automatically yield an L^p boundedness theorem. Hence we assume these conditions separately, as follows: If r is a positive integer, a kernel of type $(0, r)$ is a distribution K on G which is of class C^r on $G \backslash \{e\}$ and satisfies (7.6) with $\alpha = 0$, and which also satisfies $\| T_K f \|_2 \leq C \| f \|_2$. Thus by a standard argument (see chapter 2 of $[S2]$), we further have $\| T_k f \|_p \leq C \| f \|_p$ for all $1 < p < \infty$. The following two theorems are analogous of the results in the case of the Euclidean spaces (see for example $[FS]$ and $[TW]$).

Theorem 19 ($[F1]$) Suppose that r is a positive integer and $0 < \alpha <$ n. If $n/(n+r) < p_1 < n/\alpha$, $1/p_1 = 1/p_2 + \alpha/n$, and K is a kernel of type (α, r). Then the operator T_K extends to a bounded operator from $H^{p_1}(G)$ to $H^{p_2}(G)$.

Theorem 20 ($[F1]$) Suppose that $s = [n(1/p - 1)]$ and T is an operator with a $(0, r)$ kernel. If $r > s$, then T extends to a bounded operator on $H^p(G)$.

Sketch of the Proof: By a standard argument we only need to prove

$$\| K * a \| = \| Ta \|_{H^p} \leq C$$

with a constant C independent of any (p, ∞) atom a with supp(a) $\subseteq B(e, \rho)$. We also can assume that $\rho < \epsilon_0$ for some small ϵ_0. Let ϕ be a non-negative C^∞ function which satisfies supp $(\phi) \subseteq \{1/2 \leq |y| \leq 2\}$ and $\sum_{j=-\infty}^{\infty} \phi(2^j |y|) = 1$ for $y \neq 0$. Let $\eta(x) = 1 - \sum_{j=1} \phi(2^{-j-2}\rho^{-1} \|x\|)$. Then for any $x \in G \backslash \{e\}$,

$$K(x) = \eta(x)K(x) + \sum_{j=1} K(x)\phi(2^{-j-2}\rho^{-1} \| x \|)$$

$$= K_0(x) + \sum_{j=1}^{\log_2(\epsilon_0/\rho)} K_j(x) + K_R(x)$$

where $K_0(x) = \eta(x)K(x)$ and

$$K_R(x) = \sum_{j=\log_2(\epsilon_0/\rho)}^{\infty} K(x)\phi(2^{-j-2}\rho^{-1} \| x \|)$$

Thus

$$K{*}a(x) = K_0{*}a(x) + \sum_{j=1}^{\log_2(\epsilon_0/\rho)} K_j{*}a(x) + K_R{*}a(x)$$

Clearly, each $K_j{*}a(j = 0, 1, 2, \cdots, \log_2(\epsilon_0/\rho))$ satisfies the cancellation condition, supp $(K_0{*}a) \subseteq B(e, 16\rho)$, supp$(K_R{*}a) \subseteq G$ and supp$(K_j{*}a) \subseteq B(e, 2^{j+4}\rho)$ for $j = 1, 2, \cdots, \log_2(\epsilon_0/\rho)$. By the cancellation conditions of $a(x)$, we can prove that

$$\| K_j{*}a \|_\infty \le C(2^j\rho)^{-n/p} \sum_{|J|\le s+1} (2^j)^{-|J|-n+n/p}\rho^{-|J|+s+1}$$

and $\| K_R{*}a \|_\infty \le C$, where C is a constant independent of $a(x)$. We also can show that $K_0{*}a$ is a $(p, 2)$ atom, thus by a standard argument, we obtain that $\| K_0{*}a \|_{H^p} \le C$, $\| K_R{*}a \|_{H^p} \le C$, and

$$\| \sum_{j=1}^{\log_2(\epsilon_0/\rho)} K_j{*}a \|_{H^p} \le C \sum_{k=1}^{s+1} \rho^{-k+s+1} \sum_{j=1}^{\log_2(\epsilon_0/\rho)} (2^j)^{-k-n+n/p} \le C,$$

where C is a constant independent of atom a.

Using a simple computation, one can obtain the following two examples which imply that the boundedness of Riesz potentials and Riesz transforms in Hardy spaces are special cases of Theorem 19 and Theorem 20.

Example 21 Riesz Transforms $R_j (j = 1, 2, \cdots, n)$ are Calderón-Zygmund operators with $(0, r)$ kernels for all $r > 0$. Thus they are bounded operators in $H^p(G)$.

Example 22 The Riesz potential $(-\Delta)^{-\alpha/2+i\nu}, \nu \in \mathbf{R}$, is a Calderón-Zygmund operator with $a(\alpha, r)$ kernel for all $r > 0$.

It is also possible to consider the non-convolution Calderón-Zygmund operators, with the $(0, r)$ kernels and set up some boundedness criterions in $H^p(G)$.

Problem To set up a molecular theory on compact Lie group, then use it to study the boundedness properties of the Calderón-Zygmund operators (see [TW], [To] and [FJW] for the details of the molecular theory on $\mathbf{R}^n$).

8. Multiplier Theorems on $H^p(G)$

Given a bounded multi-sequence $\{m(\lambda)\}_{\lambda\in\Lambda}, m(\lambda) \in \mathbf{C}$, define the operator T on the space of finite linear combinations of entry functions on G by writing

$$\widehat{Tf}(\lambda) = m(\lambda)\hat{f}(\lambda), \ \lambda \in \Lambda \tag{8.1}$$

T commutes with left and right translation. If $f \in H^p(G)$ then the Hilbert-Schmidt norm of $\hat{f}$ satisfies(see [X])

$$|||\hat{f}(\lambda)||| \le C\|f\|_{H^p} d_\lambda^{n+2} \| \lambda + \beta \|^{n(1/p-1)} \tag{8.2}$$

It follows that if $m \in L^\infty(\Lambda)$ then $f \mapsto Tf$ is well defined from $H^p(G)$ to the Schwartz distribution space $\mathcal{S}'(G)$. We say that m is a Fourier multiplier of $H^p(G)$ if this mapping takes H^p continuously into H^p.

Remember that Λ can be identified with a subset of a laufm tice in $\mathbf{R}^\ell$. Hence we can speak of the partial difference operator δ^J acting on multi-sequences $\{m(\lambda)\}_{\lambda \in \hat{G}}$, with $J = (j_1, \dots, j_\ell) a$ multi-index of order $|J| = j_1 + \cdots + j_\ell$.

Theorem 23 ([$FX2$]) Let s be the smallest even integer such that $s > \frac{n}{p} - \frac{n}{2}$ where $0 < p \leq 1$. Suppose that $m \in L^\infty$ such that for all J with $|J| \leq s$ and any $R > 0$

$$\sum_{R \leq |\lambda| \leq 2R} |\delta^J m(\lambda)|^2 \leq C R^{\ell - 2|J|} \tag{8.3}$$

then m is a Fourier multiplier of $H^p(G)$.

Similar theorems in other spaces can be found in [BS], [CW], and [W].

Sketch of the Proof Consider the operator T with $\widehat{Tf}(\lambda) = m(\lambda)\hat{f}(\lambda)$. By the condition on $m(\lambda)$ we can prove $\|Ta\|_p \leq C$, where C is a constant independent of atom a(x) and $N = [n(\frac{1}{p} - 1)]$. To prove the theorem, it suffices to prove $\| Ta \|_{H^p} \leq C$ unifomly for all regular (p, ∞, N) atom a(x). In order to so we will introduce the generalized Riesz transforms R_J on compact Lie groups which were studied in [BX]. For any $p \in (0, 1]$ take L large enough so that $p > (n-1)/(n-1+L)$. By Theorem 29 in the next section, for any (p, ∞, N) atom $a(x)$ we have the atomic decomposition of $R_J(a)$: $R_J(a) = \sum \lambda_i b_i$, where $b_i's$ are atoms and $\sum |\lambda_j|^p \cong \|R_J(a)\|_{H^p}^p \leq C$. Now notice that both T and R_J are convolution operators, we have $\|Ta\|_{H^p}^p \leq C \sum_J \|T(R_J(a))\|_p^p \leq C \sum_i |\lambda_i|^p \|Tb_i\|_p^p \leq C$. Thus the theorem is proved.

We now easily obtain an application as follows:

Theorem 24 ([$FX2$]) Suppose T is the bi-invariant operator associated to the multiplier $m(\lambda) = \| \lambda + \beta \|^{ia}$ for some $a \in \mathbf{R}$. Then T is bounded in H^p spaces for any $0 < p \leq 1$.

The operators studied above are all bi-invariant (central operators). But it is also possible to study some kind of non-bi-invariant operators including the well-known Riesz transforms. For the sake of simplicity, we will study this case on the simplest compact Lie group $SU(2)$, but the idea should work on general compact Lie groups. Recall $X_i (i = 1, 2, 3)$ is an orthonormal basis on $SU(2)$. For any bi-invariant operator T associated with a multiplier $m(\lambda)$, we write $Tf = f * K$ and define a new operator $\mathcal{T}_j f = f * X_j K$. Then we have the following theorem:

Theorem 25 ([$FX2$]) Let s be the smallest even integer such that $s > 3/p - 3/2$. Suppose that $m(\lambda) \in L^\infty$ and that for all integer J with $0 \leq J \leq s$ and all $R > 0$

$$\sum_{R \leq |\lambda| \leq 2R} |\delta^J m(\lambda)|^2 \leq C R^{-2J-1} \tag{8.4}$$

Then $\mathcal{T}_j$ is a linear bounded operator on $H^p(SU(2))(0 < p \leq 1)$.

The above discussed Hörmander multiplier theorem on compact Lie group G was first set up in $L^p(G)$, $1 < p < \infty$, by N. Weiss [W]. Both Weiss' result for $p > 1$ and ours for $0 < p \leq 1$ need a restriction $\{\lambda\} \in M(2, L)$ with $L > n/p - n/2$ ($L > n/2$ for the case $p > 1$) and L being an even integer, where $\lambda \in M(2, L)$ if and only if $\{\lambda\}$ satisfies (8.3) for $s = L$. But for $SU(2)$, we have a following theorem which does not have the restriction of even integer.

Theorem 26 ([$FX3$]) Let $s > 3/p - 3/2$. Suppose $m \in L^\infty$ such that for all positive integers $J \leq s$ and any $R > 0$

$$\sum_{R \leq |\lambda| \leq 2R} |\delta^J m(\lambda)|^2 \leq C R^{1-2J},$$

where C is a constant independent of R. Then m is a Fourier multiplier on $H^p(SU(2))$.

N. Weiss' result was recently improved on the classical groups without the restriction L being an even integer. The detail will be published elsewhere [$F3$].

9. Other Characterizations Of The Hardy Spaces

We start this section with a modified Poisson kernel which is defined by

$$\psi_t(x) = e^{t\|\beta\|} \sum_{\lambda \in \Lambda} e^{-t\|\lambda+\beta\|} d_\lambda \chi_\lambda(x) \tag{9.1}$$

This kernel was introduced by S. Gong on the unitary groups [$G1$] in order to get a convergence theorem of Fourier series for both continuous and $L^p(p \geq 1)$ functions. If we defined the differential operator

$$\Delta_1 = -\Delta + d^2/dt^2 - 2d/dt - 2(\| \beta \|^2 - \| \beta \|)I \tag{9.2}$$

(here I denotes the identity map). Then a solution of the differential equation

$$\Delta_1 \Psi(g, t) = 0, \quad \Psi(g, 0) = f \tag{9.3}$$

for $f \in L^1(G)$ is given by

$$\Psi(g, t) = e^{-t\|\beta\|} \psi_t * f(g) \tag{9.4}$$

We can modify the kernel (9.1) by $\tilde{\psi}_t(x) = e^{-t\|\beta\|}\psi_t(x)$. Then $\tilde{\psi}_t * f(g)$ is a solution of the differential equation

$$\Delta_2 \Psi(g, t) = 0, \ \Psi(g, 0) = f(g) \ \text{ for } f \in L^1(G) \tag{9.5}$$

where

$$\Delta_2 = -\Delta + d^2/dt^2 - \| \beta \|^2 \, \mathrm{I} \tag{9.6}$$

Like all the kernels we discussed before, $\tilde{\psi}_t$ and Ψ_t are central functions and are determined by their restrictions to $\mathbf{T}$. Like before, after defining the maximal functions $\Psi^+ f$ and $\tilde{\psi}^+ f$, then the H^p type space $H^p_\Psi(G)$ is the collection of all distribution f on G for which $\Psi^+ f \in L^p(G)$. The H^p_Ψ norm of f is defined as $\| f \|_{H^p_\Psi(G)} = \| \Psi^+ f \|_p$. We also can use the kernel $\tilde{\Psi}_t(x)$ to define the maximal function and the H^p type space $H^p_{\tilde{\Psi}}(G)$. Let $H^1_{\tilde{\Psi}}(G) = H_{\tilde{\Psi}}(G)$, then we have the following theorem:

Theorem 27 ([$F4$]) $H^1(G) = H^1_\Psi(G) = H_{\tilde{\Psi}}(G)$.

Moreover, for any $f \in L^1(G)$, we have $\| f \|_{H^1(G)} \cong \| f \|_{H^1_\Psi(G)} \cong \| f \|_{H^1_{\tilde{\Psi}}(G)}$.

We also can use the Riesz transforms to characterize the Hardy spaces.

Theorem 28 ([X]) An $L^1(G)$ function f is in $H^1(G)$ if and only if $R_j f$ is in $L^1(G)$ for each $j = 1, 2, \cdots, n$. Moreover, we have

$$\| f \|_{H^1(G)} \cong \| f \|_1 + \sum_{j=1}^{n} \| R_j f \|_1 \tag{9.7}$$

To characterize the H^p spaces for $0 < p < 1$ by using the Riesz transforms, we need introduce the generalized Riesz transforms. For an integer $N \geq 0$ and a multi-index

$$J = (j_1, \cdots, j_n) \in \{1, 2, \cdots, n\}^N$$

let $R_J(f)$ denote the generalized Riesz transform $R_J(f) = R_{j_1} \cdots R_{j_N} f$ where $R_j f$ is the j-th Riesz transform of f if $j \neq 0$ and $R_0 f$ = f.

Theorem 29 ([BX]) Let $p > (n-1)/(n-1+N)$, and $f \in H^p(G) \cap L^2(G)$. Then there exist constants C_1, C_2 and C depending only on p, n, and N such that

(i) $C_1 \Sigma_J \| R_J(f) \|_p \leq \| f \|_{H^p} \leq C_2 \Sigma_J \| R_J(f) \|_p$,

(ii) $\| R_J(f) \|_{H\ p} \leq C \| f \|_{H\ p(G)}$, where the sums in (i) are taken over all multi-indices $J \in \{0, 1, \cdots, n\}^N$. Consider the classical domain $\mathcal{R}_n$ which consists of all $n \times n$ matrics X satisfying $I - X\bar{X}' > 0$, where X' is the transpose of X and ">" means the positive definition. This domain was studied by E. Carton [Car]. The characteristic manifold of $\mathcal{R}_n$ is the classical compact Lie group $U(n)$([Hua]). Hua [Hua] proved that there exists a completed orthogonal basis $\{\psi_m(X)\}_{m=0}^{\infty}$ on $\mathcal{R}_n$ whose restriction on $U(n)$ is also a orthonormal basis of $U(n)$. The Cauchy-Szegö kernel on $\mathcal{R}_n$ is defined by

$$H(X, \bar{\xi}) = \sum_{j=0}^{\infty} \psi_m(X) \overline{\psi_m(\xi)} \tag{9.8}$$

where $X \in \mathcal{R}_n$ and $\xi \in U(n)$. ¿From the above Cauchy-Szegö kernel, Hua defined the Poisson kernel by

$$P(X, \bar{\xi}) = H(X, \bar{\xi}) H(\xi, \bar{X}) / H(X, \bar{X}) \tag{9.9}$$

and the Poisson integral for a continuous function $f(\xi)$ on $U(n)$ by

$$\int_{U(n)} f(\xi)P(X,\xi)d\xi \tag{9.10}$$

Similar to the Hardy spaces on $\mathbf{R}^2_+$, the Hardy space $H^1(\mathcal{R}_n)$ consists of all analytic functions $F(X)$ on $\mathcal{R}_n$ such that

$$\| F \|_{H^1} = \sup_{0 \le r < 1} \int_{U(n)} |F(r\xi)|d\xi \ < \ \infty \tag{9.11}$$

Theorem 30 (see [G2]) Suppose $F(X)$ is an analytic function on $\mathcal{R}_n$, then $F(X)$ can be wriufm ten by its Poisson integral

$$F(X) = \int_{U(n)} F(\xi)P(X,\xi)d\xi$$

if and only if $F \in H^1(\mathcal{R}_n)$.

It was shown in [G1] that if $F \in H^1(\mathcal{R}_n)$, then the real part , $u(X)$ of $F(X)$ converges, as $r \to 1^-$, to a real valued function $f(\xi)$. So we can define the real H^1 space on $U(n)$ consisting of all real parts of boundary values of functions in $H^1(\mathcal{R}_n)$.

Problem Does this above defined H^1 space coincide with those H^1 spaces defined either by using atoms or by using maximal functions ?

REFERENCES

[B] B. Blank, Nontangential maximal functions over compact Riemannian manifolds, Proc. Amer. Math. Soc. **103** (1988), 999-1002.

[BF1] B. Blank and D. Fan, Hardy spaces on compact Lie groups, preprint.

[BF2] B. Blank and D. Fan, S-functions, g_λ-functions and Riesz potentials, Preprint.

[BS] A. Baernstein and E. Sawyer, Embedding and multiplier theorems on $H^p(\mathbf{R}^n)$, Memoirs of the American Math. Society, **318** (1985).

[BX] W. Bloom and Z. Xu, Approximation of H^p functions by Bochner-Riesz means on compact Lie groups, to appear, Math. Z.

[Ca] L. Carleson, Two remarks on H^1 and BMO, Advances in Math, **22** (1976), 269-275.

[Car] E. Cartan, Sur les domaines bonés homogènes de l'espace de n variables complexes, Hamburg Univ. Math. sem. Abhandl., **11** (1936), 106-162.

[C] J. L. Clerc, Bochner-Riesz means of H^p functions $(0 < p < 1)$ on compact Lie groups. Lecture Notes in Math **1243** (1987), 86-107.

[CMR] M. Cowling, A. Mantero and F. Ricci, Pointwise estimate for some kernels on compact Lie groups, Rend. Circ. Mat. Palermo Series II, **XXXI** (1982), 145-158.

[Co] L. Colzani, Hardy space on sphere, Ph.D. thesis, Washington University, *St* Louis, 1982.

[Coi] R. Coifman, A real variable characterization of H^p, Studia Math., **51**(1974), 269-274.

[CW] R. Coifman and G. Weiss, Extension of Hardy spaces and their use in analysis, Bull. Amer. Math. Soc. **83** (1977) , 569-645.

[Da] David, G., Wavelets and Singular Integrals on Curves and Surfaces, Lecture Notes in Math., **1465**, 1991.

[$F1$] D. Fan, Calderón-Zygmund operators on compact Lie groups, to appear, Math. Z.

[$F2$] D. Fan, Multiplier transformation on H^p spaces, Preprint, 1990.

[$F3$] D. Fan, L^p estimate for bi-invariant operators on classial groups, Preprint, 1993.

[$F4$] D. Fan, Hardy spaces and Gong kernel on compact Lie groups, preprint, 1993.

[$FX1$] D. Fan and Z. Xu, Characterization of Lipschitz spaces on compact Lie groups, to appear, J. of Australian Math.

[$FX2$] D. Fan and Z. Xu, H^p estimates for bi-invariant operators on compact Lie groups, to appear, Proceedings of Amer. Math. Soc.

[$FX3$] D. Fan and Z. Xu, A Hörmander multiplier theorem on $SU(2)$, to appear, Chinese Ann. Math.

[FJW] M. Frazier, B. Jawerth and G. Weiss, Liufm tlewood-Paley theory and the study of function spaces, AMS-CBMS Regional Conference Series, Vol. **79**.

[Fe] R. Fefferman, Multiprameter Fourier Analysis, Beijing Lectures in Harmonic Analysis, edited by E. Stein, Annals of Math. Studies, Princeton Univ. Press, Princeton, 1986.

[FS] C. Fefferman and E.M. Stein, H^p space of several variables, Acta Math. (1972), 137-193.

[FTW] M. Frazier, R. Torresand and G. Weiss, The boundedness of Calderón-Zygmund operators on the spaces $\dot{F}_p^{\alpha,q}$, Rev. Mat. Iberoamericana, **4** (1988) , 41-72.

[FoS] G. Folland and E. Stein, Hardy spaces on Homogeneous Groups, Princeton University Press, Princeton, New Jersey, 1982.

[$G1$] S. Gong, Harmonic Analysis on Classical Groups, Springer-Verlag, Science Press, Beijing, 1991.

[$G2$] S. Gong, Singular integral in several complex variables, Series of modern mathematics, Sciences and Technology Press, Shanghai, 1982.

[*H*] Y. Han, Certain Hardy-type spaces, Ph.D. thesis, Washington Univ. St. Louis, 1984.

[*Hua*] L. K. Hua, Harmonic analysis of functions of several complex variables in the classical domain, Vol. **6**, Translations of Math. Monographs, Amer. Math. Soc., Providence, 1963.

[*JN*] F. John and L. Nirenberg, On functions of bounded mean oscillation, Comm. Pure Appl. Math. **14** (1961), 415-426.

[*L*] R. Laufm ter, A characterization of $H^p(\mathbf{R}^n)$ in term of atoms, Studia Math. **LXII** (1978) 93-101.

[*Lu*] S. *Lu*, Decomposition of kernel and maximal generalized Bochner-Riesz means, Research Report, Center for Math. Anal., The Australian National Univ., Canberra, 1987.

[*deL*] K. deLeeuw, On L^p multipliers, Ann. of Math. **81** (1965), 364-379.

[*S*1] E.M. Stein, Topics in Harmonic Analysis, Ann. of Math. Studies **63**, Princeton University Press, 1970.

[*S*2] E.M. Stein, Singular integrals and differentiability properties of Functions, Princeton University Press, 1970.

[*Sr*] R. Strichartz, Boundary values of solutions of elliptic equations satisfying H^p conditions, Trans. Amer. Math. Soc. **176** (1973), 445-462.

[*SW*] E.M. Stein and G. Weiss, Introduction to Fourier analysis on Euclidean spaces, Princeton University Press, 1971.

[*T*] A. Tochinsky, Real variable method in harmonic analysis, Pure and Appl. Math. Vol. **123**, Academic Press, 1986.

[*To*] R. Torros, Boundedness results for operators with singular kernels on distribution spaces, Mem. Amer. Math. Soc. **442** (1991).

[*Tr*] W. Triebels, Some Fourier multiplier criteria and spherical Bochner-Riesz, Rev. Roum. Math. Pures et Appl. Tome *XX*, No. 10 (1975), 1173-1185.

[*TW*] M. Taibleson and G. Weiss, The molecular characterization of certain Hardy spaces, Astersque **77** (1980), 68-149.

[*U*] A. Uchiyama, A maximal function characterization of H^p on the space of homogeneous type, Trans. Amer. Math. Soc. **262** (1980), 579-592.

[*Wa*] J. Wang, Generalized Bochner-Riesz means on compact Lie groups, to appear, Acta Math. Sin.

[*W*] N. J. Weiss, L^p estimate for bi-invariant operators on compact Lie Group, Amer. J. of Math., **94** (1972), 103-118.

[*X*] Z. *Xu*, The generalized Abel means of H^p functions on compact Lie Groups, Chinese Ann. Math., **13A:1** (1992). 101-110.

[Z] A. Zygmund, Trigonometric Series, 2*nd* ed. Cambridge Univ. Press, 1959.

Current Address:
Department of Mathematical Sciences
University of Wisconsin-Milwaukee
Milwaukee, *WI*53201, U.S.A.
e-mail: fan@csd4.csd.uwm.edu

The Unitary Dual of the Covering Groups of GL(n) over a Local Field

Jing-Song Huang
Department of Mathematics
HKUST
Clear Water Bay, Kowloon
Hong Kong

1. Introduction

A fundamental problem in harmonic analysis on a reductive group G over a local field F is to describe the equivalence classes of the irreducible representations of G, which is well known as the unitary dual of G. In the case F is an archimedean field, the big break through in last decade are David Vogan's classification of the unitary dual of general line groups [V] and Dan Barbasch's classification of the unitary dual for complex classical Lie groups [Ba]. In the case F is a non-archimedean local field, Marko Tadić classified the unitary dual of the general linear groups [T2]. Besides these groups the only groups whose unitary duals are known are those groups of small ranks. The unitary dual problem remains to be one of the most challenging questions in noncommutative harmonic analysis.

Let **G=GL(n)** be the general linear group and F a local field. Write $G = \mathrm{GL}(n)$ for the group of F-rational points of **G**. If F is archimedean, this author classified the unitary dual of the universal covering group of G [H1]. Denote by $\widetilde{G}$ a covering group of G with a preferred section (cf. Section 2). In this paper we obtain a description of the unitary dual of $\widetilde{G}$. We show that any unitary representation is parabolically induced from a special class of unitary representations called Speh representations and the complementary series of Speh representations. Together with known results mentioned above this offers new insight of the nature of the unitary dual of reductive groups over a local field.

More precisely, we let k be a positive integer and define the k-th root of unit to be the set

$$\mu_k(F) = \{x \in F \mid x^k = 1\}. \tag{1.1}$$

Assume that $|\mu_k(F)| = k$. The covering group $\widetilde{G} = \widetilde{\mathrm{GL}}(n)$ is defined to be a central extension of G with a preferred section s:

$$1 \to \mu_k(F) \to \widetilde{G} \underset{s}{\overset{p}{\rightleftarrows}} G \to 1. \tag{1.2}$$

The set $\mu_k(F)$ may be identified as a subgroup of the center of $\widetilde{G}$.

Fix a faithful character ε of $\mu_k(F)$. Let $\mathrm{R}\widetilde{G}$ be the category of smooth representations of $\widetilde{G}$ such that $\mu_k(F)$ acts via ε. An element in $\mathrm{R}\widetilde{G}$ is called a genuine representation. A

M. Cheng et al. (eds.), Harmonic Analysis in China, 103–124.

representation of $\widetilde{G}$ which is not genuine can factor through a k_1-fold cover of G, where k_1 is a factor of k and $k_1 \neq k$. From now on, we assume that all of the representations of $\widetilde{G}$ we consider are genuine.

The main result in this paper can be formulated as follows. Let $\mathcal{D}^u$ denote the set of equivalence classes of irreducible unitary square-integrable representations and Irr^u the set of equivalence classes of irreducible unitary representations of $\widetilde{\mathrm{GL}}(n)$, $n = 1, 2, \cdots$. For $\sigma, \tau \in \mathrm{R}\widetilde{G}$, denote by $\sigma \times \tau$ the induced representation from a tensor product of σ and τ (cf. Section 2). If $n \in \mathbb{N}$ and $\delta \in \mathcal{D}^u$, we set the Speh representation $u(\delta, n)$ to be the Langlands quotient of the induced representation

$$\nu^{(n-1)/2}\delta \times \nu^{(n-3)/2}\delta \times \cdots \times \nu^{-(n-1)/2}\delta.$$

Here ν is the character of $\widetilde{G}$ such that $\nu(\tilde{g}) = |\det op(\tilde{g})|^{1/k}$. For $0 < \alpha < 1/2$ the induced representation $\nu^{\alpha}u(\delta, n) \times \nu^{-\alpha}u(\delta, n)$ is irreducible and unitary. These kind of representations are called the complementary series of Speh representations. Denote by A the set of all Speh representations and their complementary series, i.e.

$$A = \{u(\delta, n), \nu^{\alpha}u(\delta, n) \times \nu^{-\alpha}u(\delta, n) \mid \delta \in \mathcal{D}^u, n \in \mathbb{N}, 0 < \alpha < 1/2\}.$$

Then one has

(i) If $\sigma_1, \cdots, \sigma_r$ are elements in A, then $\sigma_1 \times \cdots \times \sigma_r \in \mathrm{Irr}^u$.

(ii) If $\pi \in \mathrm{Irr}^u$ then there exist $\tau_1, \cdots, \tau_s \in A$ so that

$$\pi = \tau_1 \times \cdots \times \tau_s.$$

Moreover, such τ_i's are unique up to a permutation.

Here is the organization of the paper. We first prove Kirillov's conjecture for the covering groups in Section 3 and 4 and then describe the Zelevinsky classification for the covering groups in section 5 and prove the unitarity and irreducibility results in Section 6. We make the preparation from Section 7 through Section 9 and obtain the classification of the unitary dual in Section 10. In the last section we prove the key lemma used in Section 10. We assume that the additive automorphism t (defined in Section 7) of the ring of the category of Grothendick group of all smooth representations, which relates the Langlands classification and Zelevinsky classification, is multiplicative as well in Section 10. This is the same approach used in [T2].

We have a complete different approach of the unitary dual of the covering group $\widetilde{G}$ by using Hecke algebra isomorphisms studied in [H2], [HM] and [BK]. We will discuss this approach at another paper [H3]. It is a pleasure to thank D. Miličić and M. Tadić for helpful conversations.

2. The covering groups

Let F be a non-Archimedean local field. We assume that F contains the group of all k-th roots of unit. We put p for the residual characteristic of F; R for its ring of integers; π for a uniformizer; q for the cardinality of the field $R/\pi R$.

Let $G = \mathrm{GL}(n)$ be the general linear group over F with a fixed open maximal compact subgroup K contained in $\mathrm{GL}(n, R)$. If $(n, k) = 1$, we can put $K = \mathrm{GL}(n, R)$. In [KP] the k-fold covers of G was constructed by the method of Matsumoto [M]. Each cover has a preferred section s. There are finite number of distinct covers and they are parametrized by some non-negative integers c. The representation theories of these covers are essentially the same. We will fix one cover and will not mention its dependence on the parameter c anymore. This covering group is denoted by $\widetilde{G} = \widetilde{\mathrm{GL}}(n)$ and its dependence on c should not be forgotten.

The group $\widetilde{G}$ is the central extension of G with a preferred section s:

$$1 \to \mu_k(F) \to \widetilde{G} \underset{s}{\overset{p}{\rightleftarrows}} G \to 1. \tag{2.1}$$

The cover is split over the maximal compact subgroup K. We write $K^* = s(K)$, the lift of K. Let $Z = \{xI \mid x^{n-1+2nc} \in F^{\times k}\} \cong F^{\times k/d}$ be a subgroup of G, where $d = (k, n-1+2nc)$. Then $\widetilde{Z} = p^{-1}\{Z\}$ is the center of $\widetilde{G}$ as well as the center of $\widetilde{B} = p^{-1}\{B\}$, where B is the Borel subgroup of G consisting of all upper triangular matrices (cf. [KP] Proposition 0.1.1).

Fix a faithful character ε of $\mu_k(F)$. Let $\mathrm{R}\widetilde{G}$ be the category of smooth representations of $\widetilde{G}$ such that $\mu_k(F)$ acts via ε. To define the parabolic induction it is convenient to fix a character $\widetilde{\omega}$ of $\widetilde{Z}$ whose restriction to $\mu_k(F)$ is ε. By a parabolic (resp. Levi) subgroup $\widetilde{P}$ (resp. $\widetilde{M}$) of $\widetilde{G}$ we mean the pullback via $p : \widetilde{G} \to G$ of such a subgroup P (resp. M) of G. We define the representations of $\widetilde{G}$ induced from representations of $\widetilde{M}$ same as that defined in [FK]. Suppose that $M = M_1 \times \cdots \times M_l$ with $M_i = \mathrm{GL}(n_i)$. We identify M_i as a subgroup of M. Then each $\widetilde{m} \in \widetilde{M}$ can be written (not uniquely) in the form of $\widetilde{m}_1 \times \cdots \times \widetilde{m}_l$ with $\widetilde{m}_i \in \widetilde{M}_i$.

Let B be a maximal subgroup of $F^\times$ with the property that $(b, b') = 1$ for all $b, b' \in B$, where $(\ ,\ )$ is the k-th order Hilbert symbol. The group B contains $F^{\times k}$ as a subgroup. Let $\widetilde{M}_i^B$ be the group of $\widetilde{m}_i \in \widetilde{M}_i$ with $\det \mathrm{op}(\widetilde{m}_i)$ in B, and $\widetilde{M}^B$ be the group of $\widetilde{m} \in \widetilde{M}$ with $\widetilde{m}_i \in \widetilde{M}_i^B$ for all i. Then $\widetilde{m}_i\widetilde{m}_j = \widetilde{m}_j\widetilde{m}_i$ for any $\widetilde{m}_i, \widetilde{m}_j$ in $\widetilde{M}^B$ any $i \neq j$. Let $\widetilde{\rho}_i$ be a genuine irreducible $\widetilde{M}$-module and we extend it to be $\widetilde{Z}\widetilde{M}_i$-module which transforms under $\widetilde{Z}$ by the character $\widetilde{\omega}$. Its restriction $\widetilde{\rho}_i|_{\widetilde{Z}\widetilde{M}_i^B}$ is the sum of the conjugates $\widetilde{\rho}'^m$ of some irreducible $\widetilde{\rho}'$ by $m \in \widetilde{M}_i/\widetilde{Z}\widetilde{M}_i^B$. They are all inequivalent unless $M_i = GL(1)$. Let $\widetilde{\rho} = \otimes\widetilde{\rho}'_i$ be their tensor product. The $\widetilde{M}$-module $\widetilde{\rho}$ induced from the $\widetilde{Z}\widetilde{M}^B$-module $\widetilde{\rho}'$ is irreducible and independent of the choice of each $\widetilde{\rho}'_i$, because for any $\widetilde{m} \in \widetilde{M} - \widetilde{Z}\widetilde{M}^B$ the modules $\widetilde{\rho}'$ and its conjugate $\widetilde{\rho}'^m$ are inequivalent. Hence we write $\otimes\widetilde{\rho}_i$ for $\widetilde{\rho}$ and $\mathrm{I}(\widetilde{\rho}) = \widetilde{\rho}_1 \times \cdots \times \widetilde{\rho}_l$ for the $\widetilde{G}$-module unitarily induced from $\widetilde{\rho}$.

3. Distributions on the covering groups

This section is a preparation for the proof of Kirillov's conjecture for the covering groups in the next section. We will make use Bernstein's results in [B] and extend them to the covering groups.

Definition 3.1 Let X be an l-space, that is a Hausdorff topological space which has a basis consisting of open compact subsets. Let $S(X)$ denote the space of smooth (locally constant) functions on X with compact support. A linear functional E on $S(X)$ is called a distributions on X. The set of all distributions on X is denoted by $S^*(X)$.

Let $q : X \to T$ be a continuous map of l-spaces. Then both $S(X)$ and $S^*(X)$ are $S(T)$-modules. For any $t \in T$ consider the fiber $X_t = q^{-1}\{t\}$ and identify subspace $S^*(X_t)$ with the subspace $S^*_{X_t}(X) \subset S^*(X)$ of distributions with support on the fiber X_t.

Proposition 3.2 ([B] Localization principal) *Let W be a closed subspace of $S^*(X)$ which is also a $S(T)$-submodule. Then W is generated by distributions supported on the fibers.*

Corollary 3.3 [B] *Let an l-group G act on the space X preserving each fiber X_t, and let P be a subgroup of G. Suppose that for each $t \in T$ all P-invariant distributions on X_t are also G-invariant, then any P-invariant distribution on X is G-invariant.*

Now put $G = G_n = \mathrm{GL}(n)$ and $P = P_n$ a parabolic subgroup of G defined as

$$P_n = \{g = (g_{ij} \in G_n \mid g_{ni} = \delta_{ni} \text{ for all } i\}.$$

Let $\widetilde{G}$ be the the covering group of G described in Section 2, and let $\widetilde{P}$ be the $p^{-1}(P)$ be the parabolic subgroup of $\widetilde{G}$. Denote by Ad the adjoint action.

Let $X = X_n$ denote the set of all n by n matrices with entries in F. For $x \in X_n$, let t_x denote the characteristic polynomial of x. Let $A = A_n$ denote the space of row-vectors of length n and let $e_1, \cdots, e_n$ be a basis of A. Let K_x be the subspace of A spanned by $\{e_n, e_n x, \cdots, e_n x^n\}$. Let τ_x denote the characteristic polynomial of the operator x on K_x.

A matrix $x \in X$ is called P-regular if $\tau_x(x) = 0$. It is clear that t_x is constant along G-orbit and τ_x is constant along P-orbit. We need the following geometric lemma.

Lemma 3.4 ([B] Geometric Lemma)

(a) For any polynomial t the set $X_t = \{x \in X \mid t_x = t\}$ contains a finite number of G-orbits.

(b) Each G-orbit $\mathcal{O}$ contains a finite number of P-orbits.

(c) Each G-orbit $\mathcal{O}$ contains a unique P-orbit $\mathcal{O}_P$ open and dense in $\mathcal{O}$. Namely $\mathcal{O}_P = \{x \in \mathcal{O} \mid x \text{ is } P-\text{regular}\}$.

Let $\tilde{x} \in \widetilde{G}$ then the centralizer of $\tilde{x}$ in $\widetilde{G}$ depends only on $p(\tilde{x})$ and therefore be denoted by $\widetilde{G}_x$. We say $x \in G$ is good if $p(\widetilde{G}_x) = G_x$, i.e. $yxy^{-1} = x$ and $p(\tilde{y}) = y$ imply that $\tilde{y}\tilde{x}\tilde{y}^{-1} = \tilde{x}$. If x is good then G-orbit $\mathcal{O}_x$ can be identified with the $\widetilde{G}$-orbit $\mathcal{O}_{\tilde{x}}$ via the isomorphism $G/G_x \cong \widetilde{G}/\widetilde{G}_x$. If x is not good then there exists $y \in G_x$ such that $\tilde{y}\tilde{x}\tilde{y}^{-1} = \varepsilon\tilde{x}$ for some $\varepsilon \in \mu_k(F)$. In this case $\widetilde{G}$-orbit $\mathcal{O}_{\tilde{x}}$ is a k-fold cover of the G-orbit $\mathcal{O}_x$ via the projection map p. We say $x \in G$ is P-good if $p(\widetilde{P}_x) = P_x$. It is clear that x is P-good if it is good, but conversely being P-good does not imply being good.

Remark 3.5 The statements (a) and (b) of Lemma 3.4 also hold for the covering group $\widetilde{G}$. But (c) may fail because a $\widetilde{G}$-orbit $\mathcal{O}_{\tilde{x}}$ contains k open $\widetilde{P}$-orbit instead of only one if x is P-good but not good.

Lemma 3.6 *([B] Key Lemma) Let E be a P-invariant distribution on G and S =supp E. Then S contains an open P-orbit $\mathcal{O}_P$ which consists of P-regular elements.*

Lemma 3.7 *([R],[FK]) For any $\widetilde{G}$-orbit $\mathcal{O} \subset \widetilde{G}$ there exists a $\widetilde{G}$-invariant distribution $\mu_{\mathcal{O}}$ such that supp $\mu_{\mathcal{O}} = \overline{\mathcal{O}}$.*

It is proved in [B] that any P-invariant distribution on G is actually G-invariant. The similar statement for the covering group is also true with a little modification.

Theorem 3.8 *Let E be a distribution on $\widetilde{G}$ invariant under the adjoint action of $\widetilde{P}$. If E is also quasi-invariant under the (translation) action of $\mu_k(F)$, which means $\mu_k(F)$ acting on E via a character. Then E is invariant under the adjoint action of $\widetilde{G}$.*

Proof. Put $S =$ supp $E \subset \widetilde{G}$. Let T be the space of polynomial of degree n and $\tilde{q}$ be the composition of the projection map p and the characteristic map $q : x \mapsto t_x$. Using the localization principle (Corollary 3.3) we may assume that $S \subset X_t$ for some t. Then by Lemma 3.5 (a) and (b), S contains only finite number of $\widetilde{P}$-orbits. We will prove the theorem by induction on the number of $\widetilde{P}$-orbits in S.

The projection map $p : \widetilde{G} \to G$ induces a map from Schwarz space of G to Schwarz space of $\widetilde{G}$, i.e.

$$p_I : S(G) \to S(\widetilde{G})$$

given by pull back. The adjoint of p_I is a map from the distributions on $\widetilde{G}$ to distributions on G, i.e.

$$p_I^* : S^*(\widetilde{G}) \to S^*(G).$$

Let $E_1 = p_I^*(E)$ and S_1 =supp E_1. It is clear that $S_1 = p(S)$. Since E is $\widetilde{P}$-invariant, E_1 is P-invariant. By Lemma 3.6 S_1 contains an open P-orbits $\mathcal{O}_P$ which consists of P-regular elements. Since E is also quasi-invariant under the translation action of $\mu_k(F)$, S is stable under the action of $\mu_k(F)$. Therefore S contains the preimage $\mathcal{O}_{\widetilde{P}} = p^{-1}(\mathcal{O}_P)$, which is either an open $\widetilde{P}$-orbit or a union of k open $\widetilde{P}$-orbits. Consider $\widetilde{G}$-orbit(s) $\mathcal{O}$ =Ad$(G)\mathcal{O}_{\widetilde{P}}$. Since $\mathcal{O}_P$ consists of P-regular elements, $\mathcal{O}_{\widetilde{P}}$ is open dense in $\mathcal{O}$, i.e. $\overline{\mathcal{O}} = \overline{\mathcal{O}_{\widetilde{P}}} \subset S$.

By Lemma 3.7 there exists a $\widetilde{G}$-invariant distribution $\mu_{\mathcal{O}}$ such that supp $\mu_{\mathcal{O}} = \overline{\mathcal{O}}$. Since $\mathcal{O}_{\widetilde{P}}$ is open in S, we can consider the restriction of the distribution E and $\mu_{\mathcal{O}}$ on the $\widetilde{P}$-orbits $\mathcal{O}_{\widetilde{P}}$. They are both $\widetilde{P}$-invariant and non-zero. Hence there exists a $c \in \mathbb{C}^*$ such that $E|_{\mathcal{O}_{\widetilde{P}}} = c\mu_{\mathcal{O}}|_{\mathcal{O}_{\widetilde{P}}}$. This mean that the distribution $E_0 = E - c\mu_{\mathcal{O}}$ is $\widetilde{P}$-invariant and supp E_0 contains strictly fewer $\widetilde{P}$-orbits than S. Hence an induction argument will give the desired conclusion.

Q.E.D.

4. Kirillov's conjecture for the covering groups

We retain the notations in the previous sections. Let π be an unitary irreducible representation of G, then its restriction of π to P is also irreducible. This is well-known as Kirillov's conjecture and was proved by Bernstein [B]. In this section we will prove the similar statement is also true for the covering groups.

Theorem 4.1 *Let $(\pi, \widetilde{G}, V_\pi)$ be a genuine irreducible unitary representation, then its restriction to $\widetilde{P}$ is also irreducible.*

We will follow the same line of Bernstein's proof of the case $G = \mathrm{GL}(n)$ with a slight modification.

Lemma 4.2 (cf. [B] 5.6) *For any admissible representation $(\pi, \widetilde{G}, V_\pi)$ let $(\pi^*, \widetilde{G}, V_\pi^*)$ be its contragradient. Then there exists a nonzero morphism of $\widetilde{G} \times \widetilde{G}$-modules*

$$\pi_\mu : S(\widetilde{G}) \to V_\pi \otimes V_\pi^*.$$

If π is irreducible, then π_μ is an epimorphism.

Proposition 4.3 *Let $(\pi, \widetilde{G}, V_\pi)$ be a genuine smooth irreducible representation, $(\pi^*, \widetilde{G}, V_\pi^*)$ be its contragradient, $B_0 : V_\pi \times V_\pi^* \to \mathbb{C}$ the canonical pairing $B_0(v, v^*) = \langle v, v^* \rangle$. Then any $\widetilde{P}$-invariant and $\mu_k(F)$-quasi-invariant pairing $B : V_\pi \times V_\pi^* \to \mathbb{C}$ is $\widetilde{G}$-invariant and therefore is proportional to B_0.*

Proof. It is known that π and π^* are admissible and so $(\pi \otimes \pi^*, \widetilde{G} \times \widetilde{G}, V_\pi \otimes V_\pi^*)$ is admissible and irreducible. Consider the regular representation $(R, \widetilde{G} \times \widetilde{G}, S(\widetilde{G}))$ given by $R(g_1, g_2)f(x) = f(g_1^{-1} x g_2)$, for $f \in S(\widetilde{G})$ and $g_1, g_2, x \in \widetilde{G}$. The parings B correspond to morphism $V_\pi \otimes V_\pi^* \to \mathbb{C}$, i.e. to elements of the dual of $V_\pi \times V_\pi^*$, which will be denoted by $(V_\pi \times V_\pi^*)^*$. By Lemma 4.2 there is a epimorphism of $\widetilde{G} \times \widetilde{G}$-modules

$$\pi_\mu : S(\widetilde{G}) \to V_\pi \otimes V_\pi^*.$$

Its adjoint $\pi_\mu^* : V_\pi \otimes V_\pi^* \to S^*(\widetilde{G})$ is a $\widetilde{G} \times \widetilde{G}$-invariant monomorphism. In particular, if a pairing B is g-invariant for some $g \in \widetilde{G}$ ($\widetilde{G}$ is imbedded diagonally in $\widetilde{G} \times \widetilde{G}$), then the corresponding distribution E_B is $\mathrm{Ad}(g)$-invariant. It is clear that E_B is also $\mu_k(F)$-quasi-invariant. Hence by Theorem 3.8, E_B is $\widetilde{G}$-invariant. Since π is irreducible and admissible, B is proportional to the standard pairing B_0.

Q.E.D.

Corollary 4.4 *Let $(\pi, \widetilde{G}, V_\pi)$ be a smooth irreducible representation, and let B_0 be a nonzero $\widetilde{G}$-invariant and $\mu_k(F)$-quasi-invariant bilinear (resp. Hermitian) form on V_π. Then any $\widetilde{P}$-invariant and $\mu_k(F)$-quasi-invariant bilinear (resp. Hermitian) form B on V_π is proportional to B_0.*

Proof. Since π is admissible and irreducible, the pairing B_0 defines an isomorphism $V_\pi \cong V_\pi^*$ (resp. $\bar{V}_\pi \cong V_\pi^*$, where $\bar{V}_\pi$ is the complex dual of V_π). The form B defines a $\widetilde{P}$-invariant and $\mu_k(F)$-quasi-invariant pairing of with V_π with $V_\pi \cong V_\pi^*$ (resp. $\bar{V}_\pi \cong V_\pi^*$. Hence B is proportional to B_0.

Q.E.D.

Proof of Theorem 4.1 It is sufficient to show that any $\widetilde{P}$-equivariant morphism $\alpha : V_\pi \to V_\pi$ is a scalar operator. Consider the Hermitian form B_α given by $B_\alpha(v,w) = \langle \alpha v, w \rangle$, where $\langle \ , \ \rangle$ is the scalar product on V_π. The form B_α is $\widetilde{P}$-invariant (and is also $\mu_k(F)$-quasi-invariant), by Corollary 5.4 we have $B_\alpha(v,w) = c\langle v,w \rangle$ for some $c \in \mathbb{C}$. That is $\alpha = cI$.

Q.E.D.

Lemma 4.5 *Let $(\pi, \widetilde{G}, E)$ be a genuine smooth irreducible representation. Suppose that π is Hermitian and $\pi|_{\widetilde{P}}$ is unitarizable. Then π is unitarizable.*

Proof. Let B_0 be a $\widetilde{G}$-invariant Hermitian form on E and B a $\widetilde{P}$-invariant positive definite forms on E. By Corollary 4.4 B is $\widetilde{G}$-invariant and propositional to B_0. Therefore π is unitarizable.

Q.E.D.

5. Zelevinsky classification

Let $\mathrm{R}\widetilde{G}_n$ (resp. $\mathrm{R}\widetilde{P}_n$) be the category of all genuine smooth $\widetilde{G}_n$-modules (resp. $\widetilde{Z}\widetilde{P}_n$-modules) of finite length. Let $\mathcal{R}_n = \mathcal{R}(\widetilde{G}_n)$ be the Grothendieck group of the category of $\mathrm{R}\widetilde{G}_n$ and $\mathcal{R} = \oplus \mathcal{R}_n (n = 0, 1, 2, \cdots)$. The induction functor defined in Section 2 $(\pi_1, \pi_2) \to \pi_1 \times \pi_2$ induces a bilinear morphism $\mathcal{R}_m \times \mathcal{R}_n \to \mathcal{R}_{m+n}$. Hence $\mathcal{R}$ is a ring. The set of all equivalence classes of genuine irreducible representations in $\mathrm{R}\widetilde{G}_n (n = 0, 1, 2, \cdots)$ is denoted by $\mathrm{Irr} = \cup \mathrm{Irr}\widetilde{G}_n$ and the subset of unitarizable representations in Irr is denoted by Irr^u.

The group G_{n-1} (and also P_{n-1}) is embedded into P_n in a standard way; denote by $V = V_n$ the unipotent radical of P_n:

$$V = \{(g_{ij}) \mid g_{ij} = \delta_{ij} \text{ for } j < n\}.$$

The cover is split over V. Denote by V^* the lift of V. It is clear that $\widetilde{G}_{n-1}$ normalizes V^*, $\widetilde{G}_{n-1} \cap V^* = \{e\}$ and $\widetilde{P}_n = \widetilde{G}_{n-1} \cdot V^*$. Fix a non-trivial additive character ψ of the field F and define the character θ of the group V_n (and therefore V_n^*) by $\theta((v_{ij})) = \psi(v_{n-1,n})$. $\widetilde{P}_{n-1}$ normalizes θ. The functors Ψ^-, Φ^-, Ψ^+ and Φ^+ are defined in a similar way as in [BZ2]:

$$\Psi^- : \mathrm{R}\widetilde{P}_n \to \mathrm{R}\widetilde{G}_{n-1}, \quad \Phi^- : \mathrm{R}\widetilde{P}_n \to \mathrm{R}\widetilde{P}_{n-1}$$
$$\Psi^+ : \mathrm{R}\widetilde{G}_{n-1} \to \mathrm{R}\widetilde{P}_n, \quad \Phi^+ : \mathrm{R}\widetilde{P}_{n-1} \to \mathrm{R}\widetilde{P}_n$$

with $\Psi^- = r_{V^*,1}$, $\Phi^- = r_{V^*,\theta}$, $\Psi^+ = i_{V^*,1}$ and $\Phi^+ = i_{V^*,\theta}$ (cf. [BZ2] §1.8).

For any representation $(\pi, \widetilde{P}_n, E) \in \mathrm{R}\widetilde{P}_n$ its derivatives $\pi^{(l)} \in \mathrm{R}\widetilde{G}_{n-l}$ $(l = 1, 2, \cdots, n)$ are defined by $\pi^{(l)} = \Psi^-(\Phi^-)^{l-1}\pi$. For any representation $(\pi, \widetilde{G}_n, E) \in \mathrm{R}\widetilde{G}_n$ its derivatives

$\pi^{(l)} \in \mathrm{R}\widetilde{G}_{n-l}$ $(l = 0, 1, 2, \cdots, n)$ are defined by $\pi^{(0)} = \pi$, $\pi^{(l)} = (\pi|_{\widetilde{P}})^{(l)}$. The shifted derivatives $\pi^{[l]} \in \mathrm{R}\widetilde{G}_{n-l}$ $(l = 0, 1, 2, \cdots, n)$ are defined by $\pi^{[l]} = \nu^{\frac{1}{2}}\pi^{(l)}$, where ν is the character of $\widetilde{G}$ such that $\nu(\tilde{g}) = |\det(p(\tilde{g}))|^{1/k}$.

A representation $\rho \in \mathrm{R}\widetilde{G}_n$ is called supercuspidal if its matrix coefficients have compact supports modulo the center. Let $\mathcal{C} \subset \mathrm{Irr}$ denote the set of all equivalence classes of genuine irreducible supercuspidal representations and $\mathcal{C}^u$ the subset of $\mathcal{C}$ which contains unitarizable elements. For $\rho \in \mathcal{C}^u$ and $\alpha \in \mathbb{R}$ let ρ_α be the supercuspidal representation $\nu^\alpha \rho$. Then $\mathcal{C} = \{\rho_\alpha \mid \rho \in \mathcal{C}^u, \alpha \in \mathbb{R}\}$.

A subset in $\mathcal{C}$ of the form $\Delta = \{\rho, \nu\rho, \nu^2\rho, \cdots, \nu^{m-1}\rho\}$ is called a segment in $\mathcal{C}$. The set of all segments in $\mathcal{C}$ is denoted by $\mathcal{S}$. To each segment $\Delta = \{\rho, \nu\rho, \nu^2\rho, \cdots, \nu^{m-1}\rho\}$ in $\mathcal{C}$ (i.e. $\Delta \in \mathcal{S}$) we associate the irreducible representation $Z\langle\Delta\rangle$. It may be defined as the unique irreducible subrepresentations of $\rho \times \nu\rho \times \cdots \times \nu^{m-1}\rho$ or as the unique irreducible quotient of $\nu^{m-1}\rho \times \nu^{m-2}\rho \times \cdots \times \rho$.

For each segment $\Delta = \{\rho, \nu\rho, \nu^2\rho, \cdots, \nu^{m-1}\rho\}$ in $\mathcal{C}$ set $\Delta^- = \Delta \backslash \{\nu^{m-1}\rho\}$ (in particular, if $\Delta = \{\rho\}$ then $\Delta^- = \emptyset$). Under the same observation in [Z1], there is exactly one of the derivatives $Z\langle\Delta\rangle^{(l)}$ for $l > 0$ is non-zero and this derivative equals $Z\langle\Delta^-\rangle$ (cf. [Z1] Theorem 3.5).

Let $\Delta_1 = \{\rho_1, \nu\rho_1, \cdots, \nu^{m_1-1}\rho_1\}$ and $\Delta_2 = \{\rho_2, \nu\rho_2, \cdots, \nu^{m_2-1}\rho_2\}$ be two segments in $\mathcal{C}$. We say that Δ_1 and Δ_2 are linked if $\Delta_1 \not\subset \Delta_2$, $\Delta_2 \not\subset \Delta_1$ and $\Delta_1 \cup \Delta_2$ is also a segment. If Δ_1 and Δ_2 are linked and $\Delta_1 \cap \Delta_2 = \emptyset$ then we say that Δ_1 and Δ_2 are juxtaposed (this means that either $\rho_2 = \nu^{m_1}$ or $\rho_1 = \nu^{m_2}\rho_2$). If Δ_1 and Δ_2 are linked and $\rho_2 = \nu^m \rho_1$ for some $m > 0$, then we say that Δ_1 precedes Δ_2.

Proposition 5.1 *Let $\Delta_1, \cdots, \Delta_r$ be segments in $\mathcal{C}$. Then the induced representation $Z\langle\Delta_1\rangle \times \cdots \times Z\langle\Delta_r\rangle$ is reducible if and only if Δ_i and Δ_j are linked for some $i, j \in \{1, 2, \cdots, r\}$.*

Theorem 5.2 *(a) Let $\Delta_1, \cdots, \Delta_r$ be segments in $\mathcal{C}$. Suppose that Δ_i does not precede Δ_j if $1 \le i < j \le r$. Then the induced representation $Z\langle\Delta_1\rangle \times \cdots \times Z\langle\Delta_r\rangle$ has a unique irreducible submodule, which is denoted by $Z\langle\Delta_1, \cdots, \Delta_r\rangle$*

(b) The modules $Z\langle\Delta_1, \cdots, \Delta_r\rangle$ and $Z\langle\Delta'_1, \cdots, \Delta'_s\rangle$ are isomorphic if and only if the sequences $(\Delta_1, \cdots, \Delta_r)$ and $(\Delta'_1, \cdots, \Delta'_s)$ are equal up to a permutation.

(c) Any representation $\pi \in \mathrm{Irr}\widetilde{G}_n$ is isomorphic to some representation of the form $Z\langle\Delta_1, \cdots, \Delta_r\rangle$.

Proposition 5.1 and Theorem 5.2 can be proved by the same method of Zelevinsky (cf. [Z1] Theorem 4.2 and Theorem 6.1). For a set X, denote by $\mathrm{M}(X)$ the set of all finite multisets in X. An element in $\mathrm{M}(X)$ can be regarded as a function $a : X \to \mathbb{Z}_+$ with finite support. By this notation $\mathrm{M}(\mathcal{S})$ denotes the set of finite multisets in $\mathcal{S}$. In other words, an element $a \in \mathrm{M}(\mathcal{S})$ is a sequence of segments $\Delta_1, \cdots, \Delta_r \in \mathcal{S}$ up to permutations. For each $a \in \mathrm{M}(\mathcal{S})$ one can choose an ordering $(\Delta_1, \cdots, \Delta_r)$ of a, satisfying Theorem 5.1 (a). Then the map $a \mapsto Z\langle a\rangle$ gives a one-to-one correspondence $\mathrm{M}(\mathcal{S}) \to \mathrm{Irr}$.

For $a = (\Delta_1, \cdots, \Delta_r) \in \mathrm{M}(\mathcal{S})$ we define the support of a to be supp $a \in \mathrm{M}(\mathcal{C})$ as following

$$(\mathrm{supp}\ a)(\rho) = \mathrm{card}\{i \mid \rho \in \Delta_i\}.$$

For $\pi \in \mathrm{Irr}$ there exist $\rho_1, \cdots, \rho_k \in \mathcal{C}$ such that π is a subrepresentation of $\rho_1 \times \cdots \times \rho_r$. The support of π is defined to be $\{\rho_1, \cdots, \rho_k\}$. One has supp $Z\langle a\rangle = \mathrm{supp}\ a$.

Let $a = (\Delta_1, \cdots, \Delta_r) \in \mathrm{M}(\mathcal{S})$. If Δ_i and Δ_j are linked and $i < j$ then

$$b = (\Delta_1, \cdots \Delta_{i-1}, \Delta_i \cap \Delta_j, \Delta_{i+1}, \cdots, \Delta_{j-1}, \Delta_i \cup \Delta_j, \Delta_{j+1}, \cdots, \Delta_r)$$

is also an multisegment in $\mathrm{M}(\mathcal{S})$. This is denoted by $b < a$. For $a, b \in \mathrm{M}(\mathcal{S})$ we say $b \leq a$ if there exist $a_1, \cdots, a_m \in \mathrm{M}(\mathcal{S})$ such that

$$b = a_1 < \cdots < a_m = a.$$

It is clear that $\leq$ is a partial order on $\mathrm{M}(\mathcal{S})$.

For $a = (\Delta_1, \cdots, \Delta_r) \in \mathrm{M}(\mathcal{S})$ set

$$\pi(a) = Z\langle\Delta_1\rangle \times \cdots \times Z\langle\Delta_r\rangle.$$

Denote by $m(b, a)$ the multiplicity of $Z\langle b\rangle$ in the Jordan-Hölder series of $\pi(a)$, i.e. in the ring $\mathcal{R}$ one has

$$\pi(a) = \sum_{b \in \mathrm{M}(\mathcal{S})} m(b, a) Z\langle b\rangle.$$

By the same argument in [Z1] one can prove that $m(b, a) \neq 0$ if and only if $b \leq a$ and $m(a, a) = 1$.

6. Unitarity and irreducibility

A smooth representation $(\pi, \widetilde{G}, V)$ is called Hermitian if there exists a Hermitian form $\langle\ ,\ \rangle$ on V such that

$$\langle \pi(g)w, v\rangle = \langle w, \pi(g^{-1})v\rangle \ \mathrm{for} w, v \in V, g \in \widetilde{G}.$$

A Hermitian representation is called unitarizable if the Hermitian form is positive definite. In this section we prove the following theorem.

Theorem 6.1

(a) If $\pi, \sigma \in \mathrm{Irr}^u$ then $\pi \times \sigma \in \mathrm{Irr}^u$.

(b) If $\pi, \sigma \in \mathrm{Irr}$ are Hermitian and $\pi \times \sigma$ is unitarizable, then π and σ are both unitarizable.

Before we prove this theorem, we need a few propositions.

We need Mackey theory of unitary representations of semi-direct products. The group $\widetilde{P}_n$ is a semi-direct product of $\widetilde{G}_{n-1}$ and V^*, where $V \cong F^{n-1}$ and V^* is the lift of V. The unitary dual of V^* can be identified with F^{n-1}. There are exactly two orbits on F^{n-1} under the action of $\widetilde{G}_{n-1}$, the zero orbit and the rest of the space. We choose a character χ so that the stablizer of χ is $\widetilde{G}_{n-1}$ is $\widetilde{P}_{n-1}$. The classical Mackey theory implies

Proposition 6.2 *Every irreducible unitary representations of $\widetilde{P}_n$ is obtained in the following two ways:*

(a) by trivially extending an irreducible unitary representation of $\widetilde{G}_{n-1}$.

(b) by extending an irreducible unitary representation of $\widetilde{P}_{n-1}$ by the character χ to $\widetilde{P}_{n-1}V^*$ and unitarily inducing to $\widetilde{P}_n$.

We use the convention $\widetilde{P}_1 = \widetilde{G}_0 = \mu_k(F)$. By above proposition every irreducible unitary representation τ of $\widetilde{P}_n$ is of the form $\tau = (\Phi^+)^{k-1}\Psi^+\sigma$ for some $k \in \mathbb{N}$ and $\sigma \in \mathrm{Irr}^u\widetilde{G}_{n-k}$, where k and σ are uniquely determined by τ. It is clear that σ is the highest shifted derivative of τ. In general, a unitary representation τ (not necessarily irreducible) of $\widetilde{P}_n$ is called homogeneous of depth k if $\tau = (\Phi^+)^{k-1}\psi^+\sigma$ for some unitary representation σ of $\widetilde{G}_{n-k}$. A unitary representation π of $\widetilde{G}_n$ is called adducible of depth k if $\pi|_{\widetilde{P}_n} = (\Phi^+)^{k-1}\Psi^+\sigma$ is homogeneous of depth k. We will write $\sigma = \mathrm{A}\pi$. Note that if π is adducible then π is irreducible if and only if $\mathrm{A}\pi$ is irreducible.

Proposition 6.3 *If $\pi \in \mathrm{R}\widetilde{G}_r$ and $\sigma \in \mathrm{R}\widetilde{G}_s$ are adducible representations of depth k and l, then $\rho \times \sigma$ is adducible of depth $k + l$ and*

$$\mathrm{A}(\pi \times \sigma) = \mathrm{A}\pi \times \mathrm{A}\sigma.$$

Sahi's proof of Theorem 2.1 in [Sa] provided a proof of this proposition.

Proof of (a) of Theorem 6.1 First of all, the unitarity of $\pi \times \sigma$ is well-known. We need only to prove its irreducibility. We will prove it by induction. If $n = 0$ or 1, this is trivially true. Let us assume this is true for all $n \leq m$. By Proposition 6.2 $\mathrm{A}\pi$ and $\mathrm{A}\sigma$ are irreducible unitary representations of $\widetilde{G}_{r-k}$ and $\widetilde{G}_{s-l}$. By induction hypothesis $\mathrm{A}\pi \times \mathrm{A}\sigma$ is irreducible. Hence $\mathrm{A}(\pi \times \sigma)$ is irreducible and so is $\pi \times \sigma$.

Q.E.D.

To prove Theorem 6.1 (b) we need to recall some known results due to Birgit Speh, David Vogan and others about the effect of induction on Hermitian forms. We may as well work in the context of a general reductive group G with a maximal compact subgroup K.

Definition 6.4 Let Y be an admissible representation of a reductive G, with an invariant Hermitian form $\langle \ , \ \rangle$. *The signature of $\langle \ , \ \rangle$ is a triple (p, q, z) of three functions from $\hat{K}$ to $\mathbb{N}$, defined as follows. Fix an irreducible representation (δ, V_δ) of K, and a positive invariant Hermitian form on V_δ. Then*

$$Y_\delta = Hom_K(V_\delta, Y) \cong (V_\delta)^* \otimes_K Y$$

acquires an invariant Hermitian form $\langle \ , \ \rangle_\delta$. We define $z(\delta)$ to be the dimension of the radical of $\langle \ , \ \rangle_\delta$, and $(p(\delta), q(\delta))$ to be the signature of the induced non-degenerate form on $Y_\delta / rad\langle \ , \ \rangle_\delta$. Thus the multiplicity of δ in Y is given by

$$m(\delta) = p(\delta) + q(\delta) + z(\delta).$$

For definiteness, we may sometimes write $p(\delta, Y)$, *etc.*

Suppose that Q is a parabolic subgroup of G with a Levi decomposition

$$Q = LU.$$

Each element of G can be written as a product of an element of K and an element of Q:

$$G = KQ.$$

Proposition 6.5 *Let* $Q = LU$ *be a parabolic subgroup of* G, *and* Y *an admissible* L-*module.*

(a) As representations of K,

$$\mathrm{Ind}(Q \uparrow G)(Y) \cong \mathrm{Ind}(L \cap K \uparrow K)(Y).$$

This isomorphism is defined by restricting functions in the induced representation to K. *In particular, if* $m(\tau, \sigma)$ *denotes the multiplicity of a representation* τ *in an appropriate restriction of* σ), *we have for any* μ *in* $\hat{K}$,

$$m(\mu, \mathrm{Ind}(Q \uparrow G)(Y)) = \sum_{\tau \in (L \cap K)^\wedge} m(\tau, \mu) m(\tau, Y).$$

(b) Suppose Y *admits an invariant Hermitian form* $\langle\ ,\ \rangle_L$. *If we regard elements of the induced representation as functions on* K *with values in* Y *(as is possible by (a)), then*

$$\langle v, w \rangle = \int_K \langle v(k), w(k) \rangle_L \, dk$$

defines an invariant Hermitian form on the induced representation. Its signiture (Definition 6.4) is given by the formula in (a), with the first and third m*'s replaced by* p, q, *or* z. *In particular, the induced form is non-degenerate (respectively, positive definite) if and only if* $\langle\ ,\ \rangle_L$ *is non-degenerate (respectively, positive definite).*

This proposition tells us that not only does the parabolic induction preserve the unitarity but also it preserves failure to be unitary. Hence it proves Theorem 6.1 (b).

7. Langlands classification

A smooth representation is called square-integrable if its matrix coefficients are square-integrable modulo the center. Let $\mathcal{D}$ denote the equivalence classes of irreducible square-integrable representations and $\mathcal{D}^u$ denote the subset of $\mathcal{D}$ which contains unitarizable elements. In this section we first classify all square-integrable representations of $\widetilde{G}$ and then formulate Langlands classification of all equivalence classes of irreducible admissible representations of $\widetilde{G}$.

Let $\Delta = \{\rho, \nu\rho, \cdots, \nu^{m-1}\rho\}$ be a segment in $\mathcal{C}$. The induced representation $\rho \times \nu\rho \times \cdots \nu^{m-1}\rho$ has a unique irreducible quotient which is denoted by $L\langle\Delta\rangle$. The representation $L\langle\Delta\rangle$ is also the unique irreducible subrepresentation of the induced representation $\nu^{m-1}\rho \times \nu^{m-2}\rho \times \cdots \rho$.

Theorem 7.1 (a) *For any segment Δ in $\mathcal{D}$ the representation $L\langle\Delta\rangle$ is square-integrable.*

(b) Any irreducible square-integrable representation of $\widetilde{G}$ is isomorphic to $L\langle\Delta\rangle$ for some segment Δ in $\mathcal{D}$.

Sketch of the proof. The statement (a) follows from Harish-Chandra's criterion ([C] Theorem 4.4.4 and Theorem 6.5.1).

The proof of the exhaustion part is based on the following two facts. The first is that each square-integrable representation is generic (or non-degenerate). This was proved by Jacquet for the linear group $\mathrm{GL}(n)$ and can be extended esaily to the covering group $\widetilde{\mathrm{GL}}(n)$ (cf. [J]). The second is that each non-degenerate representation is induced from a tensor product of representations of the form $L\langle\Delta\rangle$. This was proved by Zelevinsky for the linear group $\mathrm{GL}(n)$ and his argument can be extended to the covering group $\widetilde{\mathrm{GL}}(n)$ (cf. Section 9 of [Z1]). Since a square-integrable representation cannot be induced representation from a proper parabolic subgroup, it must be of the form $L\langle\Delta\rangle$. *Q.E.D.*

Remark 7.2 The idea of proof (b) of Theorem 7.1 was passed on to me by M. Tadić. There is another approach by using Zelevinsky's classification and Harish-Chandra's criterion. Suppose π is a square-integrable representation. By Theorem 5.2 there exists segments $\Delta_1, \cdots, \Delta_r$ in $\mathcal{C}$ such that π is the unique subrepresentation of

$$Z\langle\Delta_1\rangle \times \cdots \times Z\langle\Delta_r\rangle.$$

By Harish-Chandra's Criterion, each Δ_i must contain only one supercuspidal representation in order for π to be square-integrable. Now we have π is the unique subrepresentation of

$$\rho_1 \times \cdots \times \rho_r.$$

We want to show that $\rho_{i+1} = \nu\rho_i$ for $i = 1, \cdots, r-1$. By Proposition 4.2.3 of [C] we can reduce to the case $r = 2$, that is the case π is the unique subrepresentation of $\rho_1 \times \rho_2$. By Harish-Chandra's criterion again we have that $\rho_1 \times \rho_2$ contains a square-integrable subquotient if and only if $\rho_1 \times \rho_2$ is reducible. Hence $\rho_1 = \nu\rho_2$ by Proposition 5.1.

Theorem 7.3 (a) *Let $\Delta_1, \cdots, \Delta_r$ be segments in $\mathcal{C}$. Suppose that Δ_i does not precede Δ_j if $1 \leq i < j \leq r$. Then the representation $L\langle\Delta_1\rangle \times \cdots \times L\langle\Delta_r\rangle$ has a unique irreducible quotient, which is denoted by $L\langle\Delta_1, \cdots, \Delta_r\rangle$*

(b) The modules $L\langle\Delta_1,\cdots,\Delta_r\rangle$ and $L\langle\Delta'_1,\cdots,\Delta'_s\rangle$ are isomorphic if and only if the sequences $(\Delta_1,\cdots,\Delta_r)$ and $(\Delta'_1,\cdots,\Delta'_s)$ are equal up to a permutation.

(c) Any representation $\pi \in \mathrm{Irr}\widetilde{G}_n$ is isomorphic to some representation of the form $L\langle\Delta_1,\cdots,\Delta_r\rangle$.

The above theorem is Langlands classification of irreducible representations of $\widetilde{G}$ and is a dual version of Zelevinsky classification. Let $\mathrm{M}(\mathcal{S})$ denote the set of finite multisets in $\mathcal{S}$. In other words, an element $a \in \mathrm{M}(\mathcal{S})$ is a sequence of segments $\Delta_1,\cdots,\Delta_r \in \mathcal{S}$ up to permutations. For each $a \in \mathrm{M}(\mathcal{S})$ one can choose an ordering $(\Delta_1,\cdots,\Delta_r)$ of a, satisfying Theorem 7.2 (a). Then the map $a \mapsto L\langle a\rangle$ gives a one-to-one correspondence $\mathrm{M}(\mathcal{S}) \to \mathrm{Irr}$. It is clear that the support of $L\langle a\rangle$ is $\{\rho \in \mathcal{C} \mid \rho \in \Delta_i \text{ for some } i\}$.

Now we define an endomorphism t of Irr by

$$t : Z\langle a\rangle \mapsto L\langle a\rangle.$$

There exists a unique mapping

$$t : \mathrm{M}(\mathcal{S}) \to \mathrm{M}(\mathcal{S})$$

such that

$$t(Z\langle a\rangle) = Z\langle t(a)\rangle, \text{ for all } a \in \mathrm{M}(\mathcal{S}).$$

That is $L\langle a\rangle = Z\langle t(a)\rangle$. This implies $t(L\langle a\rangle) = L\langle t(a)\rangle$. We can extend t additively to an automorphism of $\mathcal{R}$. Then by the same argument as in Section 7 of [T3] we can show that the map t is involutive, i.e. $t(L\langle a\rangle) = Z\langle a\rangle$, since the proof in [T3] is essentially of combinatorial nature.

8. Complementary series

In this section we first study the complementary series and then we will show that the involutive map t defined in the previous section preserves the unitarity, i.e. if π is a unitary representation then $t(\pi)$ is also a unitary representation.

Suppose that $\sigma \in \mathrm{Irr}$ and $\alpha \in \mathbb{R}$. We write σ_α for $\nu^\alpha\sigma$. Let σ^h be the Hermitian dual of σ. The representation $\sigma_\alpha \times \sigma^h_{-\alpha}$ is denoted by $\pi(\sigma,\alpha)$. It is clear that

$$\pi(\sigma\times\tau,\alpha) = \pi(\sigma,\alpha)\times\pi(\tau,\alpha).$$

Theorem 8.1 *Let $\rho \in \mathcal{C}^u$ and $\sigma \in \mathrm{Irr}^u$. Suppose that* $\mathrm{supp}\sigma \in \rho_{\mathbb{Z}}$ *or* $\mathrm{supp}\sigma \in \rho_{\frac{1}{2}\mathbb{Z}}$. *Then the representations $\pi(\sigma,\alpha)$ are irreducible and unitarizable for $-1/2 \le \alpha \le 1/2$ and all composition factors of $\pi(\sigma,1/2)$ are unitarizable.*

The proof for the similar result for linear groups is standard and well-known. The same argument also works for the covering groups.

Now we need to introduce some notations. For $i,j \in \mathbb{Z}$ and $i \le j$ we call a set $\{k \mid i \le k \le j\}$ a segment in $\mathbb{Z}$. The set of all segments in $\mathbb{Z}$ is denoted by $\mathcal{S}(\mathbb{Z})$. If $\Delta \in \mathcal{S}(\mathbb{Z})$ we set $\Delta_\alpha = \{i+\alpha \mid i \in \Delta\}$. The set of all Δ_α, for $\Delta \in \mathcal{S}(\mathbb{Z})$ and $\alpha \in \mathbb{R}$, is denoted by $\mathcal{S}(\mathbb{R})$. The set of multisets in $\mathcal{S}(\mathbb{R})$ is denoted by $\mathrm{M}(\mathcal{S}(\mathbb{R}))$. For $a = (\Delta_1,\cdots,\Delta_r) \in$

$\mathrm{M}(\mathcal{S}(\mathbb{R}))$ and $\alpha \in \mathbb{R}$ we set $a_\alpha = ((\Delta_1)_\alpha, \cdots, (\Delta_r)_\alpha)$. For $\Delta \in \mathcal{S}(\mathbb{R})$ and $\rho \in \mathcal{C}$ we put $\Delta^{(\rho)} = \{\nu^\alpha \rho \mid \alpha \in \Delta\} \in \mathcal{S}$, where $\mathcal{S}$ is the set of all segments in $\mathcal{C}$ and is defined in Section 4. If $a = (\Delta_1, \cdots, \Delta_r) \in \mathrm{M}(\mathcal{S}(\mathbb{R}))$ and $\rho \in \mathcal{C}$ then we put $a^{(\rho)} = (\Delta_1^{(\rho)}, \cdots, \Delta_r^{(\rho)}) \in \mathrm{M}(\mathcal{S})$, where $\mathrm{M}(\mathcal{S})$ is the set of all finite multisets in $\mathcal{S}$. Let $a = (\Delta_1, \cdots, \Delta_r) \in \mathrm{M}(\mathcal{S}(\mathbb{R}))$. We define the support of a by

$$(\mathrm{supp}a)(i) = \mathrm{card}\{j \mid i \in \Delta_j\}.$$

For $\Delta \in \mathcal{S}(\mathbb{R})$ we set $-\Delta = \{-i \mid i \in \Delta\}$. If

$$a = (\Delta_1, \cdots, \Delta_r) \in \mathrm{M}(\mathcal{S}(\mathbb{R})),$$

we put

$$-a = (-\Delta_1, \cdots, -\Delta_r).$$

Suppose that $\rho \in \mathcal{C}^u$. Then the Hermitian dual of $Z\langle a^{(\rho)}\rangle$ which will be denoted by $Z\langle a^{(\rho)}\rangle^h$, is isomorphic to $Z\langle(-a)^{(\rho)}\rangle$.

For $n \in \mathbb{N}$ we set $\Delta(n) = \{-(n-1)/2, \cdots, (n-1)/2\} \in \mathcal{S}(\mathbb{Z})$. For $d, n \in \mathbb{N}$ we define

$$a(n,d) = (\Delta(d)_{-(n-1)/2}, \cdots, \Delta(d)_{(n-1)/2}) \in \mathrm{M}(\mathcal{S}(\mathbb{R})).$$

If $\rho \in \mathcal{C}$ then we have

$$Z\langle a(n,d)^{(\rho)}\rangle^h = Z\langle a(n,d)^{(\rho)}\rangle.$$

From now on, we assume that $\rho \in \mathcal{C}^u$ in this section.

Lemma 8.2 *The representation $Z\langle a(1,d)^{(\rho)}\rangle$ is unitarizable.*

Proof. We prove this lemma by induction on d. For $d = 1$ the representation $Z\langle a(1,1)^{(\rho)}\rangle = \rho$ is unitarizable. Assume that for $d \le m$ the representations $Z\langle a(1,d)^{(\rho)}\rangle$ is unitarizable. We want to show that $\pi = Z\langle a(1,m+1)^{(\rho)}\rangle = Z\langle \Delta(m+1)^{(\rho)}\rangle$ is unitarizable. By Lemma 4.5 it is enough to show that $Z\langle a(1,m+1)^{(\rho)}\rangle|_{\widetilde{P}}$ is unitarizable. Assume $\rho \in \mathrm{Irr}\widetilde{G}_k$ (we say $\deg\rho = k$). Then $Z\langle \Delta(m+1)^{(\rho)}\rangle|_{\widetilde{P}} = (\Phi^+)^{K-1}\Psi^+\sigma$, where $\sigma = Z\langle\Delta(m+1)\rangle^{[k]} = Z\langle\Delta(m)\rangle$ (cf. [Z] §3.6). By induction σ is unitarizable, hence π is unitarizable.
Q.E.D.

Lemma 8.3 *Let $d, n \in \mathbb{N}$. There exist $\mu_1^{(n,d)}, \cdots, \mu_n^{(n,d)} \in \mathbb{Z}_+$ such that with $\tau(n,d) = \mu_1^{(n,d)} a(1,d) + \cdots + \mu_n^{(n,d)} a(n,d)$:*

(i) Representation $Z\langle\tau(n,d)^{(\rho)}\rangle$ is unitarizable.

(ii) Representation $t(Z\langle\tau(n,d)^{(\rho)}\rangle)$ is unitarizable.

(iii) $\mu_n^{(n,.d)} = 1$.

(iv) $\mu_k^{(n,d)} = 0$ if $k+n$ is odd.

Proof. We prove this lemma by induction on n. The case $n = 1$ follows from Lemma 8.2. Suppose that we have $\tau(m,d)$ satisfying the lemma. Note that $\mathrm{supp}\tau(m,d)^{(\rho)} \in \mathrm{M}(\rho_{\mathbb{Z}})$ or

$\mathrm{supp}\tau(n,d)^{(\rho)} \in \mathrm{M}((\rho_{1/2})_{\mathbb{Z}})$. Now $\pi = Z\langle\tau(m,d)^{(\rho)}_{1/2} + \tau(m,d)^{(\rho)}_{-1/2}\rangle$ is a composition factor of $\pi(Z\langle\tau(m,d)\rangle, 1/2)$. Theorem 8.1 implies that π is unitarizable. Since

$$a(i,d)_{-1/2} + a(i,d)_{1/2} = a(i+1,d) + a(i-1,d),$$

we have

$$\tau(m,d)_{1/2} + \tau(m,d)_{-1/2} = \mu_2^{(m,d)}a(1,d) + \mu_m^{(m,d)}a(m+1,d) + \sum_{j=2}^{m}(\mu_{j-1}^{(m,d)} + \mu_{j+1}^{(m,d)})a(j,d)$$
$$= \sum_{i=1}^{m+1} \mu_i^{(m+1,d)}a(i,d).$$

This implies that $\mu_i^{(m+1,d)}$ satisfy (iii) and (iv) of the lemma. We can define $\tau(m+1,d) = \tau(m,d)_{-1/2}+\tau(m,d)_{1/2}$. The only thing remained to prove is that $t(Z\langle\tau(m+1,d)\rangle)$ is unitarizable. Note that $t(Z\langle\tau(m,d)^{(\rho)}_{1/2} + \tau(m,d)^{(\rho)}_{-1/2}\rangle)$ is a composition factor of the representation $\pi(t(Z\langle\tau(m,d)^{(\rho)}\rangle), 1/2)$. Since $\mathrm{supp}t(Z\langle\tau(m,d)^{(\rho)}\rangle) = \mathrm{supp}Z\langle\tau(m,d)^{(\rho)}\rangle$, we can apply Theorem 8.1 and induction hypothesis to obtain that all composition factors of $\pi(t(Z\langle\tau(m,d)^{(\rho)}\rangle), 1/2)$ is unitarizable. That gives the desired conclusion.

Q.E.D.

Lemma 8.4 [T1] *Let $a \in \mathrm{M}(\mathcal{S}(\mathbb{Z}))$.*

(i) There exist $(n_1,d_1),\cdots,(n_r,d_r),(n_1^*,d_1^*),\cdots,(n_s^*,d_s^*) \in \mathbb{N}\times\mathbb{N}$ and $\alpha_1,\cdots,\alpha_r,\beta_1,\cdots,\beta_s \in \{0,-1/2\}$ such that

(a) $a + \tau(n_1,d_1)_{\alpha_1} + \cdots + \tau(n_r,d_r)_{\alpha_r} = \tau(n_1^*,d_1^*)_{\beta_1} + \cdots + \tau(n_s^*,d_s^*)_{\beta_s}$.

(b) $\mathrm{supp}\tau(n_i,d_i)_{\alpha_i} \in \mathrm{M}(\mathbb{Z})$ and $\mathrm{supp}\tau(n_j^*,d_j^*)_{\beta_j} \in \mathrm{M}(\mathbb{Z})$ for all i,j.

(c) If $a = -a$ then we can choose α_i and β_j to be zero for all i,j.

(ii) If $a_{1/2} = -(a_{1/2})$ then there exist $(n_1,d_1),\cdots,(n_r,d_r),(n_1^*,d_1^*),\cdots,(n_s^*,d_s^*) \in \mathbb{N}\times\mathbb{N}$ such that

(a) $a_{1/2} + \tau(n_1,d_1) + \cdots + \tau(n_r,d_r) = \tau(n_1^*,d_1^*) + \cdots + \tau(n_s^*,d_s^*)$.

(b) $\mathrm{supp}\tau(n_i,d_i) \in \mathrm{M}(\mathbb{Z}+1/2)$ and $\mathrm{supp}\tau(n_j^*,d_j^*) \in \mathrm{M}(\mathbb{Z}+1/2)$ for all i,j.

Theorem 8.5 *If $\pi \in \mathrm{Irr}^u\widetilde{G}_n$ then $t(\pi) \in \mathrm{Irr}^u\widetilde{G}_n$.*

This theorem can be proved by the same argument used in proof of Theorem 5.1 in [T1] with the following needed modification:

(1) The proof is based on the Lemma 8.4 and Lemma 8.5;

(2) Theorem 6.1 plays the same role as Corollary 8.2 of [B];

(3) The same kind of statements of Proposition 8.4 and 8.6 in [Z1] are also true for the covering groups.

9. Speh representations

For $\rho \in \mathcal{C}^u$ and $n, d \in \mathbb{N}$ we call the representation $Z\langle a(n,d)^{(\rho)}\rangle$ a Speh representation. Denote (U^m) the following statement:

(U^m): If $n, d \in \mathbb{N}$ and $\rho \in \mathcal{C}^u$ such that $n \times d \times \deg\rho \leq m$, then $Z\langle a(n,d)^{(\rho)}\rangle$ is unitarizable.

Lemma 9.1 *Suppose that (U^{m-1}) holds for $m \geq 1$. Let $n, d \in \mathbb{N}$ and $\rho \in \mathcal{C}^u$ such that $n \times d \times \deg\rho = m$ and $n \leq d$. Then $Z\langle a(n,d)^{(\rho)}\rangle$ is unitarizable.*

Proof. If $n = 1$ this follows from Lemma 8.2. Now we suppose that $n \geq 2$. By (U^{m-1}) the representation $Z\langle a(n-1,d)^{(\rho)}\rangle$ is unitarizable. Consider the representation

$$Z\langle a(n,d)^{(\rho)} + a(n-2,d)^{(\rho)}\rangle = Z\langle \nu^{1/2}a(n-1,d)^{(\rho)} + \nu^{-1/2}a(n-1,d)^{(\rho)}\rangle.$$

It is a composition factor of

$$\nu^{1/2}Z\langle a(n-1,d)^{(\rho)}\rangle \times \nu^{-1/2}Z\langle a(n-1,d)^{(\rho)}\rangle.$$

If we can show that the representation

$$\pi = Z\langle a(n,d)^{(\rho)}\rangle \times Z\langle a(n-2,d)^{(\rho)}\rangle$$

is irreducible, then Theorem 8.1 implies that π is unitarizable. Hence by Theorem 6.1 (b) $Z\langle a(n,d)^{(\rho)}\rangle$ is also unitarizable.

What is remained to prove is the irreducibility of π. By (U^{m-1}) both $Z\langle a(n,d-1)^{(\rho)}\rangle$ and $Z\langle a(n-2,d)^{(\rho)}\rangle$ are unitarizable. Hence by Theorem 6.1 the reprsentation

$$\pi_1 = Z\langle a(n,d-1)^{(\rho)}\rangle \times Z\langle a(n-2,d)^{(\rho)}\rangle$$

is unitarizable and irreducible.

Consider the multiset

$$a = a(n,d)^{(\rho)} + a(n-2,d)^{(\rho)} = (\Delta_1, \cdots, \Delta_{2n-2}).$$

The inequality $n \leq d$ implies

$$\Delta_i \cap \Delta_j \neq \emptyset, \text{ for } 1 \leq i,j \leq 2n-2.$$

This implies for any $b \leq a$ the cardinal number of multiset b is $2n-2$. Therefore the highest derivative of b has degree $(2n-2) \times (d-1) \times \deg\rho$. If π is reducible then its highest nderivative π_1 is also reducible. This is a contradiction.

Q.E.D.

Lemma 9.2 *Let $n, d \in \mathbb{N}$ such that $n < d$. Then for any $\rho \in \mathcal{C}^u$,*

$$t(Z\langle a(n,d)^{(\rho)}\rangle) \neq Z\langle a(n,d)^{(\rho)}\rangle.$$

Lemma 9.3 *For $n, d \in \mathbb{N}$ and $\rho \in \mathcal{C}^u$, the module $Z\langle a(n,d)^{(\rho)}\rangle$ is a prime element in the ring $\mathcal{R}$. In other words, it cannot be written as induced representation of the tensor product of two irreducible representations.*

Both Lemmas 9.2 and 9.3 can be proved by the same arguments of Tadic in [T2] (cf. Propostion 3.8 and Lemma 6.3 of [T2]).

10. The unitary dual

In this section we classify all equivalence classes of genuine irreducible unitary representations of $\widetilde{G}_m = \widetilde{\mathrm{GL}}(m)$.

Theorem 10.1 (main theorem) *Let* $A = \{Z\langle a(n,d)^{(\rho)}\rangle, \pi(Z\langle a(n,d)^{(\rho)}\rangle, \alpha) \mid n, d \in \mathbb{N}, \rho \in \mathcal{C}^u, 0 < \alpha < 1/2\}$. *Fix* $m \in \mathbb{N}$. *Then*

(i) If $\sigma_1, \cdots, \sigma_r$ *are elements in* A *such that*

$$\deg\sigma_1 + \cdots + \deg\sigma_r = m,$$

then $\sigma_1 \times \cdots \times \sigma_r \in \mathrm{Irr}^u\widetilde{G}_m$.

(ii) If $\pi \in \mathrm{Irr}^u\widetilde{G}_m$ *then there exist* $\tau_1, \cdots, \tau_s \in A$ *so that*

$$\pi = \tau_1 \times \cdots \times \tau_s.$$

Moreover, such τ_i*'s are unique up to a permutation.*

(iii) The following identities hold

$$t(Z\langle a(n,d)^{(\rho)}\rangle) = Z\langle a(d,n)^{(\rho)}\rangle,$$
$$t(\pi(Z\langle a(n,d)^{(\rho)}\rangle, \alpha)) = \pi(Z\langle a(d,n)^{(\rho)}\rangle, \alpha)$$

for elements in A *with degree* m.

Proof. We prove the theorem by induction on m. For $m = 1$ the only possible element $Z\langle a(n,d)^{(\rho)}\rangle$ in A is for $d = n = 1$ and ρ a unitary character and t is identity. Hence (i), (ii) and (iii) are true for $m = 1$.

Assume that (i), (ii) and (iii) hold for those positive integers smaller than or equal to $m-1$. Them (U^{m-1}) holds. In order to prove (ii) we need the following key lemma which will be proved in the next section

Lemma 10.2 (key lemma) *Let* $m \leq 1$. *Assume that* (U^{m-1}) *holds. If* $\pi \in \mathrm{Irr}^u\widetilde{G}_m$ *and* π *is not a Speh representation, i.e.* π *is not a representation of the form* $Z\langle a(n,d)^{(\rho)}\rangle$ *for some* $n, d \in \mathbb{N}$ *and* $\rho \in \mathcal{C}^u$ *such that* $n \times d \times \deg\rho = m$, *then there exists* $\tau_1, \cdots, \tau_s \in A$ *so that*

$$\pi = \tau_1 \times \cdots \times \tau_s.$$

Lemma 9.1 implies that $Z\langle a(n,d)^{(\rho)}\rangle$ is unitarizable for $n \leq d$ and $n \times d \times \deg\rho = m$. By Theorem 8.6 $= t(Z\langle a(n,d)^{(\rho)}\rangle)$ is also unitarizable. The representation $\pi = t(Z\langle a(n,d)^{(\rho)}\rangle)$ must be a Speh representation. If it is not, by the key Lemma $\pi = \tau_1 \times \cdots \times \tau_s$ for some $\tau_1, \cdots, \tau_s \in A$. Now we need to use the fact that t is multiplicative as well. Hence $Z\langle a(n,d)^{(\rho)}\rangle = t(\pi) = t(\tau_1) \times \cdots \times t(\tau_s)$ cannot be a Speh representation, which is a contradiction. Now assume that $\pi = Z\langle a(n_1,d_1)^{(\rho_1)}\rangle$. The fact

$$\mathrm{supp}Z\langle a(n,d)^{(\rho)}\rangle = \mathrm{supp}Z\langle a(n_1,d_1)^{(\rho_1)}\rangle$$

implies that

$$\rho = \rho_1 \text{ and } \{n,d\} = \{n_1,d_1\}.$$

Therefore π is either

$$Z\langle a(n,d)^{(\rho)}\rangle \text{ or } Z\langle a(d,n)^{(\rho)}\rangle.$$

For $n < d$ by Lemma 9.2 $\pi \neq Z\langle a(n,d)^{(\rho)}\rangle$. Hence

$$\pi = Z\langle a(d,n)^{(\rho)}\rangle.$$

Thus (iii) holds. It is clear that (i) follows from Theorem 6.1.

Q.E.D

The above classification theorem can be formulated in different forms.

Theorem 10.3 *Let* $A = \{a(n,d)^{(\rho)}, (\nu^{\alpha}a(n,d)^{(\rho)} + \nu^{-\alpha}a(n,d)^{(\rho)}) |\ n,d \in \mathbb{N}, \rho \in \mathcal{C}^u, 0 < \alpha < 1/2\}$. *Let* $X(A)$ *be the additive subsemigroup of* $\mathrm{M}(\mathcal{S}(\mathcal{C}))$ *generated by* A. *Then the following two maps*

$$a \mapsto Z\langle a\rangle \text{ and } a \mapsto L\langle a\rangle$$

are bijections from $X(A)$ *to* Irr^u. *The mapping* $t : A \to A$ *given by*

$$a(n,d)^{(\rho)} \mapsto a(d,n)^{(\rho)};$$
$$(\nu^{\alpha}a(n,d)^{(\rho)} + \nu^{-\alpha}a(n,d)^{(\rho)}) \mapsto (\nu^{\alpha}a(d,n)^{(\rho)} + \nu^{-\alpha}a(d,n)^{(\rho)})$$

extends uniquely to a morphism of semigroup $t : X(A) \to X(A)$. *We have*

$$t(Z\langle a\rangle) = Z\langle t(a)\rangle = L\langle a\rangle.$$

We can also formulate the main theorem in terms of Langlangs classification. Let δ be a unitary square-integrable representation and $n \in \mathbb{N}$. We set $u(\delta, n)$ to be the Langlands quotient of the induced representation

$$\nu^{(n-1)/2}\delta \times \nu^{(n-3)/2}\delta \times \cdots \times \nu^{-(n-1)/2}\delta.$$

The representation $u(\delta,n)$ is a Speh representation. For $0 < \alpha < 1/2$, the induced representation $\nu^{\alpha}u(\delta,n) \times \nu^{-\alpha}u(\delta,n)$ is irreducible and called a complementary series of the Speh representation.

Theorem 10.4 *Let* $A = \{u(\delta,n), \nu^{\alpha}u(\delta,n) \times \nu^{-\alpha}u(\delta,n) \mid \delta \in \mathcal{D}^u, n \in \mathbb{N}, 0 < \alpha < 1/2\}$. *Then we have*

(i) If $\sigma_1, \cdots, \sigma_r$ *are elements in* A, *then* $\sigma_1 \times \cdots \times \sigma_r \in \mathrm{Irr}^u$.

(ii) If $\pi \in \mathrm{Irr}^u$ *then there exist* $\tau_1, \cdots, \tau_s \in A$ *so that*

$$\pi = \tau_1 \times \cdots \times \tau_s.$$

Moreover, such τ_i*'s is unique up to a permutation.*

Corollary 10.5 *Let* $a \in \mathrm{M}(\mathcal{S}(\mathbb{R}))$ *and* $\rho_1, \rho_2 \in \mathcal{C}^u$. *Then we have*

(i) The representation $Z\langle a^{(\rho_1)}\rangle$ *is unitarizable if and only if* $Z\langle a^{(\rho_2)}\rangle$ *is unitarizable.*

(ii) The representation $L\langle a^{(\rho_1)}\rangle$ *is unitarizable if and only if* $L\langle a^{(\rho_2)}\rangle$ *is unitarizable.*

11. Proof of the key Lemma

Let $\pi \in \mathrm{Irr}$. The representation π is called rigid, if

$$\mathrm{supp}\pi \in \mathrm{M}(\cup_{\rho \in \mathcal{C}^u} \rho_{\frac{1}{2}\mathbb{Z}}).$$

Otherwise we say π is nonrigid. Each $\pi \in \mathrm{Irr}$ can be uniquely decomposed

$$\pi = \pi(r) \times \pi(n)$$

where $\pi(r)$ is a rigid representation and $\pi(n)$ is a nonrigid representation. We have corresponding decomposition of irreducible unitary representations

$$\mathrm{Irr}^u = [\bigoplus_{\substack{\alpha \in (0,1/2),\\ \rho \in \mathcal{C}^u}} \mathrm{Irr}^u((\rho_\alpha)_{\mathbb{Z}} \cup (\rho_{-\alpha})_{\mathbb{Z}}] \bigoplus [\bigoplus_{\substack{\alpha \in \{0,1/2\},\\ \rho \in \mathcal{C}^u}} \mathrm{Irr}^u(\rho_\alpha)_{\mathbb{Z}}].$$

Lemma 11.1 [T (Lemma 4.5)] *Let $\phi \in \mathrm{M}(\mathcal{S}(\mathbb{R}))$ consist of one-point segments, i.e. $\phi \in \mathrm{M}(\mathbb{R})$. Suppose that* supp $\phi \in \mathrm{M}(\mathbb{Z})$ *(resp.* supp$\phi \in \mathrm{M}(1/2+\mathbb{Z})$*). Then there exist $n_1, \cdots, n_s, m_1, \cdots, m_r \in \mathbb{N}$ and $\alpha_1, \cdots, \alpha_s, \beta_1, \cdots, \beta_r \in \{-1/2, 0\}$ such that*

$$\phi + (\Delta[n_1]_{\alpha_1}, \cdots, \Delta[n_s]_{\alpha_s}) = (\Delta[m_1]_{\beta_1}, \cdots, \Delta[m_r]_{\beta_r}),$$

and supp$\Delta[n_i]_{\alpha_i} \in \mathrm{M}(\mathbb{Z})$ *(resp.* supp $\Delta[n_i]_{\alpha_i} \in \mathrm{M}(1/2+\mathbb{Z})$*) for $i = 1, \cdots, s$.*

Fix $m \in \mathbb{N}$. Let $A_m = \{Z\langle a(n,d)^{(\rho)}\rangle, \pi(Z\langle a(n,d)^{(\rho)}\rangle, \alpha) \mid n, d \in \mathbb{N}, \rho \in \mathcal{C}^u, 0 < \alpha < 1/2, \text{such that } n \times d \times \deg\rho \leq m\}$.

Lemma 11.2 *For $m \geq 1$ suppose that (U^m) holds. Let $\sigma \in \mathrm{Irr}\widetilde{G}_m$ be rigid. If*

$$\pi(\sigma, \alpha) = \sigma_\alpha \times \sigma^h_{-\alpha}$$

is irreducible and unitarizable for some $\alpha \in (0, 1/2)$, then there exist $\sigma^1, \cdots, \sigma^r \in A_m$ and $\varepsilon_1, \cdots, \varepsilon_r \in \{-1/2, 0\}$ such that

$$\sigma = \sigma^1_{\varepsilon_1} \times \cdots \times \sigma^r_{\varepsilon_r}.$$

Proof. We prove this lemma by induction on m. Let $\sigma = \rho_\beta$ for some $\rho \in \mathcal{C}^u$ and $\beta \in (1/2)\mathbb{Z}$. Then

$$\pi(\sigma, \alpha) = \sigma_\alpha \times \sigma^h_{-\alpha} = \rho_{\alpha+\beta} \times \rho_{-\alpha-\beta}$$

is irreducible and unitarizable if and only if $\alpha + \beta \in (-1/2, 1/2)$, i.e. $\beta \in \{-1/2, 0\}$. Hence the lemma is true for $m = 1$.

Let m be an integer larger than 1. Assume that the lemma is true for any positive integers smaller than or equal to $m-1$. Suppose that $\sigma \in \mathrm{Irr}\widetilde{G}_m$ such that

$$\pi(\sigma, \alpha) = \sigma_\alpha \times \sigma^h_{-\alpha}$$

is irreducible and unitarizable. This implies that σ is Hermitian. If $\sigma = \sigma_1 \times \sigma_2$ is a nontrivial decomposition then both σ_1 and σ_2 are Hermitian and therefore both $\pi(\sigma_1, \alpha)$ and $\pi(\sigma_2, \alpha)$ are Hermitian. Hence Theorem 6.1 implies that both $\pi(\sigma_1, \alpha)$ and $\pi(\sigma_2, \alpha)$ are unitarizable, since

$$\pi(\sigma, \alpha) = \pi(\sigma_1, \alpha) \times \pi(\sigma_2, \alpha).$$

Then we can apply the inductive assumption.

We may now restrict ourselves to the case when σ is supported in one $\mathbb{Z}$-orbit, i.e. we may assume that $\sigma = Z\langle a^{(\rho)}\rangle$ for some $\rho \in \mathcal{C}^u$ and $a \in \mathrm{M}(\mathcal{S}(\mathbb{R}))$ with supp $a \in \mathrm{M}(\mathbb{Z})$ or supp $a \in \mathrm{M}(1/2+\mathbb{Z})$. Let a_0 denote the multisegment obtained from a by removing all one point segments. Then $a = a_0 + \phi$ where ϕ consists of one point segments. Now consider the highest derivative of $\pi(\sigma, \alpha)$,

$$\mathrm{A}(\pi(\sigma, \alpha)) = \pi(\mathrm{A}(\sigma), \alpha),$$

which is irreducible and unitarizable. The inductive assumption implies that there exist $l_1, \cdots, l_k \in \mathbb{Z}$ and $\varepsilon_1, \cdots, \varepsilon_k \in \{-1/2, 0\}$ such that

$$\mathrm{A}(\sigma) = Z\langle a(l_1, d_1)_{\varepsilon_1}^{(\rho)} + \cdots + a(l_k, d_k)_{\varepsilon_k}^{(\rho)}\rangle.$$

This implies that

$$a_0 = a(l_1, d_1 + 1)_{\varepsilon_1} + \cdots + a(l_k, d_k + 1)_{\varepsilon_k}.$$

By (U^m) each $Z\langle a(l_i, d_i + 1)\rangle$ is unitarizable.

Now for the one point segments ϕ we apply Lemma 11.1. We have

$$\begin{aligned} &\pi(\sigma, \alpha) \times \pi(Z\langle \Delta[n_1]_{\alpha_1}^{(\rho)}\rangle, \alpha) \times \cdots, \pi(Z\langle \Delta[n_s]_{\alpha_s}^{(\rho)}\rangle, \alpha) \\ &\quad = \pi(Z\langle a(l_1, d_1 + 1)_{\varepsilon_1}^{(\rho)}\rangle, \alpha) \times \cdots \times \pi(Z\langle a(l_k, d_k + 1)_{\varepsilon_k}^{(\rho)}\rangle, \alpha) \\ &\quad \times \pi(Z\langle \Delta[m_1]_{\beta_1}^{(\rho)}\rangle, \alpha) \times \cdots \times \pi(Z\langle \Delta[m_r]_{\beta_r}^{(\rho)}\rangle, \alpha). \end{aligned}$$

This implies that σ_α divides the right hand side of the identity. Thus σ_α is a product of some $Z\langle a(l_k, d_i + 1)_{\varepsilon_i}^{(\rho)}\rangle_\alpha$, $Z\langle a(l_k, d_i + 1)_{\varepsilon_i}^{(\rho)}\rangle_{-\alpha}$, $Z\langle \Delta[m_j]_{\beta_j}^{(\rho)}\rangle_\alpha$ and $Z\langle \Delta[m_j]_{\beta_j}^{(\rho)}\rangle_{-\alpha}$, since these are prime elements in the ring $\mathcal{R}$. Considering the support of σ_α we conclude that σ is a product of some $Z\langle a(l_i, d_i + 1)^{(\rho)}\rangle_{\varepsilon_i}$ and $Z\langle \Delta[m_j]^{(\rho)}\rangle_{\beta_j}$.

Q.E.D.

Proof of the key Lemma We prove the Lemma by induction on m. For $m = 1$ the statement of the lemma is trivially true. Let $m > 1$. We assume the lemma is true for all positive integers smaller than or equal to $m - 1$. Consider a representation $\pi \in \mathrm{Irr}^u \widetilde{G}_m$. Suppose that π is not a Seph representation. If we have a nontrivial decomposition

$$\pi = \sigma_1 \times \cdots \times \sigma_r,$$

then the inductive assumption and (U^{m-1}) implies the conclusion of the lemma.

Now we may assume that supp π is contained in the $\mathbb{Z}$-orbit of ρ_α and $\rho_{-\alpha}$ for some $\rho \in \mathcal{C}^u$ and $\alpha \in [0,1/2]$. In other words, we may assume that $\pi = Z\langle a^{(\rho)}\rangle$ for some $a \in \mathrm{M}(\mathcal{S}(\mathbb{R}))$. If $0 < \alpha < 1/2$ then we can apply Lemma 11.2 to obtain the conclusion. So we only have to consider the case when $\alpha = 0$ or $\alpha = 1/2$. Since the highest derivative $\mathrm{A}(\pi)$ of π is also unitarizable and irreducible, the inductive assumption implies that

$$\mathrm{A}(\pi) = Z\langle a(n_1,d_1)^{(\rho)} + \cdots + a(n_r,d_r)^{(\rho)}\rangle.$$

This implies

$$\pi = Z\langle a(n_1,d_1+1)^{(\rho)} + \cdots + a(n_r,d_r+1)^{(\rho)} + \phi\rangle,$$

where ϕ is a one point multisegments. Since π is Hermitian, we obtain that

$$\phi = \{\nu^{p_1+\alpha}\rho, \nu^{-p_1-\alpha}\rho, \cdots, \nu^{p_k+\alpha}\rho, \nu^{-p_k-\alpha}\rho\},$$

where $p_1, \cdots, p_k$ are positive integers. Now consider

$$\begin{aligned}&\pi \times Z\langle\Delta[2p_1+2\alpha-1]^{(\rho)}\rangle \times \cdots \times Z\langle\Delta[2p_k+2\alpha-1]^{(\rho)}\rangle\\ =&Z\langle a(n_1,d_1+1)^{(\rho)}\rangle \times \cdots \times Z\langle a(n_r,d_r+1)^{(\rho)}\rangle\\ &\times Z\langle\Delta[2p_1+2\alpha+1]^{(\rho)}\rangle \times \cdots \times Z\langle\Delta[2p_k+2\alpha+1]^{(\rho)}\rangle.\end{aligned}$$

Then the same argument we used in the proof of Lemma 11.2 will give us the desired conclusion.

Q.E.D.

References

[Ba] D. Barbasch, *The unitary dual for complex classical Lie groups,* Invent. Math. **96** (1989), 103-176.

[Be] J. N. Bernstein, *P-invariant distributions on GL(N) and the classification of unitary representations of GL(N),* Lie Group Representations I, R. Herb, R. Lipsman and J. Rosenberg eds., Lecture Notes in Mathematics, Vol.1024, Springer-Verlag, Berlin-Heidelberg-New York, 1984, 50-102.

[BZ] J. N. Bernstein and A. V. Zelevinsky, *Prepresentations of the group GL(n,F), where F is a local field,* Russian Math. Surveys, Vol. 31, No. 3, 1976, 1-68.

[BZ'] J. N. Bernstein and A. V. Zelevinsky, *Induced prepresentations of reductive p-adic groups, I* Ann. scient. Èc. Norm. Sup., 4ᵉ śerie, t.10, 1977, 441-471.

[BK] C. Bushnell and P. Kutzko, *The Admissible Dual of GL(N) via Compact Open Subgroups,* Annals of Mathematics Studies, No. 129, Princeton University Press 1993.

[C] W. Casselman, *Introduction to the theory of admissible representations of p-adic reductive groups,* preprint.

[FK] Y. Flicker and D. Kazhdan, *Metaplectic correspondence*, Publ. Math. IHES. 64, 53-110 (1986).

[G] S. Gelbart, *Weil's Representation and the Spectrum of the Metaplectic Group*, Lecture Notes in Mathematics 530, Spring-Verlag, 1972.

[HM] R. Howe and A. Moy, *Heche alegbra isomorphisms for $GL(n)$ over a p-adic field*, Journal of Algebra, Vol. 131, No 2, 388-424.

[H1] J. S. Huang, *The unitary dual of the universal covering group of $GL(n, \mathbb{R})$*, Duke Mathematical Journa, Vol. 61 (1990), No. 3, 705-745.

[H2] J. S. Huang, *Metaplectic correspondences and unitary representations*, Compositio Mathematica **80** (1991): 309-322.

[H3] J. S. Huang, *Metaplectic correspondence, unitary dual and Kazhdan-Lusztig conjecture*, in preparation.

[J] H. Jacquet, *Generic representations* Non-Commutative Harmonic Analysis, edited by J. Carmona and M. Vergne Lecture Notes in Mathematics, Vol. 587, Springer-Verlag, Berlin-Heidelberg-New York, 1977, 91-101.

[KP] D. Kazhdan and S. Patterson, *Metaplectic forms*, Publ. Math. IHES 59, 35-142 (1984).

[Ma] H. Matsumoto, *Sur les sous-groupes arithmétiques des groupes semi-simple déployés*, Ann. scient. Èc. Norm. Sup., 4e śerie, t.2, 1969, 1-62.

[R] R. R. Rao, *Orbital integrals in reductive groups*, Ann. of Math., **96** (1972), 157-222.

[Sa] S. Sahi, *On Kirillov's conjecture for Archemedean fields*, Compositio Mathematica 72 (1989), 67-86.

[Si] A. J. Silberger, *Introduction to Harmonic Analysis on Reductive p-adic Groups*, Princeton Math. Note, 23 (1979).

[T1] M. Tadić, *Proof of a conjecture of Bernstein*, Math. Ann., Vol. 272, 1985, 11-16.

[T2] M. Tadić, *Classification of unitary representations in irreducible representations of general linear Ggroup (non-archimedean case)*, Ann. scient. Éc. Norm. Sup., 4e série, t. 19, 1986, 335-382.

[T3] M. Tadić, *Induced representations of GL(n,A) for p-adic division algebras A*, J. reine angew. Math. **405**(1990), 48-77.

[V] D. Vogan, *The unitary dual of $GL(n)$ over an archimedean field*, Invent. Math. **83** (1986), 445-505.

[Z1] A. V. Zelevinsky, *Induced representations of reductive p-adic groups II*, Ann. scient. Èc. Norm. Sup., 4[e] śerie, t.13, 1980, 165-210.

[Z2] A. V. Zelevinsky, *p-adic analogue of the Kazhdan-Lusztig conjecture*, Funct. Anal. Appl., Vol. 15, 1981, 165-210.

CASIMIR OPERATOR AND WAVELET TRANSFORM*

QINGTANG JIANG AND LIZHONG PENG

Deparment of mathematics, Peking University

§1 INTRODUCTION

Let D be the unit disk in the complex plane equipped with the Lebesque measure $dm(z)$. The Moebius group $G = SU(1,1)$ consists of all 2×2 complex matrices

$$g = \begin{pmatrix} a & b \\ c & d \end{pmatrix} \quad a, b, c, d \in C$$

with $c = \overline{b}, d = \overline{a}$, $ad - bc = 1$. It acts on D via the transformations

$$z \to gz := g(z) = \frac{az + b}{cz + d}$$

The group $SU(1,1)$ has a decomposition (Iwasawa decomposition): $SU(1,1) = KAN$, where

$$K := \left\{ \begin{pmatrix} e^{it/2} & 0 \\ 0 & e^{it/2} \end{pmatrix}, t \in R \right\},$$
$$A := \left\{ \begin{pmatrix} \cosh\frac{t}{2} & \sinh\frac{t}{2} \\ \sinh\frac{t}{2} & \cosh\frac{t}{2} \end{pmatrix}, t \in R \right\},$$
$$N := \left\{ \begin{pmatrix} \cos\frac{t}{2} & i\sin\frac{t}{2} \\ i\sin\frac{t}{2} & \cos\frac{t}{2} \end{pmatrix}, t \in R \right\}.$$

Its Lie algebra $su(1,1)$ is generated by

$$e_3 = \begin{pmatrix} \frac{i}{2} & 0 \\ 0 & \frac{i}{2} \end{pmatrix}, e_1 = \begin{pmatrix} 0 & \frac{1}{2} \\ \frac{1}{2} & 0 \end{pmatrix}, e_2 = \begin{pmatrix} 0 & \frac{i}{2} \\ -\frac{i}{2} & 0 \end{pmatrix}.$$

Then the Casimir operator is

$$\Delta := 4(e_1^2 + e_2^2 - e_3^2).$$

For $\alpha > -1$, consider the weighted measure on D $d\mu_\alpha(z) := (1 - |z|^2)^\alpha dm(z)$ and let $L^{\alpha 2}(D)$ be the space consisting of all functions on D square integrable with respect to the measure $d\mu_\alpha(z)$, then the group $SU(1,1)$ acts on $L^{\alpha 2}(D)$ via

$$T_g^\nu : f(z) \to f(gz)\{g'(z)\}^{\frac{\nu}{2}} = f(gz)(cz + d)^{-\nu},$$

*Research was supported in part by the National Natural Science Foundation of China

M. Cheng et al. (eds.), Harmonic Analysis in China, 125–134.

where $\nu = \alpha + 2$.

The infinitesimal actions corresponding to e_1, e_2, e_3 are given as the following

$$\begin{aligned} T^{\nu}_{e_3} &:= iz\frac{\partial}{\partial z} - i\bar{z}\frac{\partial}{\partial \bar{z}} + i\frac{\nu}{2}, \\ T^{\nu}_{e_2} &:= \frac{1}{2}(1-z^2)\frac{\partial}{\partial z} + \frac{1}{2}(1-\bar{z}^2)\frac{\partial}{\partial \bar{z}} - \frac{\nu}{2}z, \\ T^{\nu}_{e_1} &:= \frac{i}{2}(1+z^2)\frac{\partial}{\partial z} - \frac{i}{2}(1+\bar{z}^2)\frac{\partial}{\partial \bar{z}} + \frac{i\nu}{2}\bar{z}. \end{aligned} \tag{1.1}$$

The Casimir operator becomes

$$\triangle_{\nu} := -4(1-|z|^2)^2\frac{\partial^2}{\partial z \partial \bar{z}} + 4\nu\bar{z}(1-|z|^2)\frac{\partial}{\partial \bar{z}} - \nu^2 + 2\nu.$$

For $\nu = 0$, i. e. for $\alpha = -2$ the operator $\triangle_0$ has the continuous spectra and the Plancherel formula has been studied by Harish-Chandra, Helgason et al (cf. [5], [6]). For $\nu > 0$, Peetre, Peng and Zhang gave the eigenvectors of the operator $\triangle_\nu$ and established the weighted Plancherel formula in [14] (for $\nu \in R$, see [10]). They got that $L^{\alpha 2}(D)$ has some discrete components (invariant subspaces) A_k, where $k < \frac{\alpha+1}{2}$. In other word, the spectra of $\triangle_\nu$ not only consist of the continuous part but also consist of the discrete part and A_k are eigenspaces of $\triangle_\nu$ with the discrete spectra. They also gave the orthonormal basis of A_k with Romanovski polynomials. Peng and Zhang considered the invariant Hankel operators between these subspaces.

By Cayley transform: $z \to \frac{1+z}{1-z}i$, then the unit disc D becomes the upper half plane $U := \{x+iy : x \in R, y > 0\}$. The automorphism group of U is $SL(2,R)$. Let

$$L^{\alpha 2}(U) := \{f : \int_U |f(x,y)|^2 y^\alpha dxdy < \infty\}.$$

The Casimir operator $\triangle_\nu$ in this case becomes

$$\triangle_\nu := -y^2\left(\frac{\partial^2}{\partial x^2} + \frac{\partial^2}{\partial y^2}\right) + i\nu y\left(\frac{\partial}{\partial x} + i\frac{\partial}{\partial y}\right) - \frac{\nu(\nu-2)}{4}.$$

In this paper, by wavelet transform we will show that $\triangle_\nu$ is equivalent to a differential operator D_ν. This fact was pointed first by Daubiechies, Klauder and Paul in [2]. The differential operator D_ν was studied by Morse in 1929 [12] and it has discrete and continuous spectra. For the discrete spectra, we will show the eigenvectors of D_ν are admissible wavelets. The ranges of the wavelet transforms of Hardy space H^2 with these wavelets are just the eigenspaces A_k of $\triangle_\nu$. By the method in the study of wavelets (cf. [4]), we get an orthonormal basis and the reproducing kernel of A_k. We then define the Toeplitz type operators between A_k. By the similar method in [8], [9], we can establish the Schatten-von Neumann ideal class S_p properties of these operators. By Cayley transform (from U to D), we can get that our results about the orthonormal bases and reproducing kernels of A_k coincide with that in [14], and results about the S_p-properties of the Toeplitz type operators coincide with that in [14] and [16].

"Bargmann transform" is a very important concept in quantum mechanics [1], [3]. In this paper, we will introduce a similar transform associate with wavelet transform and get a formula for this transfom as that for Bargmann transform.

§2 MAIN RESULTS AND DERIVATIONS

Let $L^{\alpha 2}(U)$ be the function space defined in §1 and let $L_+^{\alpha 2}$, $L_-^{\alpha 2}$ be its subspaces defined as follows:

$$L_+^{\alpha 2} := \{f(x,y) : f \in L^{\alpha 2}(U), \hat{f}(\xi, y) = 0, \xi \leq 0\}, \tag{2.1}$$
$$L_-^{\alpha 2} := \{f(x,y) : f \in L^{\alpha 2}(U), \hat{f}(\xi, y) = 0, \xi \geq 0\}.$$

In (2.1), $\hat{f}(\xi, y)$ means the Fourier transform of $f(x,y)$ with respect to the first variable x, i. e.

$$\hat{f}(\xi, y) := \int_{-\infty}^{+\infty} e^{-i\xi x} f(x,y) dx.$$

Let ψ be an analyzing admissible wavelet, i. e. ψ is a function on R satisfies (cf [4]):

$$C_\psi := \int_0^{+\infty} \frac{|\hat{\psi}(\xi)|^2}{\xi} d\xi < \infty, \quad \text{supp}\hat{\psi} \subset [0, +\infty), \quad \hat{\psi} \quad \text{is real.}$$

The continuous wavelet transform of f with wavelet ψ is:

$$\begin{aligned} W_\psi f(b,a) : &= \frac{1}{\sqrt{a}} \int \overline{\psi}(\frac{x-b}{a}) f(x) dx \\ &= \frac{\sqrt{a}}{2\pi} \int_0^{+\infty} \hat{\psi}(a\xi) e^{i\xi b} \hat{f}(\xi) d\xi. \end{aligned}$$

The transform $f \to W_\psi f$ is an isometry (up to a constant) from H^2 (Hardy space) into $L^2(U, \frac{dadb}{a^2})$. Thus for $f \in H^2$, $a^{-\frac{\alpha+2}{2}} W_\psi f(b,a) \in L_+^{\alpha 2}$ and we denote

$$T_\psi f(b,a) := a^{-\frac{\alpha+2}{2}} W_\psi f(b,a) = \frac{1}{2\pi a^{\frac{\alpha+1}{2}}} \int_0^{+\infty} \hat{\psi}(a\xi) e^{i\xi b} \hat{f}(\xi) d\xi. \tag{2.2}$$

Now let us derivate the equivalence between Δ_ν and a two order differential operator. In the following, let Δ_ν $(\nu = \alpha + 2)$ denote

$$\Delta_\nu := -y^2(\frac{\partial^2}{\partial x^2} + \frac{\partial^2}{\partial y^2}) + i\nu y(\frac{\partial}{\partial x} + i\frac{\partial}{\partial y}).$$

Let Δ_ν^+ and Δ_ν^- be the restrict of Δ_ν on $L_+^{\alpha 2}$ and $L_-^{\alpha 2}$ respectively, i. e.

$$\Delta_\nu^+ := \Delta_\nu|_{L_+^{\alpha 2}}, \quad \Delta_\nu^- := \Delta_\nu|_{L_-^{\alpha 2}}.$$

By (2.2) and a direct calculation, we can get

$$\begin{aligned} \Delta_\nu T_\psi f(b,a) &= \Delta_\nu^+ T_\psi f(b,a) \\ &= \frac{1}{2\pi a^{\frac{\alpha+1}{2}}} \int_0^{+\infty} (D_+ \hat{\psi})(a\xi) e^{i\xi b} \hat{f}(\xi) d\xi, \end{aligned}$$

where

$$D_{+} := -x^2 \frac{d^2}{dx^2} - x\frac{d}{dx} + x^2 - \nu x + \frac{(\nu-1)^2}{4}. \tag{2.3}$$

The differential operator D_+ was studied in [12], [2] and it has (see [2]) spectra:

$$\sigma(D_+) = \{(\frac{\nu-1}{2})^2 - (\frac{\nu-1}{2} - k)^2, k \in N, k < \frac{\nu-1}{2}\} \cup \{[(\frac{\nu-1}{2})^2, +\infty)\}.$$

We now want to get the eigenvector h of D_+ with discrete spectra, i. e. to find function $\hat{\psi}$ on $R_+ = [0,\infty)$ satisfying

$$D_+ h = \lambda_k h, \tag{2.4}$$

where $\lambda_k = (\frac{\nu-1}{2})^2 - (\frac{\nu-1}{2} - k)^2, k < \frac{\nu-1}{2}$.

Let ϕ be another function defined by

$$h(x) = (2x)^{-\frac{1}{2}}\phi(2x).$$

By (2.3) and (2.4), we get

$$\phi''(t) + (-\frac{1}{4} + \frac{\nu}{2t} + \frac{1+4\lambda_k - (\nu-1)^2}{4t^2})\phi(t) = 0. \tag{2.5}$$

Above equation is just the " Whittaker's differential equation" (see [11]):

$$u''(z) + (-\frac{1}{4} + \frac{\mathcal{N}}{z} + \frac{\frac{1}{4} - \mu_k^2}{z^2})u(z) = 0, \tag{2.6}$$

with $\mathcal{N} = \frac{\nu}{2}, \mu_k = \frac{\nu-1}{2} - k$. And (2.6) has solution $M_{\mathcal{N},\mu_k}$, the Whittaker's function:

$$M_{\mathcal{N},\mu_k}(z) = e^{-\frac{z}{2}} z^{\mu_k + \frac{1}{2}} {}_1F_1(\mu_k + \frac{1}{2} - \mathcal{N}; 1 + 2\mu_k; z).$$

Thus for discrete spectra $\lambda_k = (\frac{\nu-1}{2})^2 - (\frac{\nu-1}{2} - k)^2$, $k < \frac{\nu-1}{2}$, D_+ has eigenvectors:

$$h_k(x) := (2x)^{-\frac{1}{2}} M_{\mathcal{N},\mu_k}(2x).$$

Let ψ_k be the function defined by $\hat{\psi}^k(x) := h_k(x)$, then it can be written as:

$$\begin{aligned} \hat{\psi}^k(x) = h_k(x) &= (2x)^{\frac{\nu-1}{2} - k} e^{-x} {}_1F_1(-k; \nu - 2k; 2x) \\ &= (2x)^{\frac{\alpha+1-2k}{2}} e^{-x} L_k^{(\alpha+1-2k)}(2x), \end{aligned} \tag{2.7}$$

where $L_k^{(\alpha+1-2k)}(x)$ is the Laguerre polynomial of degree of k. Since $\frac{\nu-1}{2} - k > 0$, thus $\psi^k(x)$ is an analyzing admissiable wavelet. Let H^2 denote the Hardy space, then

$$A_k := \{T_{\psi^k} f(b,a) : \hat{\psi}^k(t) = h_k(t), f \in H^2\} \tag{2.8}$$

is the eigenspace of Δ_ν with the eigenvalue $\lambda_k = (\frac{\nu-1}{2})^2 - (\frac{\nu-1}{2} - k)^2$ and it is the range of wavelet transform of functions in H^2 with wavelet ψ^k.

By (2.8), we can get easily the reproducing kernel $K^{(k)}(z,w)$ of A_k (cf.[4], [8], [9]):

$$K^{(k)}(z,w) = c_k y^{-\frac{\alpha+2}{2}} \nu^{-\frac{\alpha+2}{2}} \psi_y^k * \psi_\nu^k(x-u) \tag{2.9}$$

where $z = x+iy, w = u+iv$, $c_k = (C_{\psi^k})^{-1}$.

For $k < \frac{\nu-1}{2} = \frac{\alpha+1}{2}$, let P_k be the projection from $L^{\alpha 2}$ onto A_k, we define the Toeplitz type operators:

$$T_b^{k,k'} := P_k M_{\bar{b}} P_{k'}$$

where b is analytic on U. Let S_p denote the Schatten-von Neumann ideal class and $B_p^{\frac{1}{p}}$ be the analytic Besov space, for their properties see [7] and references therein. We can get

Theorem 1. *For $T_b^{k,k'}$ defined as above, we have*

(1) If $k < k'$, then $T_b^{k,k'} = 0$;

(2) If $k = k'$, then $T_b^{k,k'} \in L^\infty$ iff $b \in L^\infty$, and $T_b^{k,k'}$ never compact unless it is zero;

(3) If $k > k'$, $1 < p \le \infty$, then $T_b^{k,k'} \in S_p$ iff $b \in B_p^{\frac{1}{p}}$;

(4) If $k > k'$, $0 < p \le 1$, $T_b^{k,k'} \in S_p$, then $T_b^{k,k'} = 0$.

From Theorem 1, we know for all $k, k', k > k'$, $T_b^{k,k'}$ are operators having cut-off phenomenon at 1. In [8], [9], $L^{\alpha 2}(U)$ was decomposited by Laguerre polynomials to be sum of orthogonal compoments, the Toeplitz and Hankel type operators betwen the componemts were defined similarly. For $k > k'$, they are "middle" Hakel operators have cut-off at $\frac{1}{k-k'}$ or $\frac{1}{k-k'+1}$. The proof of Theorem 1 will given in section 3.

Let

$$\mathcal{F} := \{F(z) : F(z) \text{ analytic on } C, \text{ and } \quad ||F||_{\mathcal{F}}^2 := \int |F(z)|^2 e^{-\pi|z|^2} dz < \infty\}$$

denote the Fock space, or Bargmann-Fock space.

For $z \in C$, define

$$Bf(z) := 2^{\frac{1}{4}} \int f(x) e^{2\pi x z - \pi x^2 - \frac{\pi}{2} z^2} dx,$$

then Bf is called the Bargmann transform of f. The map $f \to Bf$ is an isometry from $L^2(R)$ onto the Fock space $\mathcal{F}$. Denote $B(z,x) := 2^{\frac{1}{4}} e^{2\pi x z - \pi x^2 - \frac{\pi}{2} z^2}$, then it is called the Bargmann kernel.

Denote $l_0(x) := 2^{\frac{1}{4}} e^{-\pi x^2}$, and

$$l_j(x) := \frac{2^{\frac{1}{4}}}{\sqrt{j!}} (\frac{-1}{2\sqrt{\pi}})^j e^{\pi x^2} \frac{d^j}{dx^j}(e^{-2\pi x^2}), \quad j = 1, 2, \cdots$$

be the Hermite functions, which is an orthonormal basis of $L^2(R)$. Then the Bargmann transforms of functions $l_j(x)$ are

$$\xi_j(z) = \sqrt{\frac{\pi^j}{j!}} z^j, \quad j = 0, 1, 2, \cdots$$

which is an orthonormal basis of $\mathcal{F}$. About $l_j(x), \xi_j(z)$, there is a following formula (see [3], p54).

Theorem A[3]. *Let $B(z,x)$ be the Bargmann kernel, then*

$$\sum_{j=0}^{\infty} l_j(x)\xi_j(z) = B(z,x) = 2^{\frac{1}{4}} e^{2\pi xz - \pi x^2 - \frac{\pi}{2}z^2}. \tag{2.10}$$

Let us consider the Bargmann like transform which associats with the wavelet transform. Denote $\hat{H}^2 := \{\hat{f} : f \in H^2\}$, then $\hat{H}^2 = L^2(R_+)$. From (2.2), for $\phi = \hat{f} \in \hat{H}^2 = L^2(R_+)$, we define

$$B_k\phi(z) := \frac{\tilde{c}_k}{a^{\frac{\nu-1}{2}}} \int_0^{\infty} h_k(a\xi)e^{i\xi b}\phi(\xi)d\xi$$

where $h_k(x)$ is defined in (2.7) and

$$\tilde{c}_k = \frac{(-1)^k}{\Gamma(\alpha+1-2k)} \sqrt{\frac{2\Gamma(\alpha+2-k)}{\pi(\alpha+1-2k)k!}}.$$

Then

$$B_k(z,x) := \tilde{c}_k a^{-\frac{\nu-1}{2}} h_k(ax)e^{ixb}, \tag{2.11}$$

is the kernel of this transform, where $z = b + ia$. The transform B_k is an isometry from $L^2(R_+)$ onto A_k. For $k = 0$ (A_0 is the Bergman space), it is proved in [13] by Paul.

Let

$$\phi_{kn} := c_{kn}e^{-x}L_n^{(\alpha+1-2k)}(2x)(2x)^{\frac{\alpha+1-2k}{2}}, n = 0,1,2,\cdots, \tag{2.12}$$

be the orthonormal basis of $L^2(R_+)$, here $L_n^{(\alpha+1-2k)}(x)$ are the Laguerre polynomials and

$$c_{kn} = \sqrt{\frac{n!k^2}{\Gamma(n+\alpha+2-2k)}}.$$

Thus $B_k\phi_{kn}(z)$, $n = 0,1,2,\cdots$, is an orthonormal basis of A_k. For $n \geq k$ we can get (calculating directly):

$$B_k\phi_{kn}(z) = d_{k,s}(\frac{z-i}{z+i})^{n-k} \frac{1}{(1-iz)^{\alpha+2}} {}_2F_1(-k, k-\alpha-1; k-n-\alpha-1; \frac{|z+i|^2}{4a})$$

where

$$d_{k,s} = 2^{\alpha+2} \frac{(-\alpha-1-s)_k}{k!} \sqrt{\frac{\alpha+1-2k}{\pi}} \sqrt{\frac{(\alpha+2-k)_s}{(k+1)_s}}.$$

If $k = 0$, then

$$B_0\phi_{0n}(z) = d_{0,s}(\frac{z-i}{z+i})^n \frac{1}{(1-iz)^{\alpha+2}}, n = 1,2,\cdots,$$

it is an orthonormal basis of the Bergman space on the upper half plane.

From (2.9), we can calculate directly (omitting the details) the reproducing kernel of A_k:

$$K^{(k)}(z,w) = \frac{c_k}{(i\overline{w}-iz)^{\alpha+2}} {}_2F_1(-k, k-\alpha-1; -\alpha-1; \frac{|z-\overline{w}|^2}{4yv}),$$

where $z = x + iy, w = u + iv$, and c_k is a constant.

By Cayley transform (from U to D):

$$z \to w = \frac{z-i}{z+i} \in D,$$

we get

$$B_k\phi_{kn}(w) = \frac{d_{k,n-k}}{2^{\alpha+2}} w^{n-k} {}_2F_1(-k, k-\alpha-1; k-n-\alpha-1; \frac{1}{1-|w|^2}).$$

Thus $B_k\phi_{kn}(w)$ is just the basis $p_{k,k-n}(\frac{|w|^2}{1-|w|^2})w^{n-k}$ given in [14]. And the reproducing kernel $K^{(k)}(w,z)$ coincides with that in [14]. Thus the S_p-properties of $T_b^{kk'}$ coincide with that in [14] and [16]. They all have cut-off phonomenon at 1 for $k' < k$.

Like the properties of Bargmann transform given in (2.10), for B_k, we also have the following theorem.

Theorem 2. *Let $B_k(z,x)$ be kernels by (2.11) and $\phi_{kn}(x)$ by (2.12), then we have*

$$\sum_{n=0}^{\infty} B_k\phi_{kn}(z) \cdot \phi_{kn}(x) = B_k(z,x) \tag{2.13}$$

$$= \tilde{c}_k a^{-\frac{\alpha+1}{2}} (2ax)^{\frac{\alpha+1}{2}-k} e^{-ixz} {}_1F_1(-k, \alpha+2-2k; 2ax),$$

where $z = b + ia$.

Especially, if $k = 0$, then (2.13) is

$$\sum_{n=0}^{\infty} B_0\phi_{0n}(z) \cdot \phi_{0n}(x) = \tilde{c}_0 (2x)^{\frac{\alpha+1}{2}} e^{-ixz}. \tag{2.14}$$

By Cayley transform (from U to D), (2.14) becomes

$$\sum_{m=0}^{\infty} \binom{\alpha+1+m}{m}^{\frac{1}{2}} z^m \cdot \binom{\alpha+1+m}{m}^{-\frac{1}{2}} e^{-\frac{x}{2}} L_m^{(\alpha+1)}(x) = \frac{1}{(1-z)^{\alpha+2}} exp(-\frac{x}{2}\frac{1+z}{1-z}).$$

This formula is given by Bargmann in 1961 in [1].

Theorem 2 is gotten from the definition of the kernel $B_k(z,x)$ and the normalizations of the bases $B_k\phi_{kn}(z)$ and $\phi_{kn}(x)$ for $L_+^{\alpha 2}$ and $L^2(R_+)$ respectively.

§3 The proof of Theorem 1

Let us give the proof of Theorem 1. Let τ_k be the operator from A_k onto L^2 defined by

$$\tau_k F(x) := \int_0^{\infty} \frac{1}{y^{\frac{\nu}{2}}} \psi_y^k * F(\cdot, y) y^{\alpha} dy,$$

where $\nu = \alpha + 2$ as in §1. And let $T_{k'} = T_{\psi^{k'}}$ be defined by (2.2). Then the S_p-properties of $T_b^{k,k'}$ is equivalent to $\tilde{T}_b^{k,k'} := \tau_k T_b^{k,k'} T^{k'}$. Let $K_w^{(k)}(z) := K^{(k)}(z,w)$ be the reproducing kernel of A_k given in (2.9), then

$$\hat{K}_w^{(k)}(\xi, y) = \hat{K}^{(k)}(\xi, y, w) = c_k y^{-\frac{\alpha+1}{2}} v^{-\frac{\alpha+1}{2}} \hat{\psi}^k(y\xi)\hat{\psi}^k(v\xi)e^{-i\xi u}.$$

Thus for $f \in L^2(R)$, we have

$$\begin{aligned}(T_b^{k,k'} T_{k'} f)^\wedge(\xi, y) &= \int_U \hat{K}_w^{(k)}(\xi, y)\bar{b}(w)T_{k'} f(u,v)v^\alpha du dv \\ &= c_k y^{-\frac{\alpha+1}{2}} \hat{\psi}^k(y\xi) \int_0^\infty \int_R \bar{b}(u,v)T_{k'} f(u,v)e^{-i\xi u} du \hat{\psi}^k(v\xi) v^{\frac{\alpha-1}{2}} dv \\ &= c_k y^{-\frac{\alpha+1}{2}} \hat{\psi}^k(y\xi) \int_0^\infty \frac{1}{2\pi} \int_R \hat{\bar{b}}(\xi - \eta, v)(T_{k'} f)^\wedge(\eta, v) d\eta \hat{\psi}^k(v\xi) v^{\frac{\alpha-1}{2}} dv \\ &= \frac{c_k y^{-\frac{\alpha+1}{2}} \hat{\psi}^k(y\xi)}{2\pi} \int_R \hat{\bar{b}}(\xi - \eta) a_{k,k'}(\xi, \eta) d\eta,\end{aligned}$$

where

$$a_{k,k'}(\xi, \eta) := \int_0^\infty e^{-(\eta-\xi)v} \hat{\psi}^{k'}(v\eta)\hat{\psi}^k(v\xi)\frac{dv}{v}, \tag{3.1}$$

and in the last equation we used the fact $\hat{\bar{b}}(\xi - \eta, v) = \hat{\bar{b}}(\xi - \eta)e^{-(\eta-\xi)v}$. Then we have

$$\begin{aligned}(\tilde{T}_b^{k,k'} f)^\wedge(\xi) &= \int_0^\infty \sqrt{y}\hat{\psi}^k(y\xi)(T_b^{kk'} T_{k'} f)^\wedge(\xi, y) y^{\frac{\alpha}{2}-1} dy \\ &= \frac{1}{2\pi} \int_R \hat{\bar{b}}(\xi - \eta) a_{k,k'}(\xi, \eta) d\eta.\end{aligned}$$

Thus $\tilde{T}_b^{k,k'}$ is a paracommutator and the study of its S_p-properties becomes the estimates of $a_{k,k'}(\xi, \eta)$. The paracommutator theory was established by Janson, Peetre and Peng et al, about its properties and proofs see [7], [15].

Let $a := \frac{\xi}{\eta}$. By the definition of ψ^k (by (2.6)), we have

$$\begin{aligned}a_{kk'}(\xi, \eta) =& c \int_0^\infty e^{-(\eta-\xi)t} (\eta t)^{\frac{\alpha+1}{2} - k'} e^{-\eta t} {}_1F_1(-k'; \nu - 2k'; 2\eta t) \cdot \\ & (\xi t)^{\frac{\alpha+1}{2} - k} e^{-\xi t} {}_1F_1(-k; \nu - 2k; 2\xi t)\frac{dt}{t} \\ =& c a^{\frac{\alpha+1}{2} - k} \int_0^\infty e^{-t} t^{\alpha - k - k'} {}_1F_1(-k'; \nu - 2k'; t) {}_1F_1(-k; \nu - 2k; at) dt.\end{aligned} \tag{3.2}$$

For the integral in (3.2), by direct calculating, it equals to

$$\Gamma(\alpha - k - k' + 1) \sum_{s=0}^k \frac{(-k)_s(\alpha - k - k' + 1)_s}{s!(\nu - 2k)_s} {}_2F_1(-k', \alpha - k - k' + 1 + s; \alpha + 2 - 2k'; 1)a^s.$$

By the formula (see [11])

$$ {}_2F_1(a,b;c;1) = \frac{\Gamma(c)\Gamma(c-a-b)}{\Gamma(c-a)\Gamma(c-b)}|_{b=-k} = \frac{(c-a)_k}{(c)_k}, \tag{3.3} $$

if $c \neq 0, -1, -2, \cdots, Re(a+b-c) < 0$, we have

$$ {}_2F_1(-k', \alpha - k - k' + 1 + s; \alpha + 2 - 2k'; 1) = \frac{(1-k'-s+k)_{k'}}{(\alpha+2-2k')_{k'}}. \tag{3.4} $$

If $k' > k$, or $k' \leq k$ and $s > k - k'$, then the left term in (3.4) is zero. Thus for $k' > k$, $a_{k,k'}(\xi,\eta) = 0$ and $\tilde{T}_b^{k,k'} = 0$, hence $T_b^{k,k'} = 0$, thus we get (1) of Theorem 1.

For $k = k'$, we have $a_{k,k'}(\xi,\eta) = c'(\frac{\xi}{\eta})^{\frac{\alpha+1}{2}-k}$, $c' \neq 0$, thus by the paracommutator theory (cf [7],[15]) we have (2) of Theorem 1.

For $0 \leq k' < k$, the integral in (3.2) equals to

$$ \begin{aligned} &\frac{\Gamma(\alpha-k-k'+1)}{(\alpha+2-2k')_{k'}} \sum_{s=0}^{k-k'} \frac{(-k)_s(\alpha-k-k'+1)_s}{s!(\nu-2k)_s}(1-k'-s+k)\cdot \\ &\qquad (1-k'-s+k+1)\cdots(-s+k)a^s \\ &= \frac{\Gamma(\alpha-k-k'+1)}{(\alpha+2-2k')_{k'}} {}_2F_1(-k+k', \alpha-k-k'+1; \alpha+2-2k; a). \end{aligned} \tag{3.5} $$

By the formula (see [11])

$$ {}_2F_1(a,b;c;z) = (1-z)^{c-a-b} {}_2F_1(c-a, c-b; c; z) $$

for $a = k' - k, b = \alpha - k - k' + 1, c = \alpha + 2 - 2k$, we have

$$ {}_2F_1(-k+k', \alpha-k-k'+1; \alpha+2-2k; a) = (1-a) {}_2F_1(\alpha+2-k-k', 1-k+k'; \alpha+2-2k; a). \tag{3.6} $$

Thus by (3.2), (3.5) and (3.6), for $k' < k$, we have

$$ a_{k,k'}(\xi,\eta) = c(\frac{\xi}{\eta})^{\frac{\alpha+1}{2}-k}(1-\frac{\xi}{\eta})(c_0' + c_1'(1-\frac{\xi}{\eta}) + \cdots) $$

with

$$ c_0' = \frac{(-1)^{k+k'+1}\Gamma(\alpha-k-k'+1)k!}{(\alpha+2-2k')_{k'}(\alpha+2-2k)_{k-1-k'}}. $$

Thus $\tilde{T}_b^{k,k'}$ satisfies the $A0, A1, A3(1)$ and $A4$ conditions in [7], and by the paracommutator theory (cf. [7]), we get (3) and (4) of Theorem 1.

REFERENCES

1. V. Bargmann, *On a Hilbert space of analytic functions and an associated integral transform, Part I*, Comm. Pure Appl. Math. **14** (1961), 187-214.
2. I. Daubechies, J. Klauder and T. Paul, *Wiener measures for path integrals with affine kinematic variables*, J. Math. Phys. **28** (1987).
3. G. Folland, " Harmonic analysis in phase space", Princeton Univ. Press (1989).
4. A. Grossmann and J. Morlet, *Decomposition of Hardy functions into square integrable wavelets of constant shape*, SIAM J. Math. Anal. **15** (1984), 723-736.
5. Harish-Chandra, *Plancherel formula for semi-simple Lie Groups*, Trans Amer. Math. Soc. **76** (1954), 485-528.
6. S. Helgason, "Groups and Geometric Analysis", Academic Press, New York–London (1984).
7. S. Janson and J. Peetre, *Paracommutators-boundedness and Schatten-von Neumann properties*, Trans. Amer. Math. Soc. **305** (1988), 467-504.
8. Q. Jiang and L. Peng, *Wavelet transform and Toeplitz-Hankel type operators*, Math. Scand. **70** (1992), 247-264.
9. Q. Jiang and L. Peng, *Toeplitz and Hankel type operators on the upper half-plane*, Int. Eq. Operator Theory, **15** (1992), 744-767.
10. H. Liu and L. Peng, *Weighted Plancheral formula. Irreducible unitary representations and eigenspace representations*, to appear in Math. Scand. (1993).
11. W. Magnus, F. Oberhettinger and R. Soni, "Formulas and Theorems for the Special Functions of Mathematical Physics", Springer-Verlag Berlin Heidelberg New York (1966).
12. P. Morse, *Diatomic molecules according to the wave mechanics. II. Vibrational levels*, Physical Review, **34** (1929), 57-64.
13. T. Paul, *Functions analytic on the half-plane as quantum mechanical states*, J. Math. Phys. **25** (1985), 3252-3263.
14. J. Peetre, L. Peng and K. Zhang, *A weighted Plancherel formula I. The case of the unit disk. applications to Hankel operators*, Report No. 11, Stockholm University (1990).
15. L. Peng, *Paracommutator of Schatten-von Neumann class* $S_p, 0 < p < 1$, Math. Scand. **61** (1987), 68-92.
16. L. Peng and K. Zhang, *Invariant Hankel operators*, preprint.

OSCILLATORY SINGULAR INTEGRALS WITH ROUGH KERNEL *

YINSHENG JIANG — Xinjiang University

SHANZHEN LU — Beijing Normal University

ABSTRACT. This paper is devoted to the study on the L^p-boundedness for the oscillatory singular integral defined by

$$Tf(x) = p.v. \int_{\mathbb{R}^n} e^{iP(x,y)} K(x-y) f(y) dy,$$

where $P(x,y)$ is a real polynomial on $\mathbb{R}^n \times \mathbb{R}^n$, and $K(x) = \frac{h(|x|)\Omega(x)}{|x|^n}$ with $\Omega \in L\log^+ L(S^{n-1})$ and $h \in BV(\mathbb{R}_+)$ (i.e. h is a bounded variation function on $\mathbb{R}_+$).

Let $\overline{T}$ be a singular integral operator corresponding to T, and let $\overline{T}_0$ be the truncated operator of $\overline{T}$. That means

$$\overline{T}_0 f(x) = p.v. \int_{|x-y|<1} K(x-y) f(y) dy.$$

The main result in this paper gives out a verifiable necessary and sufficient condition on $\overline{T}_0$ so that the oscillatory integral operator T is bounded on $L^p(\mathbb{R}^n), 1 < p < \infty$, for any real non-trivial polynomial $P(x,y)$.In addition, we also discuss the weighted L^p-boundedness of T.

1. Introduction.

Let us consider the oscillatory singular integral defined by

$$Tf(x) = p.v. \int_{\mathbb{R}^n} e^{iP(x,y)} K(x-y) f(y) dy,$$

where $P(x,y)$ is a real polynomial on $\mathbb{R}^n \times \mathbb{R}^n$, and $K(x) = h(|x|)\Omega(x/|x|)|x|^{-n}$ with $h(r) \in BV(\mathbb{R}_+)$ (i.e. $h(r)$ is a bounded variation function on $[0,\infty)$). It was proved by F.Ricci and E.M.Stein [5] that T was bounded on $L^p(\mathbb{R}^n), 1 < p < \infty$, if $K \in C^1(\mathbb{R}^n \setminus \{0\})$ and $h \equiv 1$.Lately, S.Z.Lu and Y.Zhang [4] proved that T was bounded on $L^p(\mathbb{R}^n), 1 < p < \infty$,provided $\Omega \in L^q(S^{n-1}), 1 < q \le \infty$,and $\int_{S^{n-1}} \Omega(x')d\sigma(x') = 0$,where $S^{n-1} = \{x \in \mathbb{R}^n : |x| = 1\}$.Meanwhile,the above result of [4] is extended by Y.S.Jiang and S.Z.Lu [3] to the case where Ω belongs to certain Block space.However,as can be seen in [1], the function class $L\log^+ L(S^{n-1})$ is the best one for L^2-boundedness of T with $h \equiv 1$ in certain sense. The purpose of this paper is to study L^p-boundedness of T in the case where $\Omega \in L\log^+ L(S^{n-1})$.

*The Project is supported by National Natural Science Foundation of China

M. Cheng et al. (eds.), Harmonic Analysis in China, 135–145.

Let us now state our results. We assume that Ω is homogeneous of degree zero throughout this paper.

Theorem 1. Suppose $\Omega \in Llog^+L(S^{n-1}), n \geq 2, \int_{S^{n-1}} \Omega(x')d\sigma(x') = 0$, and $h \in BV(\mathbb{R}_+)$. Then T is bounded on $L^2(\mathbb{R}^n)$ for any real polynomial $P(x,y)$.

Theorem 2. Suppose $\Omega \in Llog^+L(S^{n-1}), n \geq 2, \int_{S^{n-1}} \Omega(x')d\sigma(x') = 0$, and $h \equiv 1$. Then T is bounded on $L^p(\mathbb{R}^n), 1 < p < \infty$, for any real polynomial $P(x,y)$.

Let us denote the singular integral corresponding to T by

$$\overline{T}f(x) = p.v. \int_{\mathbb{R}^n} K(x-y)f(y)dy.$$

The proofs of Theorem 1 and Theorem 2 are based on the following Theorem 3.

Theorem 3. Let $1 < p < \infty, \Omega \in Llog^+L(S^{n-1})$ and $h \in BV(\mathbb{R}_+)$. If $\overline{T}$ is bounded on $L^p(\mathbb{R}^n)$, then so is T for any real polynomial $P(x,y)$.

The significance of Theorem 3 is that L^p-boundedness of oscillatory singular integrals T is restored to L^p-boundedness of singular integrals $\overline{T}$ corresponding to T. It is easy to see that Theorem 2 is a direct result of Theorem 3 and a known result of [1], and Theorem 1 is a direct consequence of Theorem 3 and a recent result of [7] where $\overline{T}$ is proved to be a bounded operator on $L^2(\mathbb{R}^n)$ provided $\Omega \in Llog^+L(S^{n-1}), \int_{S^{n-1}} \Omega(x')d\sigma(x') = 0$, and $h \in L^\infty(\mathbb{R}_+)$.

Let us now formulate the main result of this paper. If we are restricted to the case where $P(x,y)$ is a non-trivial polynomial, then we shall get a criterion on L^p-boundedness for T. Here, a real polynomial $P(x,y)$ is called non-trivial if $P(x,y)$ does not take the form of $P_1(x) + P_2(y)$ (see [4]).

Theorem 4. Let $1 < p < \infty$. If $\Omega \in Llog^+L(S^{n-1})$ and $h \in BV(\mathbb{R}_+)$, then the following three facts are equivalent:

(i) If $P(x,y)$ is a non-trivial polynomial, then T is bounded on $L^p(\mathbb{R}^n)$.

(ii) If a non-trivial polynomial $P(x,y)$ satisfies

$$P(x,y) = P(x-t, y-t) + R_1(x,t) + R_2(y,t), t \in \mathbb{R}^n, \tag{1}$$

where R_1 and R_2 are real polynimials, then T is bounded on $L^p(\mathbb{R}^n)$.

(iii) The truncated operator of $\overline{T}$,

$$\overline{T_0}f(x) = p.v. \int_{|x-y|<1} K(x-y)f(y)dy,$$

is bounded on $L^p(\mathbb{R}^n)$.

Let A_p be the Muckenhoupt's class. The definition of $L^p(\mathbb{R}^n, w(x)dx)$ is well known. We denote it by $L^p(w)$ for simplisity. To state our weighted results, we introduce a variant of the Hardy-Littlewood maximal function associated with $\Omega \in L^1(S^{n-1})$ defined by

$$M_\Omega f(x) = sup_{r>0}\frac{1}{r^n}\int_{|y|<r} |\Omega(y')||f(x-y)|dy.$$

The following theorem is the weighted form of Theorem 3 provided $\Omega \in L^q(S^{n-1}), 1 < q \leq \infty$.

Theorem 5. Let $1 < p < \infty, w \in A_p, \Omega \in L^q(S^{n-1}), 1 < q \leq \infty$, and $h \in BV(\mathbb{R}_+)$. If $\overline{T}$ and M_Ω are bounded on $L^p(w)$, then so is T for any real polynomial $P(x, y)$.

As a direct application of Theorem 5, we have

Theorem 6. Suppose $h \in BV(\mathbb{R}_+)$, $\Omega \in L^q(S^{n-1}), 1 < q \leq \infty$, and Ω satisfies

$$\int_{S^{n-1}} \Omega(x')d\sigma(x') = 0.$$

Then for any real polynomial $P(x, y)$, T is bounded on $L^p(w)$ in each of the following cases:

(A) $q' \leq p < \infty (p \neq 1)$ and $w \in A_{p/q'}$;
(B) $1 < p \leq q (p \neq \infty)$ and $w^{-1/(p-1)} \in A_{p'/q'}$,
where $1/p + 1/p' = 1$ and $1/q + 1/q' = 1$.

2. Some lemmas.

Let us begin with some lemmas that reveal the relationship between the L^p-boundedness of T and the L^p- boundedness of $\overline{T}$ in the case of $\Omega \in L^1(S^{n-1})$.

Let

$$Gf(x) = p.v. \int_{\mathbb{R}^n} K(x, y)f(y)dy,$$

and

$$G_0 f(x) = p.v. \int_{|x-y|<1} K(x, y)f(y)dy.$$

Lemma 1. Let $1 \leq p < \infty$, and $w \in A_p$. If G_0 is bounded on $L^p(w)$, then the inequality

$$\left(\int_{|x-t|<\epsilon} |G_0 f(x)|^p w(x)dx\right)^{1/p} \leq C_\epsilon\left(\int_{|y-t|<1+\epsilon} |f(y)|^p w(y)dy\right)^{1/p} \tag{2}$$

holds for any $\epsilon > 0$, where C_ϵ is independent of f and t.Conversely,if (2) holds for certain $\epsilon > 0$, then G_0 is bounded on $L^p(w)$.

Proof. We assume that (2) holds for some $\epsilon > 0$. It is easy to see that

$$\int_{\mathbb{R}^n} \int_{|x-t|<\epsilon} |G_0 f(x)|^p w(x)dxdt = |B_\epsilon| \int_{\mathbb{R}^n} |G_0 f(x)|^p w(x)dx,$$

and

$$\int_{\mathbb{R}^n}\int_{|y-t|<1+\epsilon}|f(y)|^p w(y)dydt = |B_{1+\epsilon}|\int_{\mathbb{R}^n}|f(y)|^p w(y)dy,$$

where $B_r = \{x \in \mathbb{R}^n : |x| < r\}$. Thus, it follows from (2) that

$$\Big(\int_{\mathbb{R}^n}|G_0 f(x)|^p w(x)dx\Big)^{1/p} \le C_\epsilon\Big(\int_{\mathbb{R}^n}|f(y)|^p w(y)dy\Big)^{1/p}.$$

Conversely, if $|x-t|<\epsilon$ and $|x-y|<1$,then $|y-t|<1+\epsilon$. Therefore,we have

$$\begin{aligned}\Big(\int_{|x-t|<\epsilon}|G_0 f(x)|^p w(x)dx\Big)^{1/p} &= \Big(\int_{|x-t|<\epsilon}\Big|\int_{|x-y|<1}K(x,y)f(y)\chi_{\{|y-t|<1+\epsilon\}}(y)dy\Big|^p w(x)dx\Big)^{1/p}\\ &\le \Big(\int_{\mathbb{R}^n}|G_0[f(\cdot)\chi_{\{|t-\cdot|<1+\epsilon\}}(\cdot)](x)|^p w(x)dx\Big)^{1/p}\\ &\le C_\epsilon\Big(\int_{\mathbb{R}^n}|f(y)\chi_{\{|t-y|<1+\epsilon\}}(y)|^p w(y)dy\Big)^{1/p}\\ &= C_\epsilon\Big(\int_{|y-t|<1+\epsilon}|f(y)|^p w(y)dy\Big)^{1/p}.\end{aligned}$$

Lemma 2. Let $1<p<\infty$ and $w \in A_p$. Suppose $\Omega \in L^1(S^{n-1})$ and $|K(x,y)| \le C|\Omega(x-y)|/|x-y|^n$. If G and M_Ω are bounded on $L^p(w)$, then so is G_0.

Proof. By Lemma 1, it will suffice to prove

$$\int_{|x-t|<1/4}|G_0 f(x)|^p w(x)dx \le C\int_{|y-t|<5/4}|f(y)|^p w(y)dy, t \in \mathbb{R}^n.$$

Now, we split f into three parts $f = f_1 + f_2 + f_3$ for given t, where

$$f_1(y) = f(y)\chi_{\{|y-t|<1/2\}}(y),$$
$$f_2(y) = f(y)\chi_{\{1/2\le|y-t|<5/4\}}(y),$$

and

$$f_3(y) = f(y)\chi_{\{|y-t|\ge 5/4\}}(y).$$

Note that $|x-t|<1/4$ and $|y-t|<1/2$ imply $|x-y|<1$. Thus, we have

$$G_0 f_1(x) = Gf_1(x), |x-t|<1/4.$$

Since G is bounded on $L^p(w)$, we get

$$\begin{aligned}\int_{|x-t|<1/4}|G_0 f_1(x)|^p w(x)dx &= \int_{|x-t|<1/4}|Gf_1(x)|^p w(x)dx\\ &\le C\int|f_1(y)|^p w(y)dy\\ &= C\int_{|y-t|<1/2}|f(y)|^p w(y)dy.\end{aligned}$$

By the assumption on $K(x,y)$, we have

$$|G_0 f_2(x)| \le \int_{1/4<|y|<1} \frac{|\Omega(y)|}{|y|^n} |f_2(x-y)|dy \\ \le CM_\Omega f_2(x).$$

Since M_Ω is bounded on $L^p(w)$, it follows from the above that

$$\Big(\int_{|x-t|<1/4} |G_0 f_2(x)|^p w(x)dx\Big)^{1/p} \le C\Big(\int_{|y-t|<5/4} |f(y)|^p w(y)dy\Big)^{1/p}.$$

Finally, we notice that $|x-t|<1/4$ and $|y-t|>5/4$ imply $|x-y|>1$. Thus, $G_0 f_3(x)=0$, if $|x-t|<1/4$. This completes the proof of Lemma 2.

Let T_0 and $\overline{T}_0$ be the truncated operator of T and $\overline{T}$ respectively. Then we have

Lemma 3. Let $1<p<\infty$, $w\in A_p$, $\Omega\in L^1(S^{n-1})$, and $h\in L^\infty(\mathbb{R}_+)$. If $\overline{T}_0$ and M_Ω are bounded on $L^p(w)$, then so is T_0 for any real polynomial $P(x,y)$.

Proof. We shall carry out the argument by a double induction on the degree of $P(x,y)$ in x and y as follows. Let k and l be positive integers. Note that if $\overline{T}_0$ is bounded on $L^p(w)$, then T_0 is also bounded on $L^p(w)$ for real polynomials with the form $P_1(x)+P_2(y)$. Thus, we may assume that the conclusion of Lemma 3 holds for all real polynomials which are sums of monomials of degree $<k$ in x times monomials of any degree in y, together with monomials which are of degree k in x times monomials which are of degree $<l$ in y. Put

$$P(x,y)=\sum_{|\alpha|=k,|\beta|=l} a_{\alpha,\beta}x^\alpha y^\beta + R_0(x,y),$$

where the inductive hypothesis is applied to $R_0(x,y)$. By dilation invariant, we may assume that $\sum_{|\alpha|=k,|\beta|=l}|a_{\alpha,\beta}|=1$. Now, we write

$$\begin{aligned} T_0 f(x) &= \int_{|x-y|<1} e^{iP(x,y)}\frac{h(|x-y|)\Omega(x-y)}{|x-y|^n}f(y)dy \\ &= \int_{|x-y|<1} e^{i[R_0(x,y)+\sum_{|\alpha|=k,|\beta|=l} a_{\alpha,\beta}y^{\alpha+\beta}]}\frac{h(|x-y|)\Omega(x-y)}{|x-y|^n}f(y)dy \\ &+ \int_{|x-y|<1}\Big\{e^{iP(x,y)}-e^{i[R_0(x,y)+\sum_{|\alpha|=k,|\beta|=l} a_{\alpha,\beta}y^{\alpha+\beta}]}\Big\}\frac{h(|x-y|)\Omega(x-y)}{|x-y|^n}f(y)dy \\ &:= T_{01}f(x)+T_{02}f(x). \end{aligned}$$

Applying the inductive hypothesis to T_{01}, we get

$$\|T_{01}f\|_{p,w}\le C\|f\|_{p,w}.$$

Note that if $|x| < 1$ and $|x-y| < 1$, then

$$|e^{iP(x,y)} - e^{i[R_0(x,y)+\sum_{|\alpha|=k,|\beta|=l} a_{\alpha,\beta} y^{\alpha+\beta}]}| \le C|x-y|,$$

and if $|x-t| < 1$ and $|x-y| < 1$, then $|y-t| < 2$.Thus,if $|x-t| < 1$, then

$$\begin{aligned}|T_{02}f(x)| &\le C\int_{|x-y|<1} \frac{|\Omega(x-y)|}{|x-y|^{n-1}}|f(y)|dy \\ &= C\int_{|x-y|<1} \frac{|\Omega(x-y)|}{|x-y|^{n-1}}|f(y)\chi_{B(t,2)}(y)|dy \\ &= C\int_{|y|<1} \frac{|\Omega(y)|}{|y|^{n-1}}|f(x-y)|\chi_{B(t,2)}(x-y)dy \\ &= C\sum_{k=0}^{\infty}\int_{2^{-(k+1)}\le|y|<2^{-k}} \frac{|\Omega(y)|}{|y|^{n-1}}|f(x-y)|\chi_{B(t,2)}(x-y)dy \\ &\le C\sum_{k=0}^{\infty} 2^{-k} M_\Omega\Big(f\chi_{B(t,2)}\Big)(x) \le CM_\Omega\Big(f\chi_{B(t,2)}\Big)(x).\end{aligned}$$

Since M_Ω is bounded on $L^p(w)$, we have

$$\Big(\int_{|x-t|<1} |T_{02}f(x)|^p w(x)dx\Big)^{1/p} \le C\Big(\int_{|y-t|<2} |f(y)|^p w(y)dy\Big)^{1/p}.$$

From the above and Lemma 1, it follows

$$\|T_{02}f\|_{p,w} \le C\|f\|_{p,w}.$$

Thus,we obtain

$$\|T_0 f\|_{p,w} \le C\|f\|_{p,w}.$$

Hence,the proof of Lemma 3 is complete.

We now split $P(x,y)$ as

$$\begin{aligned}P(x,y) &= \sum_{\substack{0<|\alpha|\le k' \\ 0<|\beta|\le l'}} a_{\alpha,\beta} x^\alpha y^\beta + \sum a_{\alpha,0} x^\alpha + \sum a_{0,\beta} y^\beta \\ &:= Q(x,y) + P_1(x) + P_2(y).\end{aligned}$$

Clearly,if $P(x,y)$ is non-trivial, then $0 < k' \le k$ and $0 < l' \le l$.

Lemma 4. Let $1 < p < \infty, w \in A_p, \Omega \in L^1(S^{n-1}), h \in L^\infty(\mathbb{R}_+)$, and $T_\infty = T - T_0$. If $\overline{T}$ and M_Ω are bounded on $L^p(w)$, and T_∞ are bounded on $L^p(w)$ for $Q(x,y)$,then T is bounded on $L^p(w)$ for $P(x,y)$.

Proof. If $Q(x,y)\equiv 0$, then the conclusion of lemma is trivial.If $Q(x,y)\neq 0$ and $\overline{T}$ is bounded on $L^p(w)$, then the boundedness of T on $L^p(w)$ for $P(x,y)$ is equivalent to the boundedness of T on $L^p(w)$ for $Q(x,y)$. Note that T_0 is bounded on $L^p(w)$ for $Q(x,y)$ by Lemma 2 and Lemma 3.Thus, the boundedness of T_∞ on $L^p(w)$ for $Q(x,y)$ implies the boundedness of T on $L^p(w)$ for $Q(x,y)$.This finishes the proof of Lemma 4.

Lemma 5. Let $1\le p<\infty$,$w\in A_p$, and $\overline{T}_0$ be a truncated operator defined by

$$\overline{T}_0 f(x) = p.v.\int_{|x-y|<1} K(x,y)f(y)dy.$$

If the real polynomial $P(x,y)$ satisfies (1) in Theorem 4 and the oscillatory integral T_0 defined by

$$T_0 f(x) = p.v.\int_{|x-y|<1} e^{iP(x,y)}K(x,y)f(y)dy$$

is bounded on $L^p(w)$, then $\overline{T}_0$ is also bounded on $L^p(w)$.

Proof. By (1), we write

$$\overline{T}_0 f(x) = e^{-iR_1(x,t)}p.v.\int_{|x-y|<1} e^{iP(x,y)}K(x,y)e^{-iP(x-t,y-t)}f(y)e^{-iR_2(y,t)}dy$$

for $t\in\mathbb{R}^n$.Express $e^{-iP(x-t,y-t)}$ into Taylor series:

$$\begin{aligned} e^{-iP(x-t,y-t)} &= \sum_{m=0}^{\infty}\frac{(-1)^m}{m!}[P(x-t,y-t)]^m \\ &= \sum_{m=0}^{\infty}\frac{(-i)^m}{m!}[\sum_{\alpha,\beta} a_{\alpha,\beta}(x-t)^\alpha(y-t)^\beta]^m \\ &= \sum_{m=0}^{\infty}\frac{(-i)^m}{m!}\sum_{\mu,\nu} b_{\mu,\nu}(x-t)^\mu(y-t)^\nu. \end{aligned}$$

Thus,we have

$$\begin{aligned} \Big(\int_{|x-t|<1}|\overline{T}_0 f(x)|^p w(x)dx\Big)^{1/p} &\le \sum_{m=0}^{\infty}\frac{1}{m!}\sum_{\mu,\nu}|b_{\mu,\nu}|\Big(\int_{|x-t|<1}|(x-t)^\mu\cdot \\ &\cdot\int_{|x-y|<1} e^{iP(x,y)}K(x,y)e^{-iR_2(y,t)}f(y)(y-t)^\nu dy|^p w(x)dx\Big)^{1/p} \\ &\le \sum_{m=0}^{\infty}\frac{1}{m!}\sum_{\mu,\nu}|b_{\mu,\nu}||\xi^\mu| \\ &\cdot\Big(\int_{|x-t|<1}|T_0[e^{-iR_2(\cdot,t)}f(\cdot)(\cdot-t)^\nu]|^p w(x)dx\Big)^{1/p},. \end{aligned}$$

where $\xi = (1, 1, \cdots, 1)$.By Lemma 1, we obtain

$$\begin{aligned}\Big(\int_{|x-t|<1} |\overline{T}_0 f(x)|^p w(x)dx\Big)^{1/p} &\le C\sum_{m=0}^{\infty}\frac{1}{m!}\sum_{\mu,\nu}|b_{\mu,\nu}||\xi^{\mu}|\Big(\int_{|y-t|<2}|f(y)(y-t)^{\nu}|^p w(y)dy\Big)^{1/p}\\ &\le C\sum_{m=0}^{\infty}\frac{1}{m!}\sum_{\mu,\nu}|b_{\mu,\nu}||\xi^{\mu}\eta^{\nu}|\Big(\int_{|y-t|<2}|f(y)|^p w(y)dy\Big)^{1/p}\\ &\le C\sum_{m=0}^{\infty}\frac{1}{m!}[\sum_{\alpha,\beta}|a_{\alpha,\beta}||\xi^{\alpha}\eta^{\beta}|]^m\Big(\int_{|y-t|<2}|f(y)|^p w(y)dy\Big)^{1/p}\\ &= Ce^{\sum_{\alpha,\beta}|a_{\alpha,\beta}||\xi^{\alpha}\eta^{\beta}|}\Big(\int_{|y-t|<2}|f(y)|^p w(y)dy\Big)^{1/p}\\ &\le C\Big(\int_{|y-t|<2}|f(y)|^p w(y)dy\Big)^{1/p},\end{aligned}$$

where $\eta = (2, 2, \cdots, 2)$.By Lemma 1, the above implies

$$\|\overline{T}_0 f\|_{p,w} \le C\|f\|_{p,w}.$$

3.Proofs of Theorem 3 and 4

The proofs of Theorem 3 and 4 are based on a crucial proposition.

Proposition 1. Let $1 < p < \infty$, $\Omega \in Llog^+L(S^{n-1})$,and $h \in BV(\mathbb{R}_+)$. Then $T_\infty = T - T_0$ is bounded on $L^p(\mathbb{R}^n), 1 < p < \infty$, for any real non-trivial polynomial $P(x, y)$.

Proof. We split T_∞ as follows:

$$\begin{aligned}T_\infty f(x) &= \int_{|x-y|\ge 1} e^{iP(x,y)}\frac{h(|x-y|)\Omega(x-y)}{|x-y|^n}f(y)dy\\ &= \sum_{j=1}^{\infty}\sum_{m=0}^{\infty}\int_{2^{j-1}\le|x-y|<2^j} e^{iP(x,y)}\frac{h(|x-y|)\Omega_m(x-y)}{|x-y|^n}f(y)dy\\ &:= \sum_{j=1}^{\infty}\sum_{m=0}^{\infty}T_{j,m}f(x),\end{aligned}$$

where $\Omega_m(x') = \Omega(x')\chi_{E_m}(x'), E_0 = \{x' \in S^{n-1} : |\Omega(x')| < 1\}$, and

$$E_m = \{x' \in S^{n-1} : 2^{m-1} \le |\Omega(x')| < 2^m\}, m \in \mathbb{N}.$$

Since $h \in BV(\mathbb{R}_+)$,by a method similar to that in [4],we can prove that there exists a $\delta > 0$ such that

$$\|T_{j,m}f\|_p \le C2^{-j\delta}\|\Omega_m\|_{L^\infty(S^{n-1})}\|f\|_p, \tag{3}$$

where C is independent of j and m. On the other hand, we have

$$\begin{aligned}|T_{j,m}f(x)| &\leq \int_{2^{j-1}\leq|y|<2^j} \frac{|h(|y|)\Omega_m(y)|}{|y|^n}|f(x-y)|dy \\ &\leq \|h\|_\infty \int \frac{|\Omega_m(y)|}{|y|^n}\chi_{\{2^{j-1}\leq|y|<2^j\}}(y)|f(x-y)|dy.\end{aligned}$$

Since $|\Omega_m(y)|\chi_{\{2^{j-1}\leq|y|<2^j\}}(y)/|y|^n \in L^1(\mathbb{R}^n)$, we have

$$\begin{aligned}\|T_{j,m}f\|_p &\leq C\int_{2^{j-1}\leq|y|<2^j} \frac{|\Omega_m(y)|}{|y|^n}dy\|f\|_p \\ &= C\|\Omega_m\|_{L^1(S^{n-1})}\|f\|_p. \qquad (4)\end{aligned}$$

Now we choose a positive integer $M > \delta^{-1}$. Then

$$\begin{aligned}\|T_\infty f\|_p &\leq \sum_{j=1}^\infty \sum_{m=0}^\infty \|T_{j,m}f\|_p \\ &= \sum_{j=1}^\infty \|T_{j,0}f\|_p + \sum_{m=1}^\infty \sum_{1\leq j\leq Mm} \|T_{j,m}f\|_p + \sum_{m=1}^\infty \sum_{j>Mm} \|T_{j,m}f\|_p \\ &:= I_1 + I_2 + I_3.\end{aligned}$$

Using (3), we get

$$I_1 \leq C\sum_{j=1}^\infty 2^{-j\theta\delta}\|f\|_p$$

and

$$\begin{aligned}I_3 &\leq C\sum_{m=1}^\infty \sum_{j>Mm} 2^{-j\theta\delta}2^{m\theta}\|f\|_p \\ &\leq C\|f\|_p \sum_{m=1}^\infty 2^{-\theta(M\delta-1)m} \\ &\leq C\|f\|_p.\end{aligned}$$

By (4),we obtain

$$\begin{aligned}I_2 &\leq C\sum_{m=1}^\infty \sum_{1\leq j\leq Mm} \|\Omega_m\|_{L^1(S^{n-1})}\|f\|_p \\ &\leq C\|f\|_p \sum_{m=1}^\infty m2^m|E_m| \\ &\leq C\|\Omega\|_{Llog^+L(S^{n-1})}\|f\|_p \\ &\leq C\|f\|_p.\end{aligned}$$

Thus,we have

$$\|T_\infty f\|_p \le C\|f\|_p.$$

This finishes the proof of Proposition 1.

Let us now turn to the proofs of Theorem 3 and 4.Since M_Ω is bounded on $L^p(\mathbb{R}^n)$ for $\Omega \in L^1(S^{n-1})$, Theorem 3 is directly deduced from Lemma 4 and Proposition 1.Next, $(i) \Rightarrow (ii)$ in Theorem 4 is trivial,$(ii) \Rightarrow (iii)$ is easily deduced from Lemma 2 and Lemma 5, and $(iii) \Rightarrow (i)$ is a direct result of Lemma 3 and Proposition 1.

4. Proofs of Theorem 5 and 6

The proof of Theorem 5 is based on the following proposition.

Proposition 2. Let $1 < p < \infty$, $w \in A_p$, $h \in BV(\mathbb{R}_+)$, and $\Omega \in L^q(S^{n-1}), 1 < q \le \infty$. If M_Ω is bounded on $L^p(w)$, then $T_\infty = T - T_0$ is bounded on $L^p(w)$ for any real non-trivial polynomial $P(x,y)$.

Proof. We split T_∞ as follows:

$$\begin{aligned} T_\infty f(x) &= \sum_{j=1}^{\infty} \int_{2^{j-1} \le |x-y| < 2^j} e^{iP(x,y)} \frac{h(|x-y|)\Omega(x-y)}{|x-y|^n} f(y)dy \\ &:= \sum_{j=1}^{\infty} T_j f(x). \end{aligned}$$

Note that the authors of [4] have proved that there exists a $\delta > 0$ such that

$$\|T_j f\|_p \le C2^{-j\delta}\|f\|_p, \tag{5}$$

where C is independent of j. On the other hand, we have

$$\begin{aligned} |T_j f(x)| &\le \|h\|_\infty \int_{2^{j-1} \le |y| < 2^j} \frac{\Omega(y)}{|y|^n} |f(x-y)|dy \\ &\le CM_\Omega f(x). \end{aligned}$$

Since M_Ω is bounded on $L^p(w)$, it is easy te see that

$$\|T_j f\|_{p,w} \le C\|f\|_{p,w}$$

holds for any $w \in A_p$. However,for any $w \in A_p$, it follows form the inverse Hölder's inequality that there exists a $\epsilon > 0$ such that $w^{1+\epsilon} \in A_p$. Thus, we have

$$\|T_j f\|_{p,w^{1+\epsilon}} \le C\|f\|_{p,w^{1+\epsilon}}. \tag{6}$$

Interpolating with change of measures (see[6]) between (5) and (6), we get

$$\|T_j f\|_{p,w} \le C2^{-j\theta}\|f\|_{p,w},$$

where $0 < \theta < 1$. Now,the conclusion of proposition is deduced from the above. This finishes the proof of Proposition 2.

As the same as in the proof of Theorem 3, the conclusion of Theorem 5 is a direct result of Proposition 2 and Lemma 4. Finally, we point out that Theorem 6 is a simple corollary of Theorem 5. In fact, under the conditions of Theorem 6,D.Watson [8] has proved that $\overline{T}$ is bounded on $L^p(w)$ provided $w \in A_{p/q'}$ and $q' \le p < \infty (p \neq 1)$. And J.Duoandikoetxea [2] has also proved that $w \in A_{p/q'}$ implies that M_Ω is bounded on $L^p(w)$ if $q' \le p < \infty (p \neq 1)$. Thus, the conclusion of Theorem 6 in the case (A) follows from the above two facts and Theorem 5 immediately. The conclusion of Theorem 6 in the case (B) follows from (A) and the duality. This finishes the proof of Theorem 6.

REFERENCES

1. A.P.Calderon and A.Zygmund, *On singular integrals*, Amer.J.Math. **78** (1956), 289-309.
2. J.Duoandikoetxea, *Weighted norm inequalities for homogeneous singular integrals*, Trans.Amer. Math.Soc. **336** (1993), 869-880.
3. Y.S.Jiang and S.Z.Lu, *A note on oscillatory singular integrals with rough kernel*, J.Beijing Normal Univ. **27** (1991), 398-402.
4. S.Z.Lu and Y.Zhang, *Criterion on L^p-boundedness for a class of oscillatory singular integrals with rough kernels*, Rev.Mat.Iberoamericana **8** (1992), 201-219.
5. F.Ricci and E.M.Stein, *Harmonic analysis on nilpotent groups and singular integrals,I.Oscillatory integrals*, J.Func.Analysis **73** (1987), 179-194.
6. E.M.Stein and G.Weiss, *Interpolation of operators with change of measures*, Trans.Amer.Math.Soc. **87** (1958), 159-172.
7. Q.Y.Sun, *Two problems about singular integral operator*, Approx.Theory and its Appl. **7** (1991), 83-98.
8. D.Watson, *Weighted estimates for singular integrals via Fourier transform estimates*, Duke Math.J. **60** (1990), 389-399.

The Minimal Decay of Matrix Coefficients for Classical Groups

JIAN-SHU LI*

University of Maryland

College Park MD 20742

§1. Introduction

Let G be a reductive Lie group with compact center. A unitary representation ρ of G is said to be strongly L^p if, for a dense set of vectors v in the space of ρ, the matrix coefficients $x \mapsto (\rho(x)v, v)$ all lie in $L^p(G)$. Let $p(G)$ be the smallest real number such that any irreducible infinite dimensional unitary representation is strongly L^p for any $p > p(G)$. The existence of $p(G) < \infty$ is equivalent to Kazhdan's "property (T)". Indeed, the explicit determination of $p(G)$ may be viewed as a quantitative version of Kazhdan's property (T). In this paper we compute the numbers $p(G)$ for the real classical groups. Our first main result is

Theorem A: *The number $p(G)$ for the real classical groups are given by the following table*

(1)

$G:$	$Sp(n,\mathbf{R})$	$O(p,q)$	$U(p,q)$	$Sp(p,q)$	$O^*(2n)$
$p(G):$	$2n$	$p+q-2$	$2(p+q-1)$	$2(p+q)-1$	$2n-3$
$Condition:$	$n \geq 2$	$p \geq q \geq 2$ $p+q \geq 7$	$p \geq q \geq 2$	$p \geq q \geq 1$ $p \geq 2$	$n \geq 5$

*Sloan Fellow. Supported in part by NSF grant No. DMS-9203142

M. Cheng et al. (eds.), Harmonic Analysis in China, 146–169.

with the possible exception of the five groups: $O(5,2), O(4,3), O(6,3), Sp(2,2)$ *and* $O^*(10)$.

Remark: The case with $G = Sp(n, \mathbf{R})$ was already proven by Howe [**Howc**].

In the above table, the conditions on the indices are given firstly to ensure that the groups in question have Kazhdan's property (T), and then they are further restricted to avoid the incidental local isomorphisms between various families of groups in low dimensions. For example, the groups $O(3,2), O(4,2)$ and $O(3,3)$ are locally isomorphic to $Sp(2,\mathbf{R})$, $SU(2,2)$ and $SL(4,\mathbf{R})$ respectively. The number $p(G)$ should be $4, 6$ and 6 respectively in these cases, different from what would have been indicated by their membership in the family $O(p,q)$. On the other hand, the possible exception of the five groups listed above is purely an artifact resulting from a slight deficiency in our proof. A better proof (or some ad hoc argument for these small size groups) would probably remove these exceptions.

Actually, our table only included the so called Type I classical groups. But the classical groups of Type II are of the form $GL(n, D)$ where D is either $\mathbf{R}, \mathbf{C}$ or $\mathbf{H}$ (the Hamilton quaternions). The number $p(G)$ in this case is readily known from Vogan's classification [**Vog**] of the unitary duals for these groups. Similarly, one should be able to compute $p(G)$ for the complex classical groups from Barbasch's classification [**Bar**]. For reference and comparison, we list the numbers $p(G)$ for these cases in the following table.

(2)

$G:$	$GL(n,\mathbf{R})$	$GL(\mathbf{C})$	$GL(n,\mathbf{H})$	$Sp(n,\mathbf{C})$	$O(n,\mathbf{C})$
$p(G):$	$2n-2$	$2n-2$	$2n-1$	$4n$	$n-2$
$Condition:$	$n \geq 3$	$n \geq 3$	$n \geq 3$	$n \geq 2$	$n \geq 7$

Of course, the result stated in Theorem A will be completely subsumed some day by a good classification of the unitary duals of the real classical groups. Before such a classification becomes available, however, it seems to be of sufficient interest to know the values of $p(G)$ explicitly.

The proof of Theorem A is naturally divided into two parts. For the time being let us write $r(G)$ for the numbers $p(G)$ that appear in the table (1). We first show that every irreducible infinite dimensional unitary representation of G is strongly L^p for any $p > r(G)$; hence $p(G) \leq r(G)$. This part of the proof is modeled after §7-8 of Howe's paper [**Howc**]. There is nothing particularly difficult in extending Howe's arguments to the present cases. It is the sharp results of [**CHH**], which deals with the relationship between a technical growth condition and the L^p estimates of matrix coefficients, that enables us to give a relatively quick and clean proof here.

The second part of the proof is to exhibit at least one irreducible unitary representation which is infinite dimensional and not strongly $L^{r(G)}$. This is relatively easy and will be done using the oscillator representation. This will then prove $p(G) = r(G)$.

In both steps of the proof, by far the most involved case is when $G = O(p,q)$, which we treat in great detail in the paper.

It turns out that there are unitary representations with minimal decay of matrix coefficients which are also automorphic. More precisely, for a unitary representation π of G we let $p(\pi)$ be the smallest real number such that π is strongly L^p for any $p > p(\pi)$. We have

Theorem B: *Let G and $p(G)$ be as in Theorem A. There exist infinitely many irreducible unitary representation π of G such that*

(a) *One has $p(\pi) = p(G)$.*

(b) *There exists an arithmetic lattice Γ such that π occurs as a subrepresentation of $L^2(\Gamma \backslash G)$.*

In fact, statement (b) can be made much stronger. Suppose G is neither $U(p,q)$ nor $O(p,q)$ with $p+q = 4, 8$ in the above. Then (b) can be replaced by

(b′) *For any lattice $\Gamma \subset G$, there is a finite index subgroup $\Gamma_1 \subseteq \Gamma$ such that π occurs as a subrepresentation of $L^2(\Gamma_1 \backslash G)$.*

The author would like to thank Professors Roger Howe and Peter Sarnak for their encouragement and interest in this work.

§2. Some preparation.

The real classical groups that appear in Theorem A can be described as follows. Let D be one of the division algebras $\mathbf{R}, \mathbf{C}$ or $\mathbf{H}$, with the involution $\#$ which is equal to the trivial involution, complex conjugation, and quaternionic conjugation respectively. Let V be a vector space over D, endowed with a non-degenerate sesquilinear form $(\ ,\ )$ which is either hermitian or skew-hermitian. Let G be the group of isometries of $(\ ,\ )$. By varying the data D, V and $(\ ,\ )$, we obtain the various classes of groups listed in Theorem A as our G. We write $\eta = \eta(V) = 1$ or -1 according as $(\ ,\ )$ is hermitian or skew-hermitian.

Suppose that $(V', (\ ,\ )')$ is another such "formed space" with $\eta' = \eta(V') = -\eta$. Set $W = V \otimes_D V'$ and define a symplectic form $<,>$ on W via $<,>= tr_{D/\mathbf{R}}((,) \otimes (,)'^{\#})$. Here $tr_{D/\mathbf{R}}$ is reduced trace. Let $Sp = Sp(W)$ be the symplectic group which is the isometry group of $<\ ,\ >$. Then G is a subgroup of $Sp(W)$ via its action on the first factor of $W = V \otimes_D V'$ We fix an additive character of $\mathbf{R}$ once for all and let ω be the oscillator representation of (the two-fold cover of) $Sp(W)$ attached to this character (cf.[**Howa**]).

Definition 2.1. The restriction of ω to G, to be denoted as ω or $\omega_{V'}$, is called the oscillator representation of G attached to the formed space V'.

Remark: In fact, ω is a representation of not Sp itself, but its metaplectic two-fold cover. Thus one should consider an appropriate two-fold cover $\tilde{G}$ instead of G. However, for all the cases that are of interest in what follows, it is possible to tensor $\omega|_{\tilde{G}}$ with a one dimensional unitary character of $\tilde{G}$ so that the resulting representation will factor through the covering map $\tilde{G} \longrightarrow G$. In the above definition it is to be understood that such a twist by a one dimensional character has been done. Apparently this will have no effect on the size of the matrix coefficients of $\omega_{V'}$, which is what will be of interest here.

Set

(3) $$m = dim_D V, \quad m' = dim_D V', \quad d = dim_{\mathbf{R}} D.$$

Let d_0 be the dimension of the $(-\eta)$-eigenspace of the involution $\#$ on D. The following estimate is proven in [**Howc**] in the symplectic group case; its easy extension to the general case is carried out in [**Lic**].

Lemma 2.2. *The representation $\omega_{V'}$ is strongly L^p for any*

$$p > \frac{2}{m'}\,(m - 2 + 2 \cdot \frac{d_0}{d}\,) \tag{4}$$

The right hand side of the above inequality is closely related to our number $p(G)$. Consider the case where m' is as small as possible (and thus the right hand side of (4) is as large as possible). If $D = \mathbf{R}$ and $\eta = 1$ then $(\ ,\ )'$ is a skew-symmetric form. Since it is also non-degenerate, the smallest possible dimension is $m' = 2$. If again $D = \mathbf{R}$ but $\eta = -1$ then G is a symplectic group, and $\omega_{V'}$ can only be a representation of the non-trivial two-fold cover of G (and not G itself) unless m' is even. Thus $m' = 2$ is again the minimal choice. In all other cases it is possible to take $m' = 1$ (taking the remark after Definition 2.1 into consideration). Let $m'(D)$ be this minimal m', that is

$$m'(D) = \begin{cases} 2, & D = \mathbf{R} \\ 1, & D = \mathbf{C} \text{ or } \mathbf{H} \end{cases} \tag{5}$$

The reader can easily verify that the number $p(G)$ in Theorem A is given by

$$p(G) = \frac{2}{m'(D)}\,(m - 2 + 2 \cdot \frac{d_0}{d}\,) \tag{6}$$

We next recall that the *rank* of a unitary representation of G is an integer between 0 and the split rank of G. Here the split rank of G is the same as the Witt index of V. We refer the reader to[**Howc**] and [**Lib**] for a precise discussion of the notions of rank and *pure rank.* Recall that when $D = \mathbf{R}$ and $\eta = 1$, the possible ranks are all even integers; in this case let $r(G)$ be the largest even integer less than or equal to the split rank of G. In all other cases, let $r(G)$ be the split rank of G. Thus $r(G)$ is the largest possible rank for all unitary representations of G. We note that the representation $\omega_{V'}$ is of pure rank $\min(m', r(G))$.

The following result is clear from the proof of Corollary 2.13 of [**Howc**]

Lemma 2.3. *Let ρ be a unitary representation of G of pure rank l. Fix an integer $k \leq l$ which is even when $D = \mathbf{R}, \eta = 1$. Let V_1 be a non-degenerate subspace of V. Denote by G_1 the subgroup of G leaving the orthogonal complement $V_1^\perp$ pointwise fixed. Suppose that the Witt index of $V_1^\perp$ is at least k. Then the restriction $\rho|_{G_1}$ is a finite direct sum of representations of the form $\omega_{V'} \otimes \sigma$, where $\omega_{V'}$ is the oscillator representation of G_1 attached to a formed space V' (Definition 2.1) of dimension k over D, and σ is a unitary representation of G_1 of rank $\min(r(G_1), l-k)$.*

§3. A strategy of Howe.

Notation: We will say that a representation is strongly $L^{p+\epsilon}$ if it is strongly $L^{p'}$ for all $p' > p$.

We shall make repeated use of a general strategy of Howe [**Howc**]. Let G be a reductive Lie group with compact center. Let B be a minimal parabolic subgroup of G, A a maximal split torus contained in B, $A^+ \subset A$ the positive Weyl chamber determined by B. Let H be a reductive subgroup of G with compact center, such that $B(H) = B \cap H$ is a minimal parabolic subgroup of H, and $A(H) = A \cap H$ is a maximal split torus of H. Then the corresponding positive Weyl chamber $A^+(H)$ in $A(H)$ contains A^+. Let δ and δ_H be the modulus functions for B and $B(H)$ respectively.

Suppose $H = H_1 \times \cdots \times H_m$, a product of reductive subgroups. Then in obvious notations one has $A(H) = A(H_1) \times \cdots \times A(H_m)$ and $\delta_H = \prod_{j=1}^m \delta_{H_j}$.

Theorem 3.1: *Let ρ be a unitary representation of G. Fix a real number $r > 0$. Suppose that*

(a) For $1 \leq j \leq m$, $\rho|_{H_j}$ is strongly $L^{p_j+\epsilon}$. Let k_j be a positive integer with $p_j \leq 2k_j$.

(b) For $a \in A^+$, one has

$$\prod_{j=1}^{m} \delta_{H_j}(a)^{\frac{1}{2k_j}} \geq \delta(a)^{\frac{1}{r}} \tag{7}$$

Then ρ itself is strongly $L^{r+\epsilon}$.

This result follows directly from the main corollary of [**CHH**, page 108], Proposition 6.3 of [**Howc**] and the considerations of [**Howc**, §8]. We shall also make use of the following lemma which is an obvious consequence of[**CHH**]

Lemma 3.2. *Let k be a positive integer.*

(a) If the unitary representation ρ is strongly $L^{2k+\epsilon}$ then every matrix coefficient of ρ is in $L^{2k+\epsilon}$.

(b) Suppose ρ is a direct integral of unitary representations. Then it is strongly $L^{2k+\epsilon}$ if and only if almost all the integrands are.

We now introduce some notations to be used later. Let V_1, V_1^* be two maximal totally isotropic subspaces of V such that the restriction of $(\ ,\ )$ to $V_1 + V_1^*$ is non-degenerate. Let V_0 be the orthogonal complement of $V_1 + V_1^*$. Then $(\ ,\ )$ restricts to an anisotropic form on V_0 and we have a direct sum decomposition

$$V = V_1 \oplus V_0 \oplus V_1^* \tag{8}$$

Let r be the dimension of V_1 over D. We may choose a basis of V compatible with the above decomposition, so that the matrix of $(\ ,\ )$ with respect to this basis has the form

$$J = \begin{pmatrix} & & I_r \\ & J_0 & \\ \eta I_r & & \end{pmatrix}, \tag{9}$$

where J_0 is the matrix for the restriction of $(\ ,\ )$ to V_0. Consequently G can be realized as the group of $m \times m$ matrices g with entries in D, such that ${}^t g^{\#}\, J\, g = J$.

Let A be the subgroup of G consisting of elements of the form

$$a = \begin{pmatrix} a_1 & & & & & & \\ & \ddots & & & & & \\ & & a_r & & & & \\ & & & I_{m-2r} & & & \\ & & & & a_1^{-1} & & \\ & & & & & \ddots & \\ & & & & & & a_r^{-1} \end{pmatrix} \qquad (a_1, \cdots, a_r \in \mathbf{R}^\times)$$

We will write such a matrix as $a = (a_1, \cdots, a_r)$. Then A is a maximal split torus in G. Choose a minimal parabolic subgroup B containing A. We may assume B is in a block upper triangular form compatible with (8)-(9), so that

$$(10) \qquad A^+ = \{ a \in A \mid a_1 \geq a_2 \geq \cdots \geq a_r \geq 1\}$$

is the the positive Weyl chamber of A relative to B (when $G = O(r,r)$ we must replace the last inequality by $a_{r-1}a_r \geq 1$. Let $D_0 = \{ t \in D \mid t^\# = -\eta t \}$ and $d_0 = dim_{\mathbf{R}} D_0$. The modular function δ of B, when restricted to A, is easily seen to be

$$(11) \qquad \delta(a) = \prod_{i=1}^{r} |a_i|^{d(m-2i)+2d_0}$$

§4. **The estimates.**

Notation: $\{x\}$ will denote the smallest integer $\geq x$.

We begin with $G = O(p,q), p \geq q \geq 2$. We introduce the usual coordinates for A, writing a typical element as $a = (a_1, \cdots, a_q)$. Choose the positive Weyl chamber so that for $a \in A^+$ one has

$$a_1 \geq \cdots \geq a_q$$

If $p = q$ then $a_{q-1} \geq a_q^{\pm 1}$; otherwise $a_q \geq 1$. The modulus function is given by

$$\delta(a) = a_1^{p+q-2} \cdots a_q^{p-q}$$

In the following ρ will always denote a unitary representation containing no one dimensional characters as its constituents.

Lemma 4.1. *Let ρ be a unitary representation of G which does not contain any one dimensional character of G. Then for any subgroup $O(m,n) \subset O(p,q)$ ($m < p, n < q$), embedded in a obvious way, the restriction $\rho|_{O(m,n)}$ is strongly $L^{2k+\epsilon}$ where k is any positive integer with $2k \geq m+n-2$.*

Proof. We consider $O(p,q)$ as the group of isometries of $\mathbf{R}^{p,q}$ where $\mathbf{R}^{p,q}$ denotes $\mathbf{R}^{p+q}$ endowed with a symmetric bilinear form (,) of signature (p,q). Consider the subgroup of $O(p,q)$ which leaves a fixed isotropic vector in $\mathbf{R}^{p,q}$ invariant. This subgroup is isomorphic to the semi-direct product $O(p-1,q-1)\cdot\mathbf{R}^{p-1,q-1}$, in which we may embed $O(m,n)\cdot\mathbf{R}^{m,n}$ in a natural way. By a well known result of Howe and Moore [**HoM**], there are no non-zero vectors in the space of ρ which are fixed by the abelian group $\mathbf{R}^{m,n}$. We now analyze the unitary representations of $O(m,n)\cdot\mathbf{R}^{m,n}$ with this property using Mackey's orbit method.

Fix a non-trivial additive character of the field $\mathbf{R}$. The Pontrjagin dual of the vector space $V = \mathbf{R}^{m,n}$ can be firstly identified with its linear dual V^*, and then with itself via the given symmetric bilinear form. For any real number r set

$$V_r^* = \{x \in \mathbf{R}^{m,n} | x \neq 0, (x,x) = r\}$$

By Witt's theorem, the non-zero orbits of $O(m,n)$ on $\mathbf{R}^{m,n}$ are precisely the V_r^*'s just defined.

We first take $r \neq 0$ and fix a character $\psi \in V_r^*$. For definiteness assume $r > 0$. The stabilizer of ψ in $O(m,n)$ will be $O(m-1,n)$. We extend ψ to $O(m-1,n)\cdot\mathbf{R}^{m,n}$ by making it trivial on $O(m-1,n)$. Let σ be a unitary representation of $O(m-1,n)$ and extend it to $O(m-1,n)\cdot\mathbf{R}^{m,n}$ trivial on $\mathbf{R}^{m,n}$. Consider the induction of $\psi\otimes\sigma$ from $O(m-1,n)\cdot\mathbf{R}^{m,n}$ to $O(m,n)\cdot\mathbf{R}^{m,n}$. By Mackey theory, any unitary representation of $O(m,n)\cdot\mathbf{R}^{m,n}$ with $V-$spectrum supported on the orbit V_r^* must be such an induced representation. Note that when restricted to $O(m,n)$, this is precisely the induced representation $\mathrm{Ind}_{O(m-1,n)}^{O(m,n)}(\sigma)$; we write this as $\mathrm{Ind}(\sigma)$ for simplicity. By a direct argument, the matrix coefficients of $\mathrm{Ind}(\sigma)$ are dominated by those of

$$\mathrm{Ind}(1) = L^2(O(m,n)/O(m-1,n))$$

Now either by the known spectrum analysis of the rank one symmetric space $O(m,n)/O(m-1,n)$ [**Far**], or by the estimates in [**Lia**], we see that $\mathrm{Ind}(1)$ is strongly $L^{2k+\epsilon}$.

Next consider V_0^* and take $\psi \in V_0^*$. Let P_1 be the stabilizer of ψ in $O(m,n)$. This time we have $P_1 = O(m-1,n-1)\cdot\mathbf{R}^{m-1,n-1}$. The same analysis as before shows that a unitary representation of $O(m,n)\cdot\mathbf{R}^{m,n}$ with V-spectrum supported on the orbit V_0^* must be induced, and its restriction to $O(m,n)$ must be of the form $\mathrm{Ind}_{P_1}^{O(m,n)}\sigma$ where σ is a unitary representation of P_1. As before, the matrix coefficients of the last representation are dominated by those of $\mathrm{Ind}_{P_1}^{O(m,n)}1$. Let P be the stabilizer of the line through ψ; it is a parabolic subgroup of $O(m,n)$. We have $P = GL(1)\cdot P_1$. Consider induction by stages, we see that $\mathrm{Ind}_{P_1}^{O(m,n)}1$ is the direct integral over the unitary (perhaps degenerate) principal series $\mathrm{Ind}_{P}^{O(m,n)}\chi$ with χ going through all unitary characters of $GL(1)$. It is easy to see that each $\mathrm{Ind}\chi$ is strongly $L^{m+n-2+\epsilon}$.

To summarize, we see that the restriction of ρ to $O(m,n)$ is a direct integral of unitary representations which are all strongly $L^{2k+\epsilon}$. Thus so is ρ itself by Lemma 3.2.

Proposition 4.2. *Assume that either $p+q$ is even, $p \geq q+2$, and $p \geq 5$, or $p+q$ is odd, and $p \geq 2q+2$. Then for the group $G = O(p,q)$ we have $p(G) \leq p+q-2$.*

Proof: We apply Theorem 3.1 with $H = O(3,1)\times O(p-3,q-1)$. With the notation there we may take, by the above lemma, $k_1 = 1, k_2 = \{\frac{p+q-6}{2}\}$. Also $r = p+q-2$ here. Thus the inequality (7) can be expressed as

$$a_1^2(a_2^{p+q-6}\cdots a_q^{p-q-2})^{\frac{1}{2k_2}} \geq (a_1^{p+q-2}\cdots a_q^{p-q})^{\frac{1}{p+q-2}}$$

Taking into consideration that $a \in A^+$, this is the same as

$$\frac{1}{2k_2}\sum_{j=1}^{m}(p+q-4-2j) \geq \frac{1}{p+q-2}\sum_{j=1}^{m}(p+q-2-2j)$$

for $m = 1, \cdots, q-1$. Simplifying both sides and dividing by m, we obtain

$$\frac{1}{2k_2}(p+q-m-5) \geq \frac{1}{p+q-2}(p+q-m-3)$$

This is satisfied for all m once it is for $m = q-1$. Thus letting $m = q-1$ and simplify, we are reduced to

$$(p+q-2-2k_2)(p-2) \geq 2(p+q-2) \tag{12}$$

Now suppose $p+q$ is even. Then $2k_2 = p+q-6$ and (12) is $4(p-2) \geq 2(p+q-2)$ which follows since $p \geq q+2$. On the other hand if $p+q$ is odd then we must take $2k_2 = p+q-5$ and (12) leads to the condition $p \geq 2q+2$. In any case the proposition follows.

Observe that for $q = 2$ or 3, the only cases not covered by the above proposition are $O(5,2), O(4,3)$ and $O(6,3)$ (see remarks in §1 for $O(3,2), O(4,2), O(3,3)$).

From now on assume $p \geq q \geq 4$. We first consider the special case $p = q$.

Proposition 4.3. *Suppose $n \geq 4$. Then $p(G) \leq 2n-2$ for $G = O(n,n)$.*

Proof: Apply Theorem 3.1 with $H = O(2,2) \times O(n-2, n-2)$. Here $k_1 = 1, k_2 = n-3$ and $r = 2n-2$. We must embed H in G as follows. Both factors $O(2,2)$ and $O(n-2, n-2)$ will have maximal split tori embedded in A. We assume that the split torus for $O(2,2)$ is the one consisting of elements of the form $(a_1, 1, \cdots, 1, a_n)$,and the split torus for $O(n-2, n-2)$ corresponds to the remaining coordinates of A. The inequality (7) is then

$$(a_1^2 a_n^0)^1 \cdot (a_2^{2n-6} \cdots a_{n-1}^0)^{\frac{1}{n-3}} \geq (a_1^{2n-2} \cdots a_n^0)^{\frac{2}{2n-2}}$$

This is easily seen to be valid for $a \in A^+$.

Finally assume $p > q \geq 4$. We recall the notion of rank for unitary representations of classical groups [**Howc**] [**Lib**] (see also §2). For $O(p,q)$ the ranks are all even integers $\leq q$.

Lemma 4.4. *Suppose $p > q \geq 4$. Let ρ be a unitary representation of $O(p,q)$ of pure rank at least 4. Then its restriction to a subgroup $O(3,2) \subset O(p,q)$ is tempered.*

Proof: According to Lemma 2.3, the restriction of ρ to $O(3,2)$ is a direct sum of representations of the form $\omega \otimes \sigma$, where σ is a unitary representation of $O(3,2)$ of pure rank 2 and ω is an appropriate oscillator representation attached to a symplectic form of dimension 2. Either by knowledge of the unitary dual of $O(3,2)$, or by the local isomorphism of $O(3,2)$ with $Sp(2,\mathbf{R})$ and the estimate in [**DHL**], we know that σ is $L^{4+\epsilon}$. On the other hand it is easy to see that ω is $L^{3+\epsilon}$ by Lemma 2.2. By Jordan-Hölder we conclude that $\omega \otimes \sigma$ is tempered, and hence $\rho|_{O(3,2)}$ is tempered.

Proposition 4.5. *Suppose* $p > q \geq 4$. *Let* ρ *be a unitary representation of* $O(p,q)$ *of pure rank at least* 4. *Then* ρ *is strongly* $L^{p+q-2+\epsilon}$

Proof: Apply Theorem 3.1 with $H = O(3,2) \times O(p-3,q-2)$. Here $k_1 = 1, k_2 = \{\frac{p+q-7}{2}\}$ and $r = p+q-2$. The subgroup H is embedded so that the split torus of the first factor $O(3,2)$ corresponds to the first two coordinates of A. We omit the details since it is done the same way as before.

What about representations of pure rank 2 ? By [**Lib**], all such representations come from Howe correspondence with unitary representations of $SL(2)$ (or its two fold cover if $p+q$ is odd), provided $p \geq q \geq 4$. This will be discussed next. We shall see that these representations are also $L^{p+q-2+\epsilon}$ and hence $p(G) \leq p+q-2$. At the same time we discover that $p(G) \geq p+q-2$ for all p, q. This would then conclude our study of the problem for $O(p,q)$.

We now turn to other groups, for which the argument turns out to be substantially simpler.

The group $Sp(p,q)$.

Here $(p \geq q \geq 1)$

$$\delta(a) = \prod_{j=1}^{q} a_j^{4p+4q+6-8j}$$

When $q = 1$ the desired exponent $p(G) = 2p+2q-1$ follows from classification [**Kos**]. [**Bal**]. It is also clear for rank 1 representations when $q > 1$ (see §5 below).

Proposition 4.6. *Suppose* $p > q \geq 1$. *Then* $p(G) \leq 2p+2q-1$

Proof: We may assume $q \geq 2$ and ρ is of pure rank ≥ 2. Take $H = Sp(2,1) \times Sp(p-2, q-1)$. We claim that $\rho|_{Sp(2,1)}$ is $L^{5/2+\epsilon}$. Indeed, for any $m \geq n, m < p, n < q$, the restriction of ρ to $Sp(m,n)$ is a sum of representations of the form $\omega \otimes \sigma$, where ω is an oscillator representation and σ is of pure rank at least 1. Now ω is $L^{2m+2n-1+\epsilon}$ and for $Sp(2,1)$ we also know σ is $L^{5+\epsilon}$. The claim follows.

Thus we may apply Theorem 3.1 with $k_1 = 2$, $k_2 = \{\frac{2(p+q-3)-1}{2}\} = p+q-3$ and $r = 2p + 2q - 1$. A calculation shows the inequality (7) is valid. The proposition follows.

Proposition 4.7. *Suppose $p = q = n \geq 3$ and ρ is of pure rank at least* 2. *Then it is* $L^{2(4n-1)/3+\epsilon}$

Proof: Take $H = Sp(1,1) \times \cdots \times Sp(1,1)$. We may write $\rho|_{Sp(1,1)}$ as a sum of representations of the form $\omega \otimes \sigma$ where ω is an oscillator representation attached to a skew-hermitian form of dimension 2 over $\mathbf{H}$ and σ is whatever it might be. Now ω is $L^{3/2+\epsilon}$. Hence $\rho|_{Sp(1,1)}$ is tempered. Thus we may apply Theorem 3.1 with $k_j = 1$ for $j = 1, \cdots, n$ and $r = 2(4n-1)/3$. The inequality (7) is easily verified.

Taking together, we have shown the desired estimate $p(G) \leq 2p + 2q - 1$ for all $Sp(p,q)$ with $p \geq q \geq 1, p+q \geq 3$, except for $p = q = 2$. Using $H = Sp(1,1) \times Sp(1,1)$ one may also show that $p(G) \leq 28/3$, which is slightly short of the desired $p(G) = 7$.

The group $G = U(p,q)$.

This is the easiest: just take $H = U(1,1) \times U(n-1, n-1)$ and work it out.

In fact, same for $Sp(n, \mathbf{R})$: take $H = Sp(1, \mathbf{R}) \times Sp(n-1, \mathbf{R})$ here.

The case $G = O^*(2n)$.

Let $m = [n/2]$ it would probably be more suggestive to write the group as $G(m + t, m)$, where $t = n - 2m$ $(= 0$ or $1)$. But this will not be done here.

The modulus function for $O^*(2n)$ is

$$\delta(a) = \prod_{j=1}^{m} a_j^{4n+2-8j}$$

As before, for $m \geq 2$ the rank one representations must come from Howe correspondence with $Sp(1)$, which can be easily dealt with.

Proposition 4.8. *Suppose $n \geq 6$ and ρ is of pure rank ≥ 2. Then ρ is L^{2n-3}*

Proof: We take $H = O^*(8) \times O^*(2(n-4))$, embedded in such way that the split torus of the first factor $O^*(8)$ is given as $\{(a_1, 1, \cdots, 1, a_m)\}$. Using the local isomorphism of $O^*(8)$ with $O(6,2)$, we see every non-trivial representation of it is $L^{6+\epsilon}$. The usual analysis shows $\rho|_{O^*(8)}$ is a sum of representations each of which is of the form $\omega \otimes \sigma$ with ω an oscillator representation and σ a representation of pure rank at least 1. Hence $\rho|_{O^*(8)}$ is $L^{30/11+\epsilon}$ (Jordan-Hölder used here). Thus $k_1 = 2$.

On the other hand, since the rank of the second factor is $m-2$, we see $\rho|_{O^*(2(n-4))}$ is a sum of tensor products $\omega \otimes \sigma$ with ω an oscillator representation attached to a hermitian form of dimension 2 over $\mathbf{H}$ and σ is whatever. Hence $\rho|_{O^*(2(n-4))}$ is $L^{(2(n-4)-3)/2+\epsilon}$ and we may take $k_2 = \{(2n-11)/4\}$. With $r = 2n-3$ the inequality (7) is

$$(a_1^{10}a_n^2)^{1/2}(\prod_{j=1}^{m-2} a_{j+1}^{4(n-4)+2-8j})^{1/k_2} \geq (a_1^{4n-6}a_n^{4n+2-8m}\prod_{j=1}^{m-2} a_{j+1}^{4n+2-8(j+1)})^{\frac{2}{2n-3}}$$

It suffices to get the part involving a_2 through a_{n-1}:

$$(\prod_{j=1}^{m-2} a_{j+1}^{4(n-4)+2-8j})^{1/k_2} \geq (\prod_{j=1}^{m-2} a_{j+1}^{4n+2-8(j+1)})^{\frac{2}{2n-3}}$$

This follows from

$$\frac{1}{k_2}\sum_{j=1}^{s}(4(n-4)+2-8j) \geq \frac{2}{2n-3}\sum_{j=1}^{s}(4n+2-8(j+1))$$

for $s = 1, \cdots, m-2$. This simplifies to

$$\frac{1}{k_2}(2n-2s-9) \geq \frac{2}{2n-3}(2n-2s-5)$$

for $1 \leq s \leq m-2$, which can be easily checked.

§5. Representations of minimal rank.

We now proceed to show that all irreducible unitary representations with minimal rank $m'(D)$ (cf. (5)) are strongly $L^{p(G)+\epsilon}$, yet for each G there are infinitely many minimal rank representations which are not strongly $L^{p(G)}$. Together with results from §4 this will prove Theorem A.

In the setting of §2 let G' be the group of isometry of $(V', (\ ,\ )')$. Then G, G' is a reductive dual pair in $Sp(W)$ [**Howa**]. The oscillator representation gives rise to a bijection between certain subsets of representations of G and of G' [**Howd**]. Suppose π and σ are representations of G and G' respectively which correspond to each other under the above said bijection. We will write $\pi = \theta(\sigma)$ and call π the the *local theta lift* of σ. Our grasp on the minimal rank representations comes from the following result, proven in [**Lic**].

Lemma 5.1. *Suppose $r(G) > m'(D)$ and π is an irreducible unitary representation of G with minimal rank $m'(D)$. Then there exist a formed space V' of dimension $m'(D)$ as in §4, with group of isometries G', and an irreducible unitary representation σ of (possibly a two-fold cover of) G', such that $\pi = \theta(\sigma)$.*

We will thus study the representations $\theta(\sigma)$. The case of $G = O(p,q)$, $G' = SL(2, \mathbf{R})$ is by far the most involved; the study of the local theta lift in this case is interesting in its own right. We will write $\theta_{p,q}$ for θ in this case, and allow the possibility $p < q$ here.

Let $\widetilde{SL}(2)$ be the metaplectic two-fold cover of $SL(2, \mathbf{R})$. Consider first the case where σ is a discrete series representation of $\widetilde{SL}(2)$. Then $\theta_{p,q}(\sigma)$ has been computed by Rallis and Schiffmann [**RaS**]. Suppose σ is holomorphic with lowest weight Λ, where Λ is an integer or half integer and $\Lambda > 1$. We write $\sigma = \sigma_\Lambda$. It is well known that we must have $\Lambda \equiv (p+q)/2 \ (mod\ \mathbf{Z})$ for $\theta_{p,q}(\sigma_\Lambda)$ to be non-zero; assume this is th case here. For an integer $a \geq 0$ we denote by $S_n(a)$ the representation of $O(n)$ on the spherical harmonics of degree a; for $n \geq 3$ this is the irreducible representation of highest weight $(a, 0, \cdots, 0)$, when restricted to $SO(n)$. When $n = 1$ it is to be understood that only the values $a = 0$ and 1 are allowed, which correspond to the trivial and sign characters of $O(1)$ respectively. Set $K = O(p) \times O(q)$; this is a maximal compact subgroup of $O(p,q)$.

Lemma 5.2. [**RaS**] *All the K-types of $\theta_{p,q}(\sigma_\Lambda)$ are of the form $S_p(a) \otimes S_q(b)$, which occurs if and only if*

$$a - b = \Lambda - \frac{1}{2}(p-q) + 2k \tag{13}$$

for some non-negative integer k. In such case, the K-type $S_p(a) \otimes S_q(b)$ occurs with multiplicity one.

One also knows [**Howb**] that the infinitesimal character of $\theta_{p,q}(\sigma)$ is

$$(\lambda, \frac{p+q}{2} - 2, \cdots, \frac{p+q}{2} - [\frac{p+q}{2}]) \tag{14}$$

where λ is the infinitesimal character of σ and $[x]$ denotes the largest integer $\leq x$.

In principle (13) and (14) together should determine $\theta_{p,q}(\sigma_\Lambda)$ completely. But we would like to describe it in terms of its Langlands parameters. If $\Lambda \geq \frac{p+q}{2}$ then $\theta_{p,q}(\sigma_\Lambda)$ is a unitary representation with non-zero cohomology [**Ada**] [**Lid**]; its Langlands parameters can be obtained via §6 of [**VoZ**]. In order to deal with the range $\Lambda < (p+q)/2$ we shall make use of the following "Kudla principle" which was first discovered by Kudla [**Kud**] in the p-adic case. For a proof in the real case see [**Moe**, IV.5. Corollaire]. Let $r \leq p$, q and consider a subgroup $G_1 = O(p-r, q-r) \subseteq O(p,q) = G$. Let P_1 be a parabolic subgroup of G_1 with Levi decomposition $P_1 = L_1 N_1$. Obviously there is a parabolic subgroup $P = LN$ of G containing P_1, such that its Levi component $L = GL(1)^r \cdot L_1$.

Lemma 5.3. *Let σ be an irreducible admissible representation of $\widetilde{SL}(2)$. Suppose that the theta lift of σ to G_1 is the Langlands quotient of $\mathrm{Ind}_{P_1}^{G_1} \tau_1$ where τ_1 is tempered on the semi-simple part of L_1. Then the theta lift of σ to G is the Langlands subquotient of $\mathrm{Ind}_P^G \tau$ where τ is the representation of $L = GL(1)^r \cdot L_1$ given by*

$$(a_1, \cdots, a_r) \cdot x \mapsto |a_1|^{\frac{p+q}{2}-2} \cdots |a_r|^{\frac{p+q}{2}-r-1} \cdot \tau_1(x)$$

for $a_j \in GL(1)$ and $x \in L_1$.

In general, the Langlands subquotient of a parabolicly induced representation $\mathrm{Ind}_P^G(\tau)$ will be denoted $J(P, \tau)$. To compute $\theta_{p,q}(\sigma_\Lambda)$ in general we first consider several special cases.

Lemma 5.4.

(a) If $\Lambda \geq p/2$ *then* $\theta_{p,0}(\sigma_\Lambda) = S_p(\Lambda - p/2)$, *the representation of* $O(p)$ *on the spherical harmonics of degree* $\Lambda - p/2$.

(b) Suppose $q = 1,\ \Lambda \leq (p-1)/2$ *then*

$$\theta_{p,1}(\sigma_\Lambda) = \mathrm{Ind}_P^G(sgn(\cdot)^{\Lambda - \frac{p-1}{2}}|\cdot|^{\Lambda - 1}) \tag{15}$$

where P *is the minimal parabolic with Levi component* $GL(1) \cdot O(p-1)$; *the inducing representation is trivial on* $O(p-1)$ *and takes* $a \in GL(1)$ *to* $sgn(a)^{\Lambda-(p-1)/2}|a|^{\Lambda-1}$.

(c) For $p = 2$, $\theta_{2,q}(\sigma_\Lambda)$ *is a representation of* $O(2,q)$ *of lowest* K*-type* $S_2(\Lambda - 1 + q/2) \otimes 1$ *(trivial on the second factor of* $K = O(2) \times O(q)$*). If* $q \neq 2$ *then the restriction of* $\theta_{2,q}(\sigma_\Lambda)$ *to the connected component* $G_0 = SO_0(2,q)$ *is the sum of two irreducible representations, one holomorphic with lowest weight* $[\Lambda - 1 + q/2] \otimes 1$ *and the other anti-holomorphic with highest weight* $[-(\Lambda - 1 + q/2)] \otimes 1$. *Here we used* $[x] \otimes 1$ *to denote the representation of* $SO(2) \times SO(q)$ *which is trivial on* $SO(q)$ *and has weight* x *on* $SO(2)$.

Proof: (a) is "well known"; it follows from Kashiwara-Vergne [**KaV**].

(b): By inspection of (13) we see that the representation $\theta_{p,1}(\sigma_\Lambda)$ is almost spherical (i.e. it contains the trivial representation of $SO(p) \times SO(1)$). From this and (14) we obtain part (b) easily.

(c): This follows easily from Lemma 5.2.

We write $GL(2,\mathbf{R})$ as the product $SL^{\pm}(2,\mathbf{R})A$ where A is the connected component of the center of $GL(2)$ identified with the positive reals.

Corollary 5.5. *If* $\Lambda \geq q/2 + 1$ *then* $\theta_{2,q}(\sigma_\Lambda)$ *is in the discrete series. If* $\Lambda = q/2$ *then* $\theta_{2,q}(\sigma_\Lambda)$ *is in the limit of discrete series. In both cases the lowest* K*-type is* $S_2(\Lambda - 1 + q/2) \otimes 1$. *Finally if* $\Lambda < q/2$ *then* $\theta_{2,q}(\sigma_\Lambda)$ *is the Langlands subquotient* $J(P, \delta \otimes \nu)$. *Here* P *is the maximal parabolic subgroup of* $O(2,q)$ *with Levi component* $GL(2)O(q-2) = SL^{\pm}(2,\mathbf{R})O(q-2)A$, δ *is trivial on* $O(q-2)$ *and equal to the discrete series representation with lowest* $O(2)$*-type* $S_2(q/2 + \Lambda - 1)$ *on* $SL^{\pm}(2,\mathbf{R})$, *and* ν *is given by* $\nu(a) = |a|^{q/2-\Lambda}$.

Proof: This amounts to a realization of the representation described in part (c) of the above lemma as a Langlands quotient. We omit the details of calculations.

We now return to the general case. We shall write the theta lift of σ_Λ as a Langlands quotient $\theta_{p,q}(\sigma_\Lambda) = J(P,\tau)$ and specify the Langlands data (P,τ). There are five cases; the first three have $p > q \geq 1$ and last two have $2 \leq p \leq q$. (Note that Lemma 5.2 obviously implies that $\theta_{p,q}(\sigma_\Lambda)$ is zero when $p = 1$). In each case we describe P and τ, and indicate which of the above lemmas are used.

(**A**) $\Lambda \leq (p-q)/2$ (Lemma 5.3 with $r = q-1$ and Lemma 5.4.(b)).

P is the minimal parabolic with Levi $O(p-q)\cdot GL(1)^q$. τ is trivial on $O(p-q)$ and on $GL(1)^q$ it is given by

$$\tau(a_1,\cdots a_q) = |a_1|^{\frac{p+q}{2}-2}\cdots|a_{q-1}|^{\frac{p-q}{2}}|a_q|^{\Lambda-1}sgn(a_q)^{\Lambda-\frac{p-q}{2}} \tag{16}$$

(**B**) $\Lambda \geq (p-q)/2+1$, $p \geq q+2$ (Lemma 5.3 with $r = q$ and Lemma 5.4.(a)).

P is the minimal parabolic subgroup, $\tau = \delta \otimes \nu$ where $\delta = S_{p-q}(\Lambda - \frac{p-q}{2})$, the representation of $O(p-q)$ on the spherical harmonics of degree $\Lambda - \frac{p-q}{2}$, and

$$\nu(a_1,\cdots a_q) = |a_1|^{\frac{p+q}{2}-2}\cdots|a_{q-1}|^{\frac{p-q}{2}}|a_q|^{\frac{p-q}{2}-1} \tag{17}$$

(**C**) $\Lambda \geq (p-q)/2+1$, $p = q+1$ (Lemma 5.3 with $r = q-1$ and Corollary 5.5).

P is the parabolic with Levi $O(2,1)\cdot GL(1)^{q-1}$ and $\tau = \delta\otimes\nu$. Here δ is a discrete series representation of $O(2,1)$ and

$$\nu(a_1,\cdots a_{q-1}) = |a_1|^{\frac{p+q}{2}-2}\cdots|a_{q-1}|^{\frac{p-q}{2}} \tag{18}$$

(**D**) $2 \leq p \leq q$ and $\Lambda \geq (q-p)/2+1$ (Lemma 5.3 with $r = p-2$ and Corollary 5.5).

P is the parabolic with Levi $O(2,q-p+2)\cdot GL(1)^{p-2}$, $\tau = \delta\otimes\nu$ with δ in the discrete series (if $\Lambda \geq (q-p)/2+2$) or limit of discrete series (if $\Lambda = (q-p)/2+1$), and

$$\nu(a_1,\cdots a_{p-2}) = |a_1|^{\frac{p+q}{2}-2}\cdots|a_{p-2}|^{\frac{q-p}{2}+1} \tag{19}$$

(**E**) $2 \leq p \leq q$ and $\Lambda \leq (q-p)/2$ (Lemma 5.3 with $r = p-2$ and Corollary 5.5). P is the parabolic with Levi

$$GL(2) \cdot O(q-p) \cdot GL(1)^{q-2} = (SL^{\pm}(2)O(q-p)) \cdot (GL(1)^{p-2} \cdot A),$$

and $\tau = \delta \otimes \nu$. Here δ is a discrete series representation of $SL^{\pm}(2)O(q-p)$ trivial on $O(q-p)$; on $SL^{\pm}(2)$ it is the discrete series with lowest $O(2)$-type $S_2((q-p)/2+\Lambda)$. Recall that A is the connected component of $GL(2)$ identified with the positive reals. The character ν is given by

$$\nu(a_1, \cdots a_{p-2}; a) = |a_1|^{\frac{p+q}{2}-2} \cdots |a_{p-2}|^{\frac{q-p}{2}+1} \cdot a^{\frac{q-p}{2}+1-\Lambda} \tag{20}$$

This concludes our discussion of $\theta_{p,q}(\sigma_\Lambda)$. A completely analogous analysis is valid for theta liftings of anti-holomorphic discrete series representations of $\widetilde{SL}(2)$; one only needs to switch p and q in the above discussion.

An obvious consequence is

Proposition 5.6. *Suppose $p, q > 0$. Let σ be any discrete series representation of $\widetilde{SL}(2)$. Then $\theta_{p,q}(\sigma)$ is strongly L^r for any $r > p+q-2$. If $p+q \geq 5$ there are infinitely many discrete series representations σ such that $\theta_{p,q}(\sigma)$ is not strongly L^{p+q-2}. If $p, q \geq 3$ none of the representations $\theta_{p,q}(\sigma)$ are strongly L^{p+q-2}.*

Proof: This follows from the description of the "continuous part" of the Langlands parameters given by (16)-(20), and Chapter 8 of [**Kna**] (see especially Theorem 8.48 and Proposition 8.61.).

We would like to prove a similar statement for unitary representations σ which are not in the discrete series. Assume $p \geq q$. Fix a minimal parabolic subgroup P of $O(p,q)$ with Langlands decomposition $P = MAN$; the Lie algebra version of this decomposition is written $\mathbf{p} = \mathbf{m} + \mathbf{a} + \mathbf{n}$. Let ρ be the half sum of the positive roots of $\mathbf{a}$ in $\mathbf{n}$. We recall from [**Kna**,Chapter 8] that the sizes of the matrix coefficients of $\theta_{p,q}(\theta)$ are determined by its *leading exponents* $\nu - \rho$. Identify $\mathbf{a}^*_{\mathbf{C}}$ with $\mathbf{C}^q$ as usual, and write $\nu = (\nu_1, \cdots, \nu_q)$. Let λ be the infinitesimal character of σ. Then the infinitesimal character of $\theta_{p,q}(\theta)$ is given by (14). By Theorem 8.33 of [**Kna**], we see that $\{\nu_1, \cdots, \nu_q\}$ is a subset of the set of coordinates appearing in (14) (up to "$\pm$" signs). Thus if

$$(p-q)/2 \geq 1 + \text{Re } |\lambda| \tag{21}$$

(Re λ denotes the real part of λ) then Re ν is dominated by

$$\nu_0 = (\frac{p+q}{2} - 2, \cdots, \frac{p-q}{2}, \frac{p-q}{2} - 1) \tag{22}$$

Since σ is unitary and not in the discrete series of $\widetilde{SL}(2)$, and it occurs in the theta correspondence with $O(p,q)$, we see that λ is either pure imaginary, or real with $|\lambda| \leq 1$ when $p+q$ is even and $|\lambda| \leq 1/2$ when $p+q$ is odd. Thus when $p \geq q+3$ (21) is always satisfied, and thus Re ν is dominated by Re ν_0. For $p \leq q+2$, ν will be dominated by

$$(\frac{p+q}{2} - 2, \cdots, 1, |\text{Re } \lambda|), \quad p = q+2$$
$$(\frac{p+q}{2} - 2, \cdots, \frac{1}{2}, |\text{Re } \lambda|), \quad p = q+1$$
$$(\frac{p+q}{2} - 2, \cdots, 1, |\text{Re } \lambda|, 0), \quad p = q$$

Using Theorem 8.48 of [**Kna**], one may now conclude (via a routine calculation) that if $p+q \geq 7$ then $\theta_{p,q}(\sigma)$ is strongly L^r for $r > p+q-2$, and if additionally $q > 1$ then $\theta_{p,q}(\sigma)$ is not strongly L^{p+q-2}. Putting this together with Proposition 5.6 and results of §4, we have proven Theorem A for $O(p,q)$.

We remark that when $p+q \leq 6$, taking the worst possible λ (i.e. $\lambda = 1$ or $1/2$ for $p+q$ even or odd) yields the unitary representations of $O(3,3), O(4,2)$ and $O(3,2)$ with minimal decay of matrix coefficients, namely $L^{6+\epsilon}$ for $O(3,3), O(4,2)$ and $L^{4+\epsilon}$ for $O(3,2)$.

We now turn to other groups; recall the setting of §2. If $G = Sp(n, \mathbf{R})$ then $m'(D) = 2$ and G' is $O(2)$ or $O(1,1)$. In the remaining three cases we have $m'(D) = 1$ and G' is $U(1), O^*(2)$ or $Sp(1)$ which are all compact. The theta liftings in all of these cases are much easier than the case of $G = O(p,q)$ discussed above, and are well known at any rate. See [**KaV**] [**Howe**] [**Howc**] etc. Thus we will be content to just state the following result.

Proposition 5.7 *Let $p(G)$ be as in Theorem A. Let σ be any unitary representation of G'. Then the theta lifting of σ to G is strongly L^r for $r > p(G)$. If $r(G) > m'(D)$ then it is not L^r for $r \leq p(G)$.*

§6. Automorphic forms with minimal decay of matrix coefficients.

We can now prove Theorem B easily. Let k be a totally real number field. Let D be a division algebra over k with an involution ι. We assume that D is either k itself, a quadratic extension of k, or a quaternion algebra over k, and ι is trivial, the Galois involution or the quaternionic involution respectively. Let V and V' be vector spaces over D endowed with non-degenerate sesquilinear forms $(\ ,\ )$ and $(\ ,\ )'$ respectively, such that one is hermitian and the other is skew-hermitian. Let $\mathbf{G}$ and $\mathbf{G}'$ be the isometry groups of $(\ ,\ )$ and $(\ ,\ )'$ respectively. Then $\mathbf{G}$, $\mathbf{G}'$ is a reductive dual pair over k. This is completely analogous to the situation in §2, see [**Howa**] [**Wei**] for references. Let m be the dimension of V over D and assume that the dimension of V' is $m'(D)$ where

$$m'(D) = \begin{cases} 2, & \text{if D = k} \\ 1, & \text{otherwise} \end{cases}$$

Let $v_1, \cdots, v_R$ be the real places of k. Write $G = \mathbf{G}(k_{v_1})$; then G is one of our real classical groups. Similarly write G' for $\mathbf{G}'(k_{v_1})$. Assume that the groups $\mathbf{G}(k_{v_j})$, $2 \leq j \leq R$ are all compact. Then any congruence subgroup Γ of $\mathbf{G}(k)$ will embed in G as a(n arithmetic) lattice.

Let $\eta = 1$ or -1 according as $(\ ,\ )$ is hermitian or skew-hermitian. Let d be the dimension of D over k and d' the dimension of the η-eigenspace of the involution ι. Finally set $\varepsilon = d'/d$. The following is then a special case of Theorem 1.1 of [**Lie**].

Proposition 6.1 *Let π be a unitary representation of G. Suppose $m > 2m'(D) + 4\varepsilon - 2$ and π is the local theta lift of a discrete series representation of G'. Then there is a congruence subgroup Γ of $\mathbf{G}(k)$, such that π occurs as a discrete summand in $L^2(\Gamma \backslash G)$.*

This together with Propositions 5.6, 5.7 clearly imply Theorem B. The remark after Theorem B follows from the fact that all lattices in the groups under consideration are arithmetic [**Mar**]. By inspection of the classification of algebraic groups over number fields [**Tit**], we see that except when $G = U(p,q)$ or $G = O(p,q)$ with $p+q = 4, 8$, every arithmetic lattice in G is commensurable with one of the congruence subgroups Γ in Proposition 6.1. We omit the details.

References

[**Ada**] J. ADAMS, *Discrete Spectrum of the Dual Pair (O(p,q),Sp(2m,**R**)*, Invent. Math.**74**(1983), 449–475.

[**Bal**] M.W. BALDONI-SILVA, *Unitary dual of* $Sp(n,1)$, $n \geq 2$, Duke Math. Journal **48**(1981), 549–583.

[**Bar**] D. BARBASCH , *The Unitary Dual for Complex Classical Lie Groups*, Invent. math. **96**(1989), 103–176.

[**CHH**] M. COWLING, U. HAAGERUP, R. HOWE, *Almost L^2 matrix coefficients*, J. reine angew. Math. **387**(1988),97–110.

[**DHL**] W. DUKE, R. HOWE AND J-S. LI, *Estimating Hecke eigenvalues of Siegel modular forms*, Duke Math. Journal,**67**(1992), 219–240.

[**Far**] J. FARAUT, *Distributions sphérique sur les espaces hyperboliques*, J. Math. Pures Appl.**58**,369–444.

[**Howa**] R. HOWE, *θ series and invariant theory,in Automorphic forms, representations, and L-functions*, Proc. Symp. Pure Math., vol. 33, American Math. Soc., Providence, 1979, 275–285.

[**Howb**] R. HOWE, *On some results of Strichartz and of Rallis and Schiffmann*, Journal of Functional Analysis, **32**(1979), 297–303.

[**Howc**] R. HOWE, *On a notion of rank for unitary representations of classical groups*, in C.I.M.E. Summer School on Harmonic Analysis, Cortona,ed.,1980, 223–331.

[**Howd**] R. HOWE, *Transcending Classical Invariant Theory*, Journal Amer. Math. Soc.,**2**(1989),535–552.

[**Howe**] R. HOWE, *Remarks on Classical Invariant Theory*, Trans. Amer. Math. Soc.**313**(1989),539–570.

[**HoM**] R. HOWE AND C. MOORE, *Asymptotic properties of unitary representations*, J. Func. Analysis,**32**(1979), 72–96.

[**KaV**] M. KASHIWARA AND M. VERGNE, *On the Segal-Shale-Weil Representations and Harmonic Polynomials* , Invent. Math.**44**(1978),1–47.

[**Kna**] A. W. KNAPP, *Representation Theory of Semisimple Groups: an overview based on examples*, Princeton Univ. Press, Princeton, N.J., 1986.

[**Kos**] B. KOSTANT, *On the existence and irreducibility of certain series of representations*, Bulletin of AMS,**75** (1969), 627–642.

[**Kud**] S. KUDLA, *On the Local Theta Correspondence*, Invent. Math.,**83**(1986), 229–255.

[**Lia**] J-S. LI, *On the discrete series of generalized Stiefel manifolds*, to appear in Transactions of AMS.

[**Lib**] J-S. LI, *On the classification of irreducible low rank unitary representations of classical groups*, Compositio Math.,**71**(1989),29–48.

[**Lic**] J-S. LI, *Singular Unitary Representations of Classical Groups*, Inven. Math. **97**(1989),237−255.

[**Lic**] J-S. LI, *Theta liftings for unitary representations with non-zero cohomology*, Duke Math. Journal,**61**(1990), 913–937.

[**Lic**] J-S. LI, *Non-vanishing Theorems for the Cohomology of Certain Arithmetic Quotients*, J. reine angew. Math. **428**(1992), 177–217.

[**Mar**] G. A. MARGULIS, *Discrete subgroups of semisimple Lie groups*, Ergeb. Math. Grenzeb. Springer-Verlag, 1989.

[**Moe**] C. MOEGLIN, *Correspondance de Howe pour les Paires Reductives Duales, Quelques Calculs Dans le Cas Archimedien*, J. of Func. Analysis,**85**(1990),1–85.

[**RaS**] S. RALLIS AND G. SCHIFFMANN, *Discrete Spectrum of the Weil Representation*, Bull. of Amer. Math. Soc.**83**(1977),267–170.

[**Tit**] J. TITS *Classification of algebraic semisimple groups*, in Algebraic groups and discontinuous subgroups, A. Borel and G. D. Mostow, eds., Proc. Symp. Pure Math., vol. IX, 33–62.

[**Vog**] D. VOGAN, *The Unitary Dual of GL(n) Over an Archimedean Field*, Invent. Math., **83** (1986),449–505.

[**VoZ**] D. VOGAN AND G. ZUCKERMAN, *Unitary Representations with Non-Zero Cohomology*, Compositio Math., **53**(1984), 51–90.

[**Wei**] A. WEIL, *Sur certains groupes d'operateurs unitaires*, Acta Math., **111** (1964),143–211.

APPLICATION OF HARMONIC ANALYSIS IN GEOPHYSICS

Li Shi-xiong
Department of Mathematics,Anhui University
Hefei,Anhui,230039,P.R.China

Abstract

The present report gives a brief survey of what we have done in these years on application of harmonic analysis in geophysics. It consists of two parts:
(1) Transference of potential field on curved surface.
(2) Inversion of generalized Radon transform and diffraction tomography.

I. TRANSFERENCE OF POTENTIAL FIELD ON CURVED SURFACE

1.1 Introduction

Transference of potential field on curved surface is an important problem in magnetic (or gravitative) prospecting method. In magnetic prospecting only the component of the magnetic field on the earth surface along a definite direction (in most cases it is the direction of terrestrial magnetism) is generally measured by the magnetometer. If we can calculate all the three components of the magnetic field from the observed data, then the ability of inference can be greatly improved. When the data are obtained on a plane, the above problem has been solved. But when the data are obtained on an uneven surface, this problem remains yet unsolved [1]. Using the methods of conformal mapping, Hilbert transform and singular integral equation, we solved this problem both in two dimensional case (1983) [2] and three dimensional case (1985)[3].

1.2. Two dimensional case [2]

Let $u(w) = u(\xi, \eta)$ be the potential of a magnetic field and $\vec{l} = (a, b)$ be a definite direction. We have the value of $\frac{\partial u}{\partial l} = f(\tau)$ on a given curve L(observed data), i.e.

M. Cheng et al. (eds.), Harmonic Analysis in China, 170–182.

$$a\frac{\partial u}{\partial \xi} + b\frac{\partial u}{\partial \eta} = f(\tau) \tag{1.1}$$

where τ is an arbitrary point on the curve L (see Fig.1) and $u(\xi, \eta)$ is a harmonic function in domain D (the part of w-plane above the curve L). The problem is to find the value of $\frac{\partial u}{\partial \xi}, \frac{\partial u}{\partial \eta}$ on the curve L.

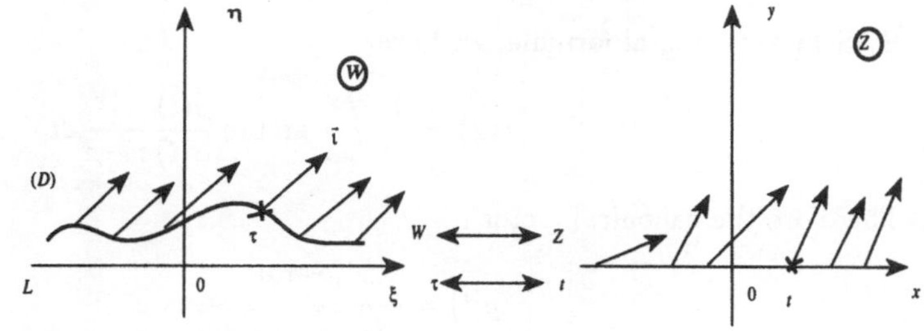

Fig. 1

Using the method of conformal mapping, we can map the domain D onto the upper half z-plane. After this transformation, the function U(z)=u(w(z))=u(x,y) is still a harmonic function, which satisfies the boundary condition:

$$\alpha(t)\frac{\partial U}{\partial x} + \beta(t)(-\frac{\partial U}{\partial y}) = c(t) \tag{1.2}$$

on the real line, with

$$\alpha(t) = \frac{a\frac{\partial x}{\partial \xi} + b\frac{\partial x}{\partial \eta}}{\sqrt{\left(a\frac{\partial x}{\partial \xi} + b\frac{\partial x}{\partial \eta}\right)^2 + \left(a\frac{\partial y}{\partial \xi} + b\frac{\partial y}{\partial \eta}\right)^2}}$$

$$\beta(t) = \frac{-\left(a\frac{\partial y}{\partial \xi} + b\frac{\partial y}{\partial \eta}\right)}{\sqrt{\left(a\frac{\partial x}{\partial \xi} + b\frac{\partial x}{\partial \eta}\right)^2 + \left(a\frac{\partial y}{\partial \xi} + b\frac{\partial y}{\partial \eta}\right)^2}}$$

$$c(t) = \frac{f(\tau(t))}{\sqrt{\left(a\frac{\partial x}{\partial \xi} + b\frac{\partial x}{\partial \eta}\right)^2 + \left(a\frac{\partial y}{\partial \xi} + b\frac{\partial y}{\partial \eta}\right)^2}}$$

Let $F(z) = \frac{\partial U}{\partial x} + i\left(-\frac{\partial U}{\partial y}\right)$, then (1.2) is equivalent to

$$Re\left[\frac{F(z)}{\alpha + i\beta}\right]_{z=t} = c(t) \tag{1.3}$$

Although c(t) is known, but in general, $\frac{F(z)}{\alpha+i\beta}$ is not an analytic function. However, we can construct a canonical factor p(t), such that $p(t)(\alpha(t) + i\beta(t))$ will be an analytic function. Let

$$p(t)\left[\alpha(t) + i\beta(t)\right] = e^{i\gamma(t)} \tag{1.4}$$

$\gamma(t) = \omega(t) + i\omega_1(t)$ be an analyic function, then

$$p(t)\sqrt{\alpha^2(t)+\beta^2(t)} = e^{-\omega_1(t)} \tag{1.5}$$

$$\arctan\frac{\beta(t)}{\alpha(t)} = \omega(t) \tag{1.6}$$

By Schwarz integral formula, we have

$$\gamma(z) = \frac{1}{\pi i}\int_{-\infty}^{\infty} \arctan\frac{\beta(t)}{\alpha(t)}\frac{1}{t-z}dt \tag{1.7}$$

Therefore the canonical factor

$$p(t) = \frac{e^{-\omega_1(t)}}{\sqrt{\alpha^2(t)+\beta^2(t)}} = e^{-\omega_1(t)} \tag{1.8}$$

Dividing both sides of (1.3) by p(t), we get

$$Re\left[\frac{F(z)}{e^{i\gamma(z)}}\right]_{z=t} = c(t)e^{\ \omega_1(t)} \tag{1.9}$$

Now, $\frac{F(z)}{e^{i\gamma(z)}}$ is an analytic function. Using Schwarz integral formula once again, we get:

$$F(z) = e^{i\gamma(z)}\left[\frac{1}{\pi i}\int_{-\infty}^{\infty}\frac{c(t)e^{\omega_1(t)}}{t-z}dt\right] \tag{1.10}$$

Because it is an integral of Cauchy type, so the boundary value of $F(z) = \frac{\partial U}{\partial x} + i\left(-\frac{\partial U}{\partial y}\right)$ on x-axis equals to

$$F^+(t_0) = \lim_{\substack{z\to t_0\\ Imz>0}} F(z) = e^{i\gamma^+(t_0)}\left[\frac{1}{\pi i}\int_{-\infty}^{\infty}\frac{e^{\omega_1(t)}c(t)-e^{\omega_1(t_0)}c(t_0)}{t-t_0}dt + e^{\omega_1(t_0)}c(t_0)\right] \tag{1.11}$$

here

$$\gamma^+(t_0) = \lim_{\substack{z\to t_0\\ Imz>0}} \gamma(z) = \frac{1}{\pi i}\int_{-\infty}^{\infty}\frac{\arctan\frac{\beta(t)}{\alpha(t)} - \arctan\frac{\beta(t_0)}{\alpha(t_0)}}{t-t_0}dt + \arctan\frac{\beta(t_0)}{\alpha(t_0)} \tag{1.12}$$

Because

$$\frac{\partial u}{\partial \xi} = \frac{\partial U}{\partial x}\frac{\partial x}{\partial \xi} + \frac{\partial U}{\partial y}\frac{\partial y}{\partial \xi}$$
$$\frac{\partial u}{\partial \eta} = \frac{\partial U}{\partial x}\frac{\partial x}{\partial \eta} + \frac{\partial U}{\partial y}\frac{\partial y}{\partial \eta}$$

it is easy to get the value of $\frac{\partial u}{\partial \xi}, \frac{\partial u}{\partial \eta}$ on curve L by expression (1.11).

In general, the observed data always involves noise. Let the accurate and noisy value be c(t) and $\tilde{c}(t)$ respectively, and let $\Delta c(t) = \tilde{c}(t) - c(t)$. We have accordingly

$$\begin{aligned} \Delta F^{+}(t_0) &= \tilde{F}^{+}(t_0) - F^{+}(t_0) \\ &= e^{i\gamma^{+}(t_0)} \left[\frac{1}{\pi i} \int_{-\infty}^{\infty} \frac{e^{\omega_1(t)}\Delta c(t) - e^{\omega_1(t_0)}\Delta c(t_0)}{t - t_0} dt + e^{\omega_1(t_0)}\Delta c(t_0) \right] \end{aligned} \tag{1.13}$$

By Riesz inequality

$$\begin{aligned} \| \Delta F^{+}(t_0) \|_2 &\le \max_{t\in(-\infty,\infty)} e^{-\omega_1(t)} \| e^{\omega_1(t)}\Delta c(t) \|_2 + \| \Delta c(t) \|_2 \\ &\le \left(\max_{t\in(-\infty,\infty)} e^{-\omega_1(t)} \max_{t\in(-\infty,\infty)} e^{\omega_1(t)} + 1 \right) \| \Delta c(t) \|_2 \\ &= K \| \Delta c(t) \|_2 \end{aligned}$$

So the error of calculated values $\frac{\partial u}{\partial \xi}, \frac{\partial u}{\partial \eta}$ are small if the noise involved in the observed data are weak. That means the algorithms of transference of potential field is stable.

1.3 The three dimensional case [3]

In 1.2, using the method of function of a complex variable, we have solved the problem of transference of potential field in two dimensional case. But this method can not be used to solve the three dimensional case. Essentially, it is the oblique derivative problem of Laplace equation. Using the single layer potential theory, it is not difficult to transfer it as a problem of solving a singular integral equation.

Let

$$(\textstyle\sum): \quad z = f(x, y) \tag{1.14}$$

be a given surface, $\vec{l}$ be a fixed direction (see Fig.2) and u(P) be the magenetic potential. Suppose the value of $\frac{\partial u}{\partial l}$ on the surface $(\sum)$ is known.

$$\frac{\partial u}{\partial l} = g(P), \quad P \in (\textstyle\sum) \tag{1.15}$$

The problem is to find the value of $\frac{\partial u}{\partial x}, \frac{\partial u}{\partial y}$ and $\frac{\partial u}{\partial z}$ on the surface $(\sum)$.

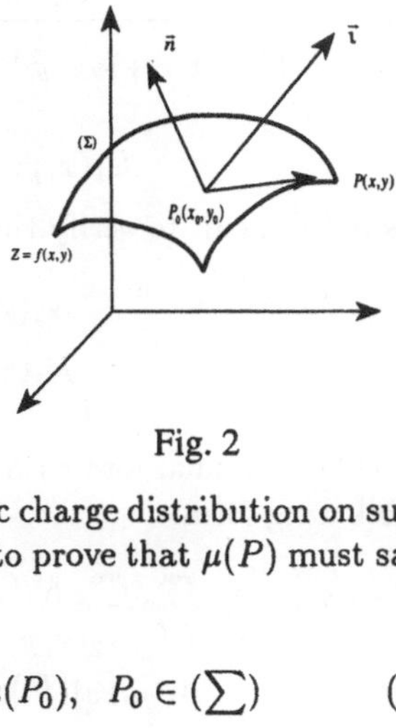

Fig. 2

Suppose $\mu(P)$ be the surface density of single layer magnetic charge distribution on surface $(\sum)$. By equation (1.15) and theory of potential, it is easy to prove that $\mu(P)$ must satisfy the integral equation:

$$g(P_0) = P.V. \iint_{(\sum)} \vec{K}(P, P_0)\mu(P)d\sigma_P - 2\pi(\vec{n}\cdot\vec{l})\mu(P_0), \quad P_0 \in (\textstyle\sum) \tag{1.16}$$

where $\vec{n}$ is the normal of surface $(\sum)$ at point P and the kernel of the integral equation is

$$\bar{K}(P,P_0)=\frac{(x-x_0)\cos(\vec{l},\vec{x})+(y-y_0)\cos(\vec{l},\vec{y})+(z-z_0)\cos(\vec{l},\vec{z})}{|P-P_0|^3}$$

If we can solve equation (1.16) to get $\mu(P)$, it is easy to calculate $\frac{\partial u}{\partial x},\frac{\partial u}{\partial y},\frac{\partial u}{\partial z}$ from $\mu(P)$. But (1.16) is a singular integral equation, so the Fredholm-Riesz-Schauder theory about completely continuous operator is not suitable for this case and the ordinary numerical calculating method such as, iteration or discretized by approximate linear algebraic equations are not applicable. We suggest a method of seperating singularities to solve this singular integral equation. Because in rectangular coordinates, $d\sigma_P=\sqrt{1+f_x^2+f_y^2}dxdy$, the integral equation (1.16) may be rewritten in the following form:

$$g(x_0,y_0)=P.V.\int_{-\infty}^{\infty}\int_{-\infty}^{\infty}K(x,y;x_0,y_0)\mu(x,y)dxdy-2\pi(\vec{n}\cdot\vec{l})\mu(x_0,y_0) \tag{1.17}$$

The kernel

$$\begin{aligned} K(x,y;x_0,y_0) &= \bar{K}(P,P_0)\sqrt{1+f_x^2+f_y^2} \\ &= \frac{(x-x_0)\cos(\vec{l},\vec{x})+(y-y_0)\cos(\vec{l},\vec{y})+(z-z_0)\cos(\vec{l},\vec{z})}{[(x-x_0)^2+(y-y_0)^2+(z-z_0)^2]^{\frac{3}{2}}}\sqrt{1+f_x^2+f_y^2} \end{aligned}$$

has a singular point at $P_0(x_0,y_0)$ (here z=f(x,y), $z_0=f(x_0,y_0)$)

On purpose for seperating of singularities, let

$$K_2(x,y;x_0,y_0)$$

$$=\frac{(x-x_0)\cos(\vec{l},\vec{x})+(y-y_0)\cos(\vec{l},\vec{y})+\left[f_x^0\cdot(x-x_0)+f_y^0\cdot(y-y_0)\right]\cos(\vec{l},\vec{z})}{\left\{(x-x_0)^2+(y-y_0)^2+\left[f_x^0\cdot(x-x_0)+f_y^0\cdot(y-y_0)\right]^2\right\}^{\frac{3}{2}}}\sqrt{1+(f_x^0)^2+(f_y^0)^2}$$

$$K_1(x,y;x_0,y_0)=K(x,y;x_0,y_0)-K_2(x,y;x_0,y_0)$$

It is not difficult to verify that

$$\begin{aligned} K_2(x,y;x_0,y_0) &= O\left(\frac{1}{(x-x_0)^2+(y-y_0)^2}\right) \\ K_1(x,y;x_0,y_0) &= O\left(\frac{1}{\sqrt{(x-x_0)^2+(y-y_0)^2}}\right) \end{aligned}$$

Let D be a circular disc with radius ρ and centre at (x_0,y_0), then (1.17) would be written in following form:

$$\begin{aligned} g(x_0,y_0) &= \textstyle\int_{-\infty}^{\infty}\int_{-\infty}^{\infty}K_1(x,y;x_0,y_0)\mu(x,y)dxdy+\int\int_{R^2\setminus D}K_2(x,y;x_0,y_0)\mu(x,y)dxdy \\ &+ \textstyle\int\int_D K_2(x,y;x_0,y_0)[\mu(x,y)-\mu(x_0,y_0)]dxdy \\ &+ \textstyle\mu(x_0,y_0)(P.V.)\int\int_D K_2(x,y;x_0,y_0)dxdy-2\pi(\vec{n}\cdot\vec{l})\mu(x_0,y_0) \end{aligned} \tag{1.18}$$

Because $K_2(x,y;x_0,y_0)$ is a known function, so the principle value of the singular integral (P.V.)$\int\int_D K_2(x,y;x_0,y_0)dxdy$ can be calculated concretely. The remaining integrals of equation (1.18) are ordinary Riemann integral, therefore using mechenical quadrature it can be discreted as a system of linear algebraic equations. The solution of these linear algebraic equations is the approximate solution of singular integral equation (1.16), which approaches to the accurate solution of (1.16) when the nodal distance approaches to zero. Make use of the solution $\mu(P)$ and let $\vec{l} = (1,0,0)$, it is easy to get $\frac{\partial u}{\partial x}$ from expression (1.16) and similar for $\frac{\partial u}{\partial y}$ and $\frac{\partial u}{\partial z}$.

II. INVERSION OF GENERALIAZED RADON TRANSFORM AND DIFFRACTION TOMOGRAPHY [6]

2.1 Introduction

The classical Radon transform Rf of a function f is defined by integration on a family of hyperplanes (straight lines and planes in 2 and 3 dimensional cases). It was proved by J.Radon in 1917[4] that a differentiable function on R^3 can be determined explicitly by means of its integrals over planes in R^3. The inverse Radon transform to recover a function from its integrals along straight lines in the plane has received considerable attention in recent years because of its many practical applications. But in many practical problems, we have to generalize Radon transform such that instead the family of hyperplanes (straight lines) one has a family of hypersurfaces (curves). A.M.Cormack[5] and S.X.Li[7] have got exact inversion formulae for generalized Radon transform. But the kind of curves they can treat is restricted, and the Cormack's formula has further serious defect — it is unstable for noisy data. It is notable that G.Beylkin [9] using the theory of Fourier integral operators has got a pseudo-inverse for the generalized Radon transform. The pseudo-inverse allows the discrepancy with the exact inverse to be a smooth function. So using the pseudo- inverse one can accurately reconstruct the discontinuities of the exact solution. Seismic exploration, medical applications, crack and void detection are examples of variety of fields where these problems are of interest. Unfortunately, Beylkin's formula is not appropriate for numerical computation. By analyzing the Fourier integral operators appeared in the Beylkin's paper carefully we established a simple and effective method to determine the position of the discontinuities of the function from its generalized Radon transform. We call it "Method of Envelope"[7]. The key of this method is to find the location of the points where generalized Radon transform $Rf(\theta,t)$ has infinite derivative, i.e. $\frac{\partial Rf(\theta,t)}{\partial t} = \infty$. Beylkin's results can be used to solve the inverse scattering problem [10],[11]. But they have the same defect — not appropriate for numerical computation. After slight improvement of our "Method of Envelope", we show that this method can be used to solve the inverse scattering problem (diffraction tomography) also. It avoids the defects of Beylkin and Cohen's formulae.

2.2 Generalized Radon transform and Fourier transform

(1) Let $L_{\theta,t}$ be a family of curves in R^2, which depend on two parameters t and θ. We define the generalized Radon transform of function f(x,y) along $L_{\theta,t}$ as follows:

$$(Rf)_L(\theta,t) = \int_{L_{\theta,t}} f(x,y)ds \tag{2.1}$$

where the ds is the differential of the arc length of the curve $L_{\theta,t}$ and f is compactly supported. Furthermore, we define the generalized Radon transform with a weight function w(x,y) as follows:

$$(Rf)_{L,W}(\theta,t) = \int_{L_{\theta,t}} f(x,y)w(x,y)ds \tag{2.2}$$

where w(x,y) is a prescribed positive function with continuous second derivatives.

If function f(x,y) has discontinous jump along curve Γ and curve $L_{\theta,t_0(\theta)}$ tangent to Γ. Using similar argument as in [7] (satisfying similar conditions also), we can prove that in the neigborhood of $(\theta, t_0(\theta)$ $(Rf)_{L,W}(\theta,t)$ has the following expression:

$$(Rf)_{L,W}(\theta,t) = \begin{cases} \sqrt{t_0(\theta)-t}h(\theta,t) + g(\theta,t), & t \le t_0(\theta) \\ g(\theta,t), & t > t_0(\theta) \end{cases} \tag{2.3}$$

where functions h(θ,t) and g(θ,t) have continuious second derivatives in some neighborhood of $(\theta, t_0(\theta))$.

(2) Let $\hat{f}(\omega)$ be the Fourier transform of f(x):

$$\hat{f}(\omega) = \int_{-\infty}^{\infty} f(x)e^{-i\omega x}dx$$

If f(x) has compact support and continuous second derivatives, then

$$\hat{f}(\omega) = O\left(\frac{1}{\omega^2}\right) \tag{2.4}$$

If f(x) has compact support and

$$f(x) = \begin{cases} \sqrt{x_0 - x}f_1(x) + f_2(x), & x \le x_0 \\ f_2(x), & x > x_0 \end{cases} \tag{2.5}$$

where $f_1(x)$ and $f_2(x)$ have compact support and continuous second derivatives, then [12]

$$\hat{f}(\omega) = \frac{ce^{-i\omega x_0}}{\omega\sqrt{\omega}} + O\left(\frac{1}{\omega^2}\right) \tag{2.6}$$

2.3 Inverse problem of the wave equation, diffraction tomography

We consider the wave equation in the plane region:

$$\nabla^2 u(x,y,t) = \frac{1}{v^2(x,y)} \frac{\partial^2 u(x,y,t)}{\partial t^2} \tag{2.7}$$

(For simplicity, we only consider the two dimensional case, although the method is appropriate also for the three dimensional case), where v(x,y) is propagation velocity of the wave, it depends on the physical properties of the medium.
Direct problem: Given v(x,y), to solve (2.7) (under suitable initial and boundary conditions).
Inverse problem: From the partial solution of (2.7) (that is the records of the detector) to recover v(x,y).

It is too difficult to solve the inverse problem accurately, we can only treat linearized inverse problem. Let

$$\frac{1}{v^2(x,y)} = \frac{1+\alpha(x,y)}{c^2(x,y)} \tag{2.8}$$

where c(x,y) is a prior known smooth function, $\alpha(x,y)$ is unknown and small relative to 1. Take Fourier transform with respect to t:

$$\hat{u}(x,y,\omega) = \int_{-\infty}^{\infty} u(x,y,t)e^{-i\omega t}dt \tag{2.9}$$

Then (2.7) will transform to:

$$\nabla^2 \hat{u} = -\frac{\omega^2}{v^2(x,y)}\hat{u} \tag{2.10}$$

Substituting (2.8) in above equation, we obtain:

$$\nabla^2 \hat{u} + \frac{\omega^2}{c^2(x,y)}[1+\alpha(x,y)]\hat{u} = 0 \tag{2.11}$$

Let $\hat{u} = \hat{u}_0 + \hat{u}_1$, and $\hat{u}_0$ satisfies:

$$\nabla^2 \hat{u}_0 + \frac{\omega^2}{c^2(x,y)}\hat{u}_0 = 0$$

Substituting $\hat{u} = \hat{u}_0 + \hat{u}_1$ in (2.11), we obtain:

$$\nabla^2 \hat{u}_1 + \frac{\omega^2 \alpha(x,y)}{c^2(x,y)}\hat{u}_0 + \frac{\omega^2}{c^2(x,y)}[1+\alpha(x,y)]\hat{u}_1 = 0 \tag{2.12}$$

taking the approximation:

$$\alpha(x,y)\hat{u}_1 = 0 \tag{2.13}$$

we obtain the linearized approximate equation:

$$\nabla^2\hat{u}_1 + \frac{\omega^2}{c^2(x,y)}\hat{u}_1 = -\frac{\omega^2\alpha(x,y)}{c^2(x,y)}\hat{u}_0 \tag{2.14}$$

2.3.1 *The case c(x,y)=c(constant)*

First, we consider the case c(x,y)=c(constant). Let $\frac{\omega^2}{c^2} = k^2$ and $\hat{u}_0$ be the cylindrical wave due to a point source located at the point $\vec{x}_s = (x_s, y_s)$:

$$\hat{u}_0(x,y,x_s,y_s,k) = \hat{u}_0(\vec{x},\vec{x}_s,k) \approx \frac{Ae^{ik|\vec{x}-\vec{x}_s|}}{\sqrt{k}\sqrt{|\vec{x}-\vec{x}_s|}} \tag{2.15}$$

where A is a constant and $|\vec{x}-\vec{x}_s| = \sqrt{(x-x_s)^2+(y-y_s)^2}$. Then we have the integral representation of the solution of (2.14) as:

$$\hat{u}_1(\vec{x}_r,\vec{x}_s,k) = A_1 k \int_X \frac{e^{ik|\vec{x}-\vec{x}_r|}e^{ik|\vec{x}-\vec{x}_s|}}{\sqrt{|\vec{x}-\vec{x}_r|\,|\vec{x}-\vec{x}_s|}}\alpha(\vec{x})d\vec{x} \tag{2.16}$$

where $d\vec{x} = dxdy$, X is the domain where $\alpha(\vec{x}) \neq 0$, $\vec{x}_r = \vec{x}_r(\xi) = (x_r(\xi), y_r(\xi))$, $\vec{x}_s = \vec{x}_s(\xi) = (x_s(\xi), y_s(\xi))$ are the position of the receivers and sources which depend on a parameter ξ. Because

$$e^{ik|\vec{x}-\vec{x}_r|}e^{ik|\vec{x}-\vec{x}_s|} = e^{ik(|\vec{x}-\vec{x}_r|+|\vec{x}-\vec{x}_s|)} \tag{2.17}$$

and $|\vec{x}-\vec{x}_r|+|\vec{x}-\vec{x}_s|$ is the sum of the distance from point $\vec{x}$ to receiver point $\vec{x}_r$ and source point $\vec{x}_s$. For a fixed value ξ, take the coordinate system (x,y) such that $\vec{x}_s = (-1,0)$, $\vec{x}_r = (1,0)$. Furthermore, take elliptic coordinates (ρ,ϕ)($\vec{x}_r, \vec{x}_s$ are the focuses of the ellipses):

$$\begin{aligned} x &= \tfrac{1}{2}(\rho+\tfrac{1}{\rho})\cos\phi, \quad \rho > 1 \\ y &= \tfrac{1}{2}(\rho-\tfrac{1}{\rho})\sin\phi, \quad 0 \le \phi < 2\pi \end{aligned} \tag{2.18}$$

Then

$$\begin{aligned} |\vec{x}-\vec{x}_r| + |\vec{x}-\vec{x}_s| &= \rho + \tfrac{1}{\rho} \\ \sqrt{|\vec{x}-\vec{x}_s||\vec{x}-\vec{x}_r|} &= \sqrt{\tfrac{1}{4}(\rho+\tfrac{1}{\rho})^2 - \cos^2\phi} \\ \left|\frac{\partial(x,y)}{\partial(\rho,\phi)}\right| &= \tfrac{1}{\rho}\left[\tfrac{1}{4}(\rho+\tfrac{1}{\rho})^2 - \cos^2\phi\right] \end{aligned}$$

Equation (2.16) can now be written as:

$$\hat{u}_1(\vec{x}_r,\vec{x}_s,k) = A_1 k \int \frac{e^{ik(\rho+\frac{1}{\rho})}}{\rho}d\rho \cdot \int \tilde{\alpha}(\rho,\phi)\sqrt{\frac{1}{4}(\rho+\frac{1}{\rho})^2 - \cos^2\phi d\phi} \tag{2.19}$$

where

$$\tilde{\alpha}(\rho,\phi)=\alpha\left(\frac{1}{2}\left(\rho+\frac{1}{\rho}\right)\cos\phi,\frac{1}{2}\left(\rho-\frac{1}{\rho}\right)\sin\phi\right)$$

But $\sqrt{\frac{1}{4}(\rho+\frac{1}{\rho})^2-\cos^2\phi}\,d\phi = ds$ is the differential of the arc length of ellipse L_ρ,

$$L_\rho:\quad \begin{cases} x = & \frac{1}{2}(\rho+\frac{1}{\rho})\cos\phi \\ y = & \frac{1}{2}(\rho-\frac{1}{\rho})\sin\phi \end{cases} \quad (for\ fixed\ \rho)$$

So $\int\tilde{\alpha}(\rho,\phi)\sqrt{\frac{1}{4}(\rho+\frac{1}{\rho})^2-\cos^2\phi}\,d\phi$ is the generalized Radon transform $(R\alpha)_L(\rho,\xi)$ of the function $\tilde{\alpha}(\rho,\phi)$. Change the variables once again:

$$u=\rho+\frac{1}{\rho},\qquad \rho=\frac{u+\sqrt{u^2-4}}{2},\qquad d\rho=\frac{1}{2}\left(\frac{u+\sqrt{u^2-4}}{\sqrt{u^2-4}}\right)du,$$

equation (2.19) can be written as:

$$\hat{u}_1(\vec{x}_r(\xi),\vec{x}_s(\xi),k)=A_1k\int\frac{e^{iku}}{\sqrt{u^2-4}}\cdot(R\alpha)_L\left(\frac{u+\sqrt{u^2-4}}{2},\xi\right)du \tag{2.20}$$

Therefore, when there are m ellipses:

$$L_{\rho_j}:\quad \begin{cases} x &= \frac{1}{2}(\rho_j+\frac{1}{\rho_j})\cos\phi \\ y &= \frac{1}{2}(\rho_j-\frac{1}{\rho_j})\sin\phi \end{cases} \quad (j=1,2,\cdots,m)$$

tangent to the curves of discontinuous points of $\alpha(\vec{x})$, then by virtue of (2.3),(2.6), we have:

$$\hat{u}_1(\vec{x}_r(\xi),\vec{x}_s(\xi),k)=\sum_{j=1}^{m}\frac{c_je^{ik(\rho_j+\frac{1}{\rho_j})}}{\sqrt{k}}+O\left(\frac{1}{k}\right)$$

or

$$\sqrt{k}\hat{u}_1(\vec{x}_r(\xi),\vec{x}_s(\xi),k)=\sum_{j=1}^{m}c_je^{ik(\rho_j+\frac{1}{\rho_j})}+O\left(\frac{1}{\sqrt{k}}\right) \tag{2.21}$$

Let

$$\begin{aligned} D(\omega,K) = & \int_{k_0}^{K}\sqrt{k}\hat{u}_1(\vec{x}_r(\xi),\vec{x}_s(\xi),k)e^{-ik\omega}dk \\ = & \sum_{j=1}^{m}\int_{k_0}^{K}c_je^{ik[(\rho_j+\frac{1}{\rho_j})-\omega]}dk+O\left(\int_{k_0}^{K}\frac{e^{-ik\omega}}{\sqrt{k}}dk\right) \end{aligned} \tag{2.22}$$

Obviously, if $\omega\neq\rho_j+\frac{1}{\rho_j}$, $j=1,2,\cdots,m$, D(ω,K) is bounded when K $\to\infty$. If ω equals to one of $\rho_j+\frac{1}{\rho_j}$, then D(ω,K) is unbounded, when $K\to\infty$. So we can determine $\rho_j+\frac{1}{\rho_j}$, (j=1,2,...,m) from $\hat{u}_1(\vec{x}_r(\xi),\vec{x}_s(\xi),k)$. Therefore, from the received wave, we can determine which ellipse tangents to the line of discontinuous points of $\alpha(\vec{x})$. When the value of the parameter ξ is changed (i.e. change the position of source and receiver), we get a series of ellipses, all of them tangent to the line of discontinuous points of $\alpha(\vec{x})$. Therefore, the envelope of these ellipses is the line of discontinuous points of $\alpha(\vec{x})$.

2.3.2 *The case* $c(\vec{x}) \neq$ *constant*

Now consider the case $c(x) \neq const.$ At this time, we have following integral representation of solution:

$$\hat{u}_1(\vec{x}_r(\xi), \vec{x}_s(\xi), \omega) = \omega \int_X \frac{G(\vec{x}, \vec{x}_r(\xi), \omega) G(\vec{x}, \vec{x}_s(\xi), \omega)}{c^2(\vec{x})} \alpha(\vec{x}) d\vec{x} \tag{2.23}$$

where G is the Green function. Using WKBJ approximation:

$$G(\vec{x}, \vec{x}_0, \omega) \approx A(\vec{x}, \vec{x}_0) e^{i\omega\tau(\vec{x}, \vec{x}_0)} \tag{2.24}$$

where $\tau(\vec{x}, \vec{x}_0)$ is the phase function, which satisfies the eikonal equation:

$$\nabla\tau \cdot \nabla\tau = \frac{1}{c^2(\vec{x})} \tag{2.25}$$

and $A(\vec{x}, \vec{x}_0)$ is the amplitude function, which satisfies the transport equation:

$$2\nabla\tau \cdot \nabla A + A\nabla^2\tau = 0 \tag{2.26}$$

Equation (2.23) can now be written as:

$$\hat{u}_1(\vec{x}_r, \vec{x}_s, \omega) \approx \omega \int_X \frac{a(\vec{x}, \vec{x}_r(\xi), \vec{x}_s(\xi))}{c^2(\vec{x})} \cdot e^{i\omega\phi(\vec{x}, \vec{x}_r(\xi), \vec{x}_s(\xi))} \alpha(\vec{x}) d\vec{x} \tag{2.27}$$

where

$$\begin{aligned} \phi(\vec{x}, \vec{x}_r(\xi), \vec{x}_s(\xi)) &= \tau(\vec{x}, \vec{x}_r(\xi)) + \tau(\vec{x}, \vec{x}_s(\xi)) \\ a(\vec{x}, \vec{x}_r(\xi), \vec{x}_s(\xi)) &= A(\vec{x}, \vec{x}_r(\xi)) A(\vec{x}, \vec{x}_s(\xi)) \end{aligned} \tag{2.28}$$

Take curvilinear coordinates (μ, ν) on (x,y) plane, such that

$$\phi(\vec{x}, \vec{x}_r(\xi), \vec{x}_s(\xi)) = \psi(\mu) \qquad (forfixed \xi)$$

Then (2.27) will transform to:

$$\hat{u}_1(\vec{x}_r(\xi), \vec{x}_s(\xi), \omega) \approx \omega \int e^{i\omega\psi(\mu)} \cdot \left\{ \int \frac{\tilde{a}(\mu, \nu, \xi)}{\tilde{c}^2(\mu, \nu)} \left| \frac{\partial(x, y)}{\partial(\mu, \nu)} \right| \tilde{\alpha}(\mu, \nu) d\nu \right\} d\mu \tag{2.29}$$

The expression in the bracket $\{\cdots\}$ is the generalized Radon transform with weight function $\frac{\tilde{a}(\mu,\nu,\xi)}{\tilde{c}^2(\mu,\nu)} \left| \frac{\partial(x,y)}{\partial(\mu,\nu)} \right|$. Taking similar argument as in $c(\vec{x}) = const.$ case, we obtain the following conclusion:

If there are m curves:

$$\phi(\vec{x}, \vec{x}_r(\xi), \vec{x}_s(\xi)) = \psi(\mu_j), \qquad (for\ fixed\ \xi)$$
$$j = 1, 2, \cdots, m.$$

tangent to the line of discontinuous points of $\alpha(\vec{x})$, then

$$\sqrt{\omega}\hat{u}_1(\vec{x}_r(\xi),\vec{x}_s(\xi),\omega)=\sum_{j=1}^{m}c_je^{i\omega\psi(\mu_j)}+O\left(\frac{1}{\sqrt{\omega}}\right) \tag{2.30}$$

Therefore, we can determine $\psi(\mu_j), j = 1,2,\cdots,m$, from $\hat{u}_1(\vec{x}_r(\xi),\vec{x}_s(\xi),\omega)$, and the curves $\phi(\vec{x},\vec{x}_r(\xi),\vec{x}_s(\xi)) = \psi(\mu_j)$, $j = 1,2,\cdots,m$, which tangent to the line of discontinuous points of $\alpha(\vec{x})$. When the value of the parameter ξ is changed, we get a series of curves, their envelope is the line of discontinuous points of $\alpha(\vec{x})$, as same as case 2.3.1, but the computation is more complicate.

References

[1] Bhattacharyya,B.K. and Chan,K.C., Reduction of magnetic and gravity data on an arbitrary surface acquired in a region of high topographic relief, Geophysics, Vol.42, No.7, 1977.

[2] Li,S.X., Ye,Y.M., and Lin,Q.P. Transference of potential field on uneven surface, Acta Geophysics Sinica, Vol.26, No.6, Nov. 1983.

[3] Li,S.X., Lin,Q.P. and Ye,Y.M., Transformation of potential field on curved surface, Acta Geophysics Sinica, vol.28, No.5, Sept.1985.

[4] Radon,J., Über die Bestimmung von Funktionen durch ihre Integralwerte längs gewisser Mannigfaltigkeiten. Beriche Sächsische Akademie der Wissenschaften, Leipzig, Math.-Phys. Kl., 69, 262-267, 1917.

[5] Cormack,A.M., The Radon Transform on a family of curves in the plane, Proc. of Am. Math. Soc., Vol.83, No.2, Oct. 1981.

[6] Li,S.X., Ye,Y.M. and Lin,Q.P., Generalized Radon transform and Diffraction tomography, Acta Geophysics Sinica, Vol.33, Supp.II, 1990.

[7] Li,S.X., Lin,Q.P. and Ye,Y.M., Radon transform on family of curves, Acta Geophysics Sinica, Vol.33, Supp.II, 1990.

[8] Li,S.X., Lin,Q.P. and Ye,Y.M., Application of the "envelope method" to the generalized Radon transform with noise, Acta Geophysics Sinica, Vol.33, Supp.II, 1990.

[9] Beylkin,G., The inverse problem and applications of generalized Radon transform, Comm. of Pure and Appl. Math., 37, 579-599, 1984.

[10] Beylkin,G., Imaging of discontinuities in the inverse scattering problem by inversion of a causal generalized Radon transform, Math. Phys., 26, 99-108, 1985.

[11] Cohen,J.K., Hagin,F.G. and Blestein,N., Three-dimensional born inversion with arbitrary reference, Geophysics, 51, 1552-1558, 1986.

[12] Erdélyi,A., Asymptotic Expansions, Ch. II, Dover Pub. Inc., 1956.

BIVARIATE BOX-SPLINE WAVELETS

X. Z. LIANG

Institute of Mathematics, Jilin University,
Changchun 130023, *P. R. China*

G. R. JIN

Institute of Mathematics, Jilin University,
Changchun 130023, *P. R. China*

H. L. CHEN

Institute of Mathematics, Academia Sinica,
Beijing 100080, *P. R. China*

Abstract: In this paper, two kinds of orthonormal periodic wavelets are constructed from the multiresolution analysis which is generated by the bivariate 3-directional box-splines. The explicit expressions are given, the decomposition and reconstruction algorithms are established.

1. Introduction

It is well known that some multivariate box-splines have been used to construct the multivariate wavelets. For example, in [5],[7] and [9], a number of box-splines have been applied to yield orthonormal wavelet bases of $L^2(R^n)$. In [2] and [10], some non-orthogonal box-spline wavelet bases of $L^2(R^n)$ have also been constructed.

In this paper, we will apply the bivariate 3-directional box-splines to yield two kinds of orthogonal wavelets. The periodic wavelet bases have explicit expressions, and they are convenient for the practical computation.

M. Cheng et al. (eds.), Harmonic Analysis in China, 183–196.

This paper consists of three sections. In section 2, we construct the periodic box-spline wavelets in the sense of Meyer[8] and Daubechies [4]. In section 3, we construct another kind of periodic box-spline wavelets. The latter is influenced by the works of Kamada et al [6] and Chen [1].

Now we introduce some notations to facilitate our presentation.

Let R^2 denote the bivariate Euclidean space, and $\mathbb{Z}^2$ the set of all two-dimensional integer points on R^2. Let $T=(T_1, T_2)^T$, where T_1 and T_2 are positive integers. Let $\dot{L}^2(T)$ denote the linear space of T-periodic functions with finite L^2 norm on $[0, T_1]\times[0, T_2]$. For $f, g\in \dot{L}^2(T)$, we define the inner product as $\langle f, g\rangle = \int_{[0,T_1]\times[0,T_2]} f\bar{g}$. Let m_1, m_2, m_3 be positive integers, and $m=(m_1, m_2, m_3)$. We denote by $B_m(X)$ the bivariate 3-directional box-spline associated with the unimodular matrix

$$(\underbrace{e_1,\cdots,e_1}_{m_1}, \underbrace{e_2,\cdots,e_2}_{m_2}, \underbrace{e_3,\cdots,e_3}_{m_3}),$$

where

$$e_1 = (1, 0)^T, e_2 = (0, 1)^T, e_3 = (1, 1)^T.$$

Then the Fourier transform of $B_m(X)$ is

$$\begin{aligned}\hat{B}_m(\xi) &= \int_{R^2} B_m(X)e^{-i\xi\cdot X}dX \\ &= \prod_{j=1}^{3}\left(\frac{1-e^{-i\xi\cdot e_j}}{i\xi\cdot e_j}\right)^{m_j}.\end{aligned} \tag{1.1}$$

So $\hat{B}_m(\xi)$ satisfies the following relation

$$\frac{\hat{B}_m(2\xi)}{\hat{B}_m(\xi)} = \prod_{j=1}^{3}\left(\frac{1+e^{-i\xi\cdot e_j}}{2}\right)^{m_j}. \tag{1.2}$$

The symmetric centre of $B_m(X)$ is

$$\sigma = \left(\frac{m_1+m_3}{2}, \frac{m_2+m_3}{2}\right)^T. \tag{1.3}$$

If we denote the trigonometric polynomial in the right-hand side of (1.2) by

$$C(\xi) = \sum_{k\in\mathbb{Z}^2} C_k e^{-ik\cdot\xi}, \text{ where supp}\{C_k\} \text{ is finite}, \tag{1.4}$$

then we can obtain the refinement equation

$$B_m(X) = \sum_{k\in\mathbb{Z}^2} 4C_k B_m(2X - k). \tag{1.5}$$

Let

$$V_n = \mathrm{clos}_{L_2}\{B_m(2^n X - k);\ k \in \mathbb{Z}^2\},\ n \in \mathbb{Z} \tag{1.6}$$

then $V_n (n\in\mathbb{Z})$ forms a multiresolution analysis of $L^2(R^2)$.

Now we can define $\varphi(X)$, which is the orthonormal scaling function of V_0, by means of its Fourier transform

$$\hat{\varphi}(\xi) = \hat{B}_m(\xi) \cdot \Big(\sum_{k\in\mathbb{Z}^2} |\hat{B}_m(\xi + 2k\pi)|^2\Big)^{-\frac{1}{2}}$$

$$\triangleq \hat{B}_m(\xi) \cdot (\Omega_m(\xi))^{-\frac{1}{2}}. \tag{1.7}$$

By Poisson summation formula and the property of box-spline, we know that

$$\Omega_m(\xi) = \sum_{j\in\mathbb{Z}^2} e^{-ij\cdot\xi} \cdot \int_{R^2} B_m(X) \cdot B_m(X - j)\mathrm{d}X$$

$$= \sum_{j\in\mathbb{Z}^2} B_{2m}(2\sigma - j)e^{-ij\cdot\xi}. \tag{1.8}$$

Let $m_0(\xi) = \hat{\varphi}(2\xi)/\hat{\varphi}(\xi)$. If $\sum_{k=1}^{3} m_k e_k \equiv (p, q)^T \bmod 2$, then the associated wavelet functions $\psi_\mu(X)(\mu=1, 2, 3)$ can be defined as

$$\begin{cases} \hat{\psi}_1(2\xi) = e^{-i\xi\cdot e_1}\,\overline{m_0(\xi + \pi e_1)}\hat{\varphi}(\xi) \\ \hat{\psi}_2(2\xi) = e^{-i\xi\cdot[e_3+(p-q)e_1]}\,\overline{m_0(\xi + \pi e_2)}\hat{\varphi}(\xi) \\ \hat{\psi}_3(2\xi) = e^{-i\xi\cdot[e_2+qe_3]}\,\overline{m_0(\xi + \pi e_3)}\hat{\varphi}(\xi) \end{cases} \tag{1.9}$$

or

$$\begin{cases} \hat{\psi}_1(2\xi) = e^{-i\xi\cdot e_3}\,\overline{m_0(\xi + \pi e_1)}\hat{\varphi}(\xi) \\ \hat{\psi}_2(2\xi) = e^{-i\xi\cdot[e_2+(p-q)e_1]}\,\overline{m_0(\xi + \pi e_2)}\hat{\varphi}(\xi) \\ \hat{\psi}_3(2\xi) = e^{-i\xi\cdot[e_1+qe_3]}\,\overline{m_0(\xi + \pi e_3)}\hat{\varphi}(\xi). \end{cases} \tag{1.10}$$

Moreover, we define

$$m_\mu(\xi) = \hat{\psi}_\mu(2\xi)/\hat{\varphi}(\xi),\ \mu = 1,2,3. \tag{1.11}$$

The above wavelet bases will be applied to yield the periodized wavelet bases in section 2.

2. Periodized Box-spline Wavelet Bases

In this section, we will construct the periodized wavelet bases (cf[8]) by means of the orthonormal box-spline wavelet bases which were introduced in section 1. We will give the explicit expressions of the periodic wavelets.

Let $\varphi(X)$ and $\psi_\mu(X)$ ($\mu=1,2,3$) denote the scaling function and wavelets which were defined in section 1, respectively. Let

$$\dot{\varphi}_{n,j}(X) = 2^n \sum_{s_1\in\mathbb{Z}}\sum_{s_2\in\mathbb{Z}} \varphi(2^n X - 2^n s_1 T_1 e_1 - 2^n s_2 T_2 e_2 - j),$$
$$n \geqslant 0,\ j \in \mathbb{Z}^2, \tag{2.1}$$

$$\dot{\psi}^\mu_{n,j}(X) = 2^n \sum_{s_1\in\mathbb{Z}}\sum_{s_2\in\mathbb{Z}} \psi_\mu(2^n X - 2^n s_1 T_1 e_1 - 2^n s_2 T_2 e_2 - j),$$
$$n \geqslant 0,\ j \in \mathbb{Z}^2, \tag{2.2}$$

and define

$$\dot{V}_n = \operatorname{span}\{\dot{\varphi}_{n,j}(X);\ j \in \mathbb{Z}^2\},\ n \geqslant 0, \tag{2.3}$$
$$\dot{W}^\mu_n = \operatorname{span}\{\dot{\psi}^\mu_{n,j}(X);\ j \in \mathbb{Z}^2\},\ n \geqslant 0. \tag{2.4}$$

It is evident that $\dot{V}_0 \subset \dot{V}_1 \subset \cdots$, and

$$\dim \dot{V}_n = 4^n T_1 T_2,\ \text{for } n \geqslant 0. \tag{2.5}$$

Moreover, we can prove

Theorem 2. 1

$$\dot{V}_n \perp \dot{W}^\mu_n,\ \text{for } n \geqslant 0,\ \mu = 1,2,3. \tag{2.6}$$
$$\dot{W}^\mu_n \perp \dot{W}^\eta_n,\ \text{for } n \geqslant 0,\ \mu,\ \eta = 1,2,3,\ \mu \neq \eta. \tag{2.7}$$

Furthermore, $\dot{W}^\mu_n$ and $\dot{V}_0$ constitute an orthogonal decomposition of $\dot{L}^2(T)$, i. e.

$$\dot{L}^2(T) = \dot{V} \bigoplus_{n\geqslant 0} \bigoplus_{\mu=1}^{3} \dot{W}^\mu_n. \tag{2.8}$$

By (2. 5), we may restrict the subscripts j's in (2. 3)—(2. 4) to

$$\mathbb{Z}(n) \triangleq \mathbb{Z}^2 / 2^n T \mathbb{Z}^2, \tag{2.9}$$

here and throughout, we define

$$T\mathbb{Z}^2 = \{s_1 T_1 e_1 + s_2 T_2 e_2;\ s_1,\ s_2 \in \mathbb{Z}\}. \tag{2.10}$$

Then we have

Theorem 2. 2 $\{\dot{\varphi}_{n,j}(X);\ j \in \mathbb{Z}(n)\}$ constitute an orthonormal basis of $\dot{V}_n (n \geqslant 0)$; and $\{\dot{\psi}^\mu_{n,j}(X);\ j \in \mathbb{Z}(n)\}$ constitute an orthonormal basis of $\dot{W}^\mu_n$.

In the following part of this section, we will give the explicit expressions of $\dot{\varphi}_{n,j}(X)$ and $\dot{\psi}^{\mu}_{n,j}(X)$. To this end, we define the periodized box-splines as

$$\dot{B}^{m}_{n,j}(X) \triangleq 2^{n}\sum_{s_1\in\mathbb{Z}}\sum_{s_2\in\mathbb{Z}} B_m(2^n X - 2^n s_1 T_1 e_1 - 2^n s_2 T_2 e_2 - j),$$

$$n \geqslant 0,\ j \in \mathbb{Z}^2. \tag{2.11}$$

First, we derive the explicit expressions of $\dot{\varphi}_{n,j}(X)$. By (1.7)—(1.8) and (2.11), we have

$$\begin{aligned}\dot{\varphi}_{n,j}(X) &= 2^n\sum_{s_1\in\mathbb{Z}}\sum_{s_2\in\mathbb{Z}}\varphi(2^n X - 2^n s_1 T_1 e_1 - 2^n s_2 T_2 e_2 - j)\\ &= 2^n\sum_{s_1\in\mathbb{Z}}\sum_{s_2\in\mathbb{Z}}\sum_{p\in\mathbb{Z}^2} a_p B_m(2^n X - 2^n s_1 T_1 e_1 - 2^n s_2 T_2 e_2 - j - p)\\ &= \sum_{p\in\mathbb{Z}^2} a_{p-j}\dot{B}^{m}_{n,p}(X)\\ &\triangleq \sum_{q\in\mathbb{Z}(n)} b_{j,q}\dot{B}^{m}_{n,q}(X),\end{aligned} \tag{2.12}$$

where

$$a_p = (2\pi)^{-2}\int_{[0,2\pi]^2}(\Omega_m(\xi))^{-\frac{1}{2}}e^{ip\cdot\xi}\mathrm{d}\xi \tag{2.13}$$

and

$$b_{j,q} = \sum_{s_1\in\mathbb{Z}}\sum_{s_2\in\mathbb{Z}} a_{q+2^n s_1 T_1 e_1 + 2^n s_2 T_2 e_2 - j}. \tag{2.14}$$

Now we compute $b_{j,q}$. By (2.13)—(2.14),

$$\begin{aligned}b_{j,q} &= (2\pi)^{-2}\sum_{s_1\in\mathbb{Z}}\sum_{s_2\in\mathbb{Z}}\int_{[0,2\pi)^2}(\Omega_m(\xi))^{-\frac{1}{2}}e^{i\xi\cdot(q-j)}e^{i\xi\cdot 2^n(s_1T_1e_1+s_2T_2e_2)}\mathrm{d}\xi\\ &= \lim_{\substack{N_1\to\infty\\ N_2\to\infty}}(2\pi)^{-2}\int_0^{2\pi}\int_0^{2\pi}(\Omega_m(\xi))^{-\frac{1}{2}}e^{i\xi\cdot(q-j)}\\ &\qquad\cdot\frac{\sin\left(N_1+\frac{1}{2}\right)2^nT_1\xi_1}{\sin 2^{n-1}T_1\xi_1}\cdot\frac{\sin\left(N_2+\frac{1}{2}\right)2^nT_2\xi_2}{\sin 2^{n-1}T_2\xi_2}\mathrm{d}\xi_1\mathrm{d}\xi_2\\ &= 4^{-n}(T_1T_2)^{-1}\lim_{\substack{N_1\to\infty\\ N_2\to\infty}}(2\pi)^{-2}\int_0^{2^{n+1}T_1\pi}\int_0^{2^{n+1}T_2\pi}\left(\Omega_m\left(\frac{\eta_1}{2^nT_1},\frac{\eta_2}{2^nT_2}\right)\right)^{-\frac{1}{2}}\end{aligned}$$

$$\cdot\, e^{i2^{-n}\left(\frac{\eta_1}{T_1},\frac{\eta_2}{T_2}\right)\cdot(q-j)} \cdot \frac{\sin\left(N_1+\frac{1}{2}\right)\eta_1}{\sin\frac{\eta_1}{2}} \cdot \frac{\sin\left(N_2+\frac{1}{2}\right)\eta_2}{\sin\frac{\eta_2}{2}} \mathrm{d}\eta_1\eta_2$$

$$= 4^{-n}(T_1T_2)^{-1}\sum_{\mu=1}^{2^nT_1}\sum_{\nu=1}^{2^nT_2}(\Omega_m(\xi_{\mu,\nu}))^{-\frac{1}{2}}e^{i(q-j)\cdot\xi_{\mu,\nu}}, \tag{2.15}$$

where

$$\xi_{\mu,\nu} = (2^{1-n}\mu T_1^{-1}\pi,\ 2^{1-n}\nu T_2^{-1}\pi)^T,\ \mu,\ \nu\in\mathbb{Z}. \tag{2.16}$$

So we have

Proposition 2.3 $\dot{\varphi}_{n,j}(X)$ has its explicit expression as

$$\dot{\varphi}_{n,j}(X) = 4^{-n}(T_1T_2)^{-1}\sum_{q\in\mathbb{Z}(n)}\sum_{\mu=1}^{2^nT_1}\sum_{\nu=1}^{2^nT_2}(\Omega_m(\xi_{\mu,\nu}))^{-\frac{1}{2}}e^{i(q-j)\cdot\xi_{\mu,\nu}}\cdot \dot{B}^m_{n,q}(X).$$

Now, it is our turn to give explicit expressions of $\dot{\psi}^\mu_{n,j}$. By (1.9)—(1.11), we have

$$\psi_\mu(\xi) = 4\sum_{p\in\mathbb{Z}^2}C^\mu_pB_m(2X-p), \tag{2.17}$$

where

$$C^\mu_p = (2\pi)^{-2}\int_{[0,2\pi)^2}m_\mu(\xi)(\Omega_m(\xi))^{-\frac{1}{2}}e^{ip\cdot\xi}\mathrm{d}\xi. \tag{2.18}$$

So

$$\begin{aligned}
&\dot{\psi}^\mu_{n,j}(X)\\
&= 2^n\sum_{s_1\in\mathbb{Z}}\sum_{s_2\in\mathbb{Z}}\sum_{p\in\mathbb{Z}^2}4C^\mu_pB_m(2^{n+1}X-2^{n+1}s_1T_1e_1-2^{n+1}s_2T_2e_2-2j-p)\\
&= 2\sum_{p\in\mathbb{Z}^2}C^\mu_p\dot{B}^m_{n+1,\,p+2j}(X)\\
&\triangleq 2\sum_{q\in\mathbb{Z}(n+1)}d^\mu_{n+1,q}\dot{B}^m_{n+1,q}(X),
\end{aligned} \tag{2.19}$$

where

$$d^\mu_{n+1,q} = \sum_{s_1\in\mathbb{Z}}\sum_{s_2\in\mathbb{Z}}C^\mu_{q-2j+2^{n+1}s_1T_1e_1+2^{n+1}s_2T_2e_2}. \tag{2.20}$$

We can compute $d^\mu_{n+1,q}$ as we did in the computation of $b_{j,q}$. So we have

Proposition 2.4 $\dot{\psi}^\mu_{n,j}(X)$ has its explicit expression as

$$\dot{\psi}^{\mu}_{n,j}(X) = 2^{-2n-1} \sum_{q \in \mathbb{Z}(n+1)} \dot{B}^{m}_{n+1,q}(X) \sum_{s=1}^{2^{n+1}T_1} \sum_{t=1}^{2^{n+1}T_2} (\Omega_m(\xi_{s,t}))^{-\frac{1}{2}} \cdot m_\mu(\xi_{s,t}) e^{i(q-2j)\cdot\xi_{s,t}}, \tag{2.21}$$

where

$$\xi_{s,t} = (2^{-n} s T_1^{-1}\pi,\ 2^{-n} t T_2^{-1}\pi)^T, \qquad s,\ t,\ \in \mathbb{Z}. \tag{2.22}$$

3. Periodic Box-spline Orthonormal Bases and Wavelets

In this section, we construct another kind of periodic box-spline wavelet. This work is an extension of its univariate case (see Chen [1]).

In [6], Kamada et al gave a method to construct periodic B-spline orthonormal basis. Here we generalize the method to the bivariate case to construct the bivariate periodic box-spline orthonormal bases which are non-tensor-product type, and apply the bases to yield the periodic wavelet bases.

As in section 2, we define the periodized box-spline as follows,

$$\dot{B}^{m}_{n,j}(X) = 2^n \sum_{s_1 \in \mathbb{Z}} \sum_{s_2 \in \mathbb{Z}} B_m(2^n X - 2^n s_1 T_1 e_1 + 2^n s_2 T_2 e_2 - j),\ n \geq 0, \tag{3.1}$$

where $B_m(X)$ denotes the box-spline introduced in section 1 and 2, and $T = (T_1, T_2)^T$ defined as before. Let

$$\dot{V}_n \triangleq \text{span}\{\dot{B}^{m}_{n,j}(X);\ j \in \mathbb{Z}(n)\}. \tag{3.2}$$

Then we have

Proposition 3.1 For $n \geq 0$, $\{\dot{B}^{m}_{n,j}(X);\ j \in \mathbb{Z}(n)\}$ are linearly independent, and $\dim \dot{V}_n = 4^n T_1 T_2$.

Proof We only need to prove the conclusion in the case $n = 0$, the proof for the case $n \geq 1$ is similar.

Assume that there exists $\{C_k\} \in l^2(\mathbb{Z}^2)$ such that

$$\sum_{j \in \mathbb{Z}(0)} C_j \dot{B}^{m}_{n,j}(X) \equiv 0,\ \forall\ X \in [0, T_1] \times [0, T_2], \tag{3.3}$$

this implies that

$$\sum_{s_1\in\mathbb{Z}}\sum_{s_2\in\mathbb{Z}}\sum_{j\in\mathbb{Z}(0)} d_{j+s_1T_1e_1+s_2T_2e_2}B_m(X - s_1T_1e_1 - s_2T_2e_2 - j) \equiv 0,\ X \in R^2, \tag{3.4}$$

where

$$d_{j+s_1T_1e_1+s_2T_2e_2} = C_j,\ j \in \mathbb{Z}(0). \tag{3.5}$$

Because $B_m(X)$ has linearly independent integer translates (see [3]), $C_j \equiv 0$ for $j \in \mathbb{Z}(0)$. This completes the proof of the first conclusion. Moreover, the second conclusion is its corollary.

Proposition 3.2 $\dot{V}_n\,(n \geqslant 0)$ forms a nested subspace sequence of $\dot{L}^2(T)$.

Proof By (1.4), $B_m(X) = 4\sum_{k\in\mathbb{Z}^2} C_kB_m(2X - k)$, here the mask of $\{C_k\}$ is finitely supported. So, for any $j\in\mathbb{Z}(n)$, $n\geqslant 0$,

$$\begin{aligned}\dot{B}^m_{n,j}(X) &= 2^n\sum_{s_1\in\mathbb{Z}}\sum_{s_2\in\mathbb{Z}} B_m(2^nX - 2^ns_1T_1e_1 - 2^ns_2T_2e_2 - j)\\ &= 2^n\sum_{s_1\in\mathbb{Z}}\sum_{s_2\in\mathbb{Z}}\sum_{k\in\mathbb{Z}^2} 4C_kB_m(2^{n+1}X - 2^{n+1}s_1T_1e_1 - 2^{n+1}s_2T_2e_2 - 2j - k)\\ &= \sum_{k\in\mathbb{Z}^2} 2C_k\dot{B}^m_{n+1,\,2j+k}(X).\end{aligned} \tag{3.6}$$

Proposition 3.3 $\mathrm{dist}_2(\dot{L}^2(T),\ \dot{V}_n) \to 0,\ n \to +\infty$.

Proof For $\forall\ f(X) \in \dot{L}^2(T)$, let

$$f_n(X) \triangleq \sum_{k\in\mathbb{Z}(n)} a_n(k)\dot{B}^m_{n,k}(X), \tag{3.7}$$

where

$$a_n(k) = 2^n\int_{2^{-n}k+[0,\,2^{-n})^2} f(X)\mathrm{d}X = 2^{-n}\int_{[0,1)^2} f(2^{-n}(Y + k))\mathrm{d}Y. \tag{3.8}$$

Then by theorem of Jia and Micchelli (see [5], Th 2.5),

$$\lim_{n\to+\infty} \|f(\cdot) - f_n(\cdot)\|_{L^2(T)} = 0. \tag{3.9}$$

Now we sum the above three propositions up as

Theorem 3.4 $\dot{V}_n(n \geqslant 0)$ form a multiresolution analysis of $\dot{L}^2(T)$

Based on the multiresolution analysis, we can construct the orthonormal bases of $\dot{V}_n$ and the related orthonormal wavelet bases. To this end, we need the following three lemmas.

Lemma 3.5 The Fourier series of $\dot{B}^m_{n,j}(X)$ is

$$2^{-n}(T_1T_2)^{-1}\sum_{t_1\in\mathbb{Z}}\sum_{t_2\in\mathbb{Z}}\hat{B}_m\left(\frac{2\pi t_1}{2^nT_1},\frac{2\pi t_2}{2^nT_2}\right)\cdot e^{2\pi i\left(\frac{t_1}{T_1},\frac{t_2}{T_2}\right)\cdot(X-2^{-n}j)}. \tag{3.10}$$

Proof Because $\dot{B}^m_{n,j}(X)=\dot{B}^m_{n,0}(X-2^{-n}j)$, it suffices to give the Fourier series of $\dot{B}^m_{n,0}(X)$.

$$\dot{B}^m_{n,0}(X)=\sum_{t_1\in\mathbb{Z}}\sum_{t_2\in\mathbb{Z}}(T_1T_2)^{-\frac{1}{2}}e^{2\pi i(t_1/T_1,\,t_2/T_2)\cdot X}\cdot\int_{[0,T_1]\times[0,T_2]}\dot{B}^m_{n,0}(Y)(T_1T_2)^{-\frac{1}{2}}e^{-2\pi i(t_1/T_1,\,t_2/T_2)\cdot Y}dY$$

$$=\sum_{t_1\in\mathbb{Z}}\sum_{t_2\in\mathbb{Z}}2^n(T_1T_2)^{-1}e^{2\pi i(t_1/T_1,\,t_2/T_2)\cdot X}\cdot\int_{R^2}B_m(2^nY)e^{-2\pi i(t_1/T_1,\,t_2/T_2)\cdot Y}dY$$

$$=(2^{-n})(T_1T_2)^{-1}\sum_{t_1\in\mathbb{Z}}\sum_{t_2\in\mathbb{Z}}\hat{B}_m\left(\frac{2\pi t_1}{2^nT_1},\frac{2\pi t_2}{2^nT_2}\right)e^{2\pi i(t_1/T_1,\,t_2/T_2)\cdot X}.$$

By (3.10), we can prove the following result.

Lemma 3.6 For $j,k\in\mathbb{Z}(n)$,

$$\langle\dot{B}^m_{n,j}(X),\dot{B}^m_{n,k}(X)\rangle=4^{-n}(T_1T_2)^{-1}\sum_{t_1\in\mathbb{Z}}\sum_{t_2\in\mathbb{Z}}\left|\hat{B}_m\left(\frac{2\pi t_1}{2^nT_1},\frac{2\pi t_2}{2^nT_2}\right)\right|^2e^{2\pi i(t_1/T_1,\,t_2/T_2)\cdot2^{-n}(k-j)}. \tag{3.11}$$

In this paper, we also need the following familiar fact.

Lemma 3.7 For any $l\in\mathbb{Z}$, $m\in N$,

$$\frac{1}{m}\sum_{k=0}^{m-1}e^{2\pi il\cdot k/m}=\begin{cases}1 & \text{if } l \text{ is a multiple of } m,\\ 0 & \text{otherwise.}\end{cases} \tag{3.12}$$

We are now in a position to state one of the main results of this paper.

Theorem 3.8 For $p\in\mathbb{Z}(n)$, $(n\geqslant0)$, let $j=(j_1,j_2)^T$, and

$$\varphi_{n,p}^{m}(X) = C_{n,p}^{m} \sum_{j \in \mathbb{Z}(n)} e^{2\pi i\left(\frac{p_1 j_1}{2^n T_1} + \frac{p_2 j_2}{w^n T_2}\right)} \cdot \mathring{B}_{n,j}^{m}(X), \tag{3.13}$$

where

$$C_{n,p}^{m} = 2^{-n}(T_1T_2)^{-\frac{1}{2}} \cdot e^{2\pi i\left(\frac{m_1+m_3}{2^{n+1}T_2}, \frac{m_2+m_3}{2^{n+1}T_2}\right)\cdot p}\left(\Omega_m\left(\frac{2p_1\pi}{2^nT_1}, \frac{2p_2\pi}{2^nT_2}\right)\right)^{-\frac{1}{2}}. \tag{3.14}$$

Then $\{\varphi_{n,p}^{m}(z);\ p \in Z(n)\}$ constitute an orthonormal basis of $\dot{V}_n$.

Proof It is evident that $\varphi_{n,p}^{m}(Z) \in \dot{V}_n$, so we only need to prove that $\langle \varphi_{n,p}^{m}, \varphi_{n,q}^{m} \rangle = \delta_{p,q}$. For $p, q \in \mathbb{Z}(n)$,

$$\langle \varphi_{n,p}^{m}, \varphi_{n,q}^{m} \rangle$$

$$= \Big\langle C_{n,p}^{m} \sum_{r \in \mathbb{Z}(n)} e^{2\pi i\left(\frac{p_1 r_1}{2^n T_1} + \frac{p_2 r_2}{2^n T_2}\right)} \mathring{B}_{n,r}^{m}(X),$$

$$C_{n,q}^{m} \sum_{s \in \mathbb{Z}(n)} e^{2\pi i\left(\frac{q_1 s_1}{2^n T_1} + \frac{q_2 s_2}{2^n T_2}\right)} \mathring{B}_{n,s}^{m}(X) \Big\rangle$$

$$= C_{n,p}^{m} \overline{C_{n,q}^{m}} \sum_{r,s \in \mathbb{Z}(n)} e^{2\pi i(p_1 r_1 - q_1 s_1)/2^n T_1} e^{2\pi i(p_2 r_2 - q_2 s_2)/2^n T_2} \langle \mathring{B}_{n,r}^{m}(X), \mathring{B}_{n,s}^{m}(X) \rangle.$$

By (3.11), we know that

$$\langle \mathring{B}_{n,r}^{m}(X), \mathring{B}_{n,s}^{m}(X) \rangle = 4^{-n}(T_1T_2)^{-1} \sum_{\mu \in \mathbb{Z}(n)} \sum_{t_1, t_2 \in \mathbb{Z}}$$

$$\left|\hat{B}_m\left(\frac{2\mu_1\pi}{2^nT_1} + 2t_1\pi, \frac{2\mu_2\pi}{2^nT_2} + 2t_2\pi\right)\right|^2 e^{2\pi i(\mu_1/T_1, \mu_2/T_2)\cdot 2^{-n}(s-r)}. \tag{3.15}$$

So

$$\langle \varphi_{n,p}^{m}(X), \varphi_{n,q}^{m}(X) \rangle = \frac{C_{n,p}^{m} \overline{C_{n,q}^{m}}}{4^n T_1 T_2} \sum_{r \in \mathbb{Z}(n)} e^{2\pi i(p_1 - q_1) r_1/2^n T_1} e^{2\pi i(p_2 - q_2) r_2/2^n T_2}.$$

$$\sum_{\mu \in \mathbb{Z}(n)} \sum_{t_1, t_2 \in \mathbb{Z}} \sum_{\eta \in \mathbb{Z}(n)} \left|\hat{B}_m\left(\frac{2\mu_1\pi}{2^nT_1} + 2t_1\pi, \frac{2\mu_2\pi}{2^nT_2} + 2t_2\pi\right)\right|^2$$

$$\cdot\ e^{2\pi i(\mu_1 - q_1)\eta_1/2^n T_1} \cdot e^{2\pi i(\mu_2 - q_2)\eta_2/2^n T_2}$$

$$= C_{n,p}^{m} \overline{C_{n,q}^{m}} \cdot \sum_{r \in \mathbb{Z}(n)} e^{2\pi i(p_1 - q_1) r_1/2^n T_1} e^{2\pi i(p_2 - q_2) r_2/2^n T_2}.$$

$$\cdot \sum_{\mu \in \mathbb{Z}(n)} \sum_{t_1, t_2 \in \mathbb{Z}} \left|\hat{B}_m\left(\frac{2\mu_1\pi}{2^nT_1} + 2t_1\pi, \frac{2\mu_2\pi}{2^nT_2} + 2t_2\pi\right)\right|_2 \delta_{\mu,q}$$

$$= 4^n T_1 T_2 C_{n,p}^{m} \overline{C_{n,q}^{m}} \delta_{p,q} \cdot \Omega_m\left(\frac{2q_1\pi}{2^nT_1}, \frac{2q_2\pi}{2^nT_2}\right)$$

$$= \delta_{p,q}.$$

This completes the proof.

Proposition 3.9 For $q \in \mathbb{Z}(n)$, the following equality holds:

$$\dot{B}^m_{n,q}(X) = 4^{-n}(T_1T_2)^{-1} \sum_{p\in\mathbb{Z}(n)} (C^m_{n,p})^{-1} e^{-2\pi i\left(\frac{p_1q_1}{2^nT_1}+\frac{p_2q_2}{2^nT_2}\right)} \dot{\varphi}^m_{n,p}. \quad (3.16)$$

Proof By (3.13), the right-hand side of (3.16) equals

$$4^{-n}(T_1T_2)^{-1} \sum_{p\in\mathbb{Z}(n)} \sum_{j\in\mathbb{Z}(n)} e^{2\pi i\left(\frac{p_1(j_1-q_1)}{2^nT_1}+\frac{p_2(j_2-q_2)}{2^nT_2}\right)} \dot{B}^m_{n,j}(X)$$

$$=4^{-n}(T_1T_2)^{-1} \sum_{j\in\mathbb{Z}(n)} \dot{B}^m_{n,j}(X) \sum_{p\in\mathbb{Z}(n)} e^{2\pi i\left(\frac{p_1(j_1-q_1)}{2^nT_1}+\frac{p_2(j_2-q_2)}{2^nT_2}\right)}$$

$$=\dot{B}^m_{n,q}(X).$$

By the above proposition, we can obtain the refinement equation which $\varphi^{m,n}_p(X)$ satisfies. In fact, for any $p \in \mathbb{Z}(n)$,

$$\varphi^m_{n,p}(X) = C^m_{n,p} \sum_{j\in\mathbb{Z}(n)} e^{2\pi i\left(\frac{p_1j_1}{2^nT_1}+\frac{p_2j_2}{2^nT_2}\right)} \sum_{k\in\mathbb{Z}^2} 2C_k \dot{B}^m_{n+1,2j+k}(X)$$

$$= C^m_{n,p} \sum_{j\in\mathbb{Z}(n)} e^{2\pi i\left(\frac{p_1j_1}{2^nT_1}+\frac{p_2j_2}{2^nT_2}\right)} \sum_{k\in\mathbb{Z}^2} \frac{2C_k}{4^{n+1}T_1T_2}$$

$$\cdot \sum_{q\in\mathbb{Z}(n+1)} (C^m_{n+1,q})^{-1} e^{-2\pi i\left(\frac{q_1(2j_1+k_1)}{2^{n+1}T_1}+\frac{q_2(2j_2+k_2)}{2^{n+1}T_2}\right)} \varphi^m_{n+1,q}(X)$$

$$= C^m_{n,p} \sum_{k\in\mathbb{Z}^2} \frac{2C_k}{4^{n+1}T_1T_2} \sum_{q\in\mathbb{Z}(n+1)} e^{-2\pi i\left(\frac{q_1k_1}{2^{n+1}T_1}+\frac{q_2k_2}{2^{n+1}T_2}\right)} \varphi^m_{n+1,q}(X)(C^m_{n+1,q})^{-1}$$

$$\cdot \sum_{j\in\mathbb{Z}(n)} e^{2\pi i\left(\frac{j_1(p_1-q_1)}{2^nT_1}+\frac{j_2(p_2-q_2)}{2^nT_2}\right)}$$

$$= \frac{1}{2}C^m_{n,p}\left[C\left(\frac{p_1\pi}{2^nT_1}, \frac{p_2\pi}{2^nT_2}\right)(C^m_{n+1,p})^{-1}\varphi^m_{n+1,p}(X)\right.$$

$$+ C\left(\frac{p_1\pi}{2^nT_1}+\pi, \frac{p_2\pi}{2^nT_2}\right)(C^m_{n+1,p+2^nT_1e_1})^{-1}\varphi^m_{n+1,p+2^nT_1e_1}(X)$$

$$+ C\left(\frac{p_1\pi}{2^nT_1}, \frac{p_2\pi}{2^nT_2}+\pi\right)(C^m_{n+1,p+2^nT_2e_2})^{-1}\varphi^m_{n+1,p+2^nT_2e_2}(X)$$

$$\left. + C\left(\frac{p_1\pi}{2^nT_1}+\pi, \frac{p_2\pi}{2^nT_2}+\pi\right)(C^m_{n+1,p+2^nT})^{-1}\varphi^m_{n+1,p+2^nT}(X)\right]$$

$$\triangleq a(n,p,1)\varphi^m_{n+1,p}(X) + a(n,p,2)\varphi^m_{n+1,p+2^nT_1e_1}$$

$$+ a(n,p,3)\varphi^m_{n+1,p+2^nT_2e_2} + a(n,p,4)\varphi^m_{n+1,p+2^nT}. \quad (3.17)$$

Here $C(X)$ is defined as in (1.2) and (1.4). By (3.14), we know that $a(n,p,s)$ $(s=1,2,3,4)$ are real numbers. So we get the following results.

Theorem 3. 10 $\varphi^m_{n,p}(X)$ satisfies the refinement equation (3. 17). And

Lemma 3. 11 For $p\in\mathbb{Z}(n)$, $n\geqslant 0$, we have

$$\sum_{s=1}^{4}(a(n,p,s))^2 = 1. \tag{3.18}$$

Moreover, the following matrix is orthogonal:

$$A(n,p) = \begin{Bmatrix} a(n,p,1) & a(n,p,2) & a(n,p,3) & a(n,p,4) \\ a(n,p,2) & -a(n,p,1) & -a(n,p,4) & a(n,p,3) \\ a(n,p,3) & a(n,p,4) & -a(n,p,1) & -a(n,p,2) \\ a(n,p,4) & -a(n,p,3) & a(n,p,2) & -a(n,p,1) \end{Bmatrix}$$

$$\triangleq (A^{k,l}_{n,p})_{1\leqslant k,l\leqslant 4}. \tag{3.19}$$

From the above lemma, we can construct the orthonormal periodic wavelets. For $p\in\mathbb{Z}(n)$, let

$$\begin{aligned}\psi^{m,\mu}_{n,p}(X) &= A^{\mu+1,1}_{n,p}\varphi^m_{n+1,p}(X) + A^{\mu+1,2}_{n,p}\varphi^m_{n+1,p+2^nT_1e_1}(X) \\ &\quad + A^{\mu+1,3}_{n,p}\varphi^m_{n+1,p+2^nT_2e_2}(X) + A^{\mu+1,4}_{n,p}\varphi^m_{n+1,p+2^nT}(X), \\ &\quad \mu = 1,2,3, \end{aligned} \tag{3.20}$$

$$\dot{W}^\mu_n \triangleq \text{span}\{\psi^{m,\mu}_{n,p}(X);\ p\in Z(n)\},\ \mu=1,2,3,\ n\geqslant 0, \tag{3.21}$$

$$\psi^{m,0}_{n,p}(X) = \varphi^m_{n,p}(X), \tag{3.22}$$

$$\dot{W}^0_n = \dot{V}_n. \tag{3.23}$$

Then we have

Theorem 3. 12 Let $\dot{W}_n$ denote the orthogonal complementary space of $\dot{V}_n$ in $\dot{V}_{n+1}$. Then $\dot{W}^\mu_n$ ($\mu=1,2,3$) form an orthogonal decomposition of $\dot{W}_n$, i. e.

$$\dot{W}_n = \dot{W}^1_n \oplus \dot{W}^2_n \oplus \dot{W}^3_n, \tag{3.24}$$

and the wavelets $\{\psi^{m,\mu}_{n,p}(X);\ p\in\mathbb{Z}(n)\}$ constitute an orthonormal basis of $\dot{W}^\mu_n$ ($\mu=1,2,3$).

Proof Because $\dot{W}^\mu_n\subset\dot{V}_{n+1}$ ($\mu=1,2,3$), it suffices to prove that

$$<\psi^{m,\mu}_{n,p}(X),\ \psi^{m,\eta}_{n,q}(X)> = \delta_{p,q}\delta_{\mu,\eta}. \tag{3.25}$$

(3. 25) can be easily proved from the definition of $\psi^{m,\mu}_{n,p}(X)$ and the fact that $\mathbb{Z}(n+1)$ is the union of $\mathbb{Z}(n)$, $\mathbb{Z}(n)+2^nT_1e_1$, $\mathbb{Z}(n)+2^nT_2e_2$ and $\mathbb{Z}(n)+2^nT$, which are disjoint to each other.

Now we consider the orthogonal projective operators.

$$E_n: \dot{L}^2(T) \to \dot{V}_n,\ n \geqslant 0,$$

$$E_n f(X) = \sum_{p \in \mathbb{Z}(n)} < f,\ \varphi^m_{n,p} > \varphi^m_{n,p}(X)$$

$$\triangleq \sum_{p \in \mathbb{Z}(n)} \alpha_{n,p} \varphi^m_{n,p}(X). \tag{3.26}$$

$$D_n: \dot{L}^2(T) \to \dot{W}_n,\ n \geqslant 0,$$

$$D_n f(X) = \sum_{\mu=1}^{3} \sum_{p \in \mathbb{Z}(n)} < f,\ \psi^{m,\mu}_{n,p} > \psi^{m,\mu}_{n,p}(X)$$

$$= \sum_{\mu=1}^{3} \sum_{p \in \mathbb{Z}(n)} \beta^{\mu}_{n,p} \psi^{m,\mu}_{n,p}(X). \tag{3.27}$$

We have the decomposition and reconstruction algorithms as follows.

Theorem 3. 13 If E_n and D_n are defined as in (3. 26)—(3. 27), then the following algorithms hold:

(1) Decomposition algorithm:

$$\alpha_{n,p} = A^{1,1}_{n,p} \alpha_{n+1,p} + A^{1,2}_{n,p} \alpha_{n+1,p+2^n T_1 e_1} + A^{1,3}_{n,p} \alpha_{n+1,p+2^n T_2 e_2} + A^{1,4}_{n,p} \alpha_{n+1,p+2^n T}, \tag{3.28}$$

$$\beta^{\mu}_{n,p} = A^{\mu+1,1}_{n,p} \alpha_{n+1,p} + A^{\mu+1,2}_{n,p} \alpha_{n+1,p+2^n T_1 e_1} + A^{\mu+1,3}_{n,p} \alpha_{n+1,p+2^n T_2 e_2} + A^{\mu+1,4}_{n,p} \alpha_{n+1,p+2^n T}. \tag{3.29}$$

(2) Reconstruction algorithm: for $p \in \mathbb{Z}(n)$,

$$\alpha_{n+1,p} = A^{1,1}_{n,p} \alpha_{n,p} + A^{2,1}_{n,p} \beta^1_{n,p} + A^{3,1}_{n,p} \beta^2_{n,p} + A^{4,1}_{n,p} \beta^3_{n,p}, \tag{3.30}$$

$$\alpha_{n+1,p+2^n T_1 e_1} = A^{1,2}_{n,p} \alpha_{n,p} + A^{2,2}_{n,p} \beta^1_{n,p} + A^{3,2}_{n,p} \beta^2_{n,p} + A^{4,2}_{n,p} \beta^3_{n,p}, \tag{3.31}$$

$$\alpha_{n+1,p+2^n T_2 e_2} = A^{1,3}_{n,p} \alpha_{n,p} + A^{2,3}_{n,p} \beta^1_{n,p} + A^{3,3}_{n,p} \beta^2_{n,p} + A^{4,3}_{n,p} \beta^3_{n,p}, \tag{3.32}$$

$$\alpha_{n+1,p+2^n T} = A^{1,4}_{n,p} \alpha_{n,p} + A^{2,4}_{n,p} \beta^1_{n,p} + A^{3,4}_{n,p} \beta^2_{n,p} + A^{4,4}_{n,p} \beta^3_{n,p}. \tag{3.33}$$

References

[1] Chen, H. L. (1993) Construction of orthonormal wavelets in the periodic case, to apper in *J. Comp. Math.*

[2] Chui, C. K., Stöckler, J. and Ward, J. (1992) Compactly

supported box-spline wavelets, *Approx. Theory and its Appl.* **8**, 77—100.

[3] Dahmen, W. and Micchelli, C. A. (1983) Recent progress in multivariate splines, in C. K. Chui et al (eds), *Approximation Theory* IV, Academic Press, 27—121.

[4] Daubechies, I. (1992) *Ten lectures on wavelet*, CBMS—NSF series in Appl. Math., SIAM Publ.

[5] Jia, R. Q. and Micchelli, C. A. (1991) Using the refinement equations for the construction of pre-wavelets II: Powers of two, in P. J. Laurent et al (eds.), *Curves and Surfaces*, 209—246.

[6] Kamada, M., Toraichi, K. and Mori, R. (1988) Periodic spline orthonormal bases, *J. Approx. Theory* **55**, 27—34.

[7] Liang, X. Z. and Jin, G. R. (1991) Bivariate spline wavelets, in "*Proceeding of Conference of Computational Mathematics*", Tianjin.

[8] Meyer, Y. (1990) *Ondelettes*, Hermann, Editeurs des Sciences et des Arts.

[9] Riemenschneider, S. and Shen, Z. W. (1990) Box-spline, cardinal series and wavelets, in C. K. Chui (ed), *Approximation Theory and Functional Analysis, Academic Press.*

[10] Riemenschneider, S. and Shen, Z. W. (1990) Wavelets and pre-wavelets in low dimensions, preprint.

On Martingale Spaces and Inequalities

R. Long

(Institute of Mathematics, Academia Sinica, Beijing)

1. Introduction

In the past twenty years, the H_p-BMO theory on $\mathbb{R}^n$ has undergone a flourishing development, which should partly give the credit to the applications of some martingale ideas and methods. A number of examples can be taken to show this. The famous Calderón-Zygmund decomposition, one of the key parts of Calderón-Zygmund's real method, may be said to be a counterpart of stopping time argument in Probability Theory; the atomic decomposition of H_p spaces, the constructive proof of the Fefferman-Stein decomposition of BMO spaces, and the good λ-inequality technique etc. were all first germinated in martingale setting. In addition, there ard also many applications of Martingale Theory to Harmonic Function Theory and many recent applications to Analysis and especially to Harmonic Analysis. Among them, two examples are worth to be mentioned. One is that D. Burtholder described an important kind of Banach spaces (called UMD spaces) by using martingales, another is that by using martingales as a tool, a much more simplified proof of the important $T(b)$ theorem in Calderón-Zygmund Singular Integral Theory was given. From the above-Cited examples we can see what an important role Martingale Theory has played in the development of Analysis, especially of Harmonic Analysis. In this topic, i.e. Martngale Spaces and Inequalities, our Chinese Mathematicians made some contributions too. In what follows, we will list some of them without proofs.

Let $(\Omega, \mathcal{F}, \mu, \{\mathcal{F}_n\}_{n\geq 0})$ be a probability space endowed with a nondecreasing family $\{\mathcal{F}_n\}_{n\geq 0}$ of sub-σ-fields, such that $V\mathcal{F}_n = \mathcal{F}$, and $(\Omega, \mathcal{F}_n, \mu)$ is complete for all $n \geq 0$. process $(\gamma_n)_{n\geq 0}$ is said to be adapted, if γ_n is $\mathcal{F}_n$-measurable, $n \geq 0$. A

Supported by National Science Foundation of China

Key words: Martingale, Singular integral operator, Geometry of Banach Spaces

Subject classification: 60GXX, 60HXX.

M. Cheng et al. (eds.), Harmonic Analysis in China, 197–209.

process $(f_n)_{n\geq 0}$ is said to be a martingale if it is adapted, and f_n is integrable, and $E(f_{n+1}|\mathcal{F}_n) = f_n$, for all n. For any process $(\gamma_n)_{n\geq 0}$, γ_{-1} is assumed as 0 unless otherwise is stated. For any process $(\gamma_n)_{n\geq 0}$, we set

$$M_n(\gamma) = \sup_{k\leq n} |\gamma_k|, M(\gamma) = M_\infty(\gamma), \tag{1.1}$$

$$S_n(\gamma) = \left(\sum_{k=0}^{n} | \Delta_k \gamma|^2\right)^{\frac{1}{2}}, S(\gamma) = S_\infty(\gamma), \Delta_k = \gamma_k - \gamma_{k-1}, k \geq 0, \tag{1.2}$$

$$\sigma_n(\gamma) = \sum_{k=0}^{n} E(| \Delta_k \gamma|^2|\mathcal{F}_{k-1}) = |\gamma_0|^2 + \sum_{1}^{n} E(| \Delta_k \gamma|^2)|\mathcal{F}_{k-1}), \sigma(\gamma) = \sigma_\infty(\gamma). \tag{1.3}$$

A martingale $f = (f_n)_{n\geq 0}$ is said to be in $L^p, 1 \leq p \leq \infty$, if

$$||f||_p = \sup_n ||f_n||_p < \infty, \tag{1.4}$$

is said to be in L^1_u, if there exists $f \in L^1 = L^1(\Omega, \mathcal{F}, \mu)$ such that

$$f_n = E(f|\mathcal{F}_n), \quad \forall n \geq 0. \tag{1.5}$$

Define

$$H_p^S = \{\text{martiugale } f = (f_n)_{n\geq 0} : ||f||_{H_p^S} = ||S(f)||_p < \infty\}, 0 < p \leq \infty, \tag{1.6}$$

$$H_p^* = \{\text{martingale} f = (f_n)_{n\geq 0} : ||f||_{H_p^*} = ||M(f)||_p < \infty\}, 0 < p \leq \infty, \tag{1.7}$$

$$h_p = \{\text{martingale} f = (f_n)_{n\geq 0} : ||f||_{h_p} = ||\sigma(f)||_p < \infty\}, 0 < p \leq \infty. \tag{1.8}$$

2. Regular maringales

In [17, 18], Long introduced a kind of regularity for martingales and studied H_p martingales $(0 < p \leq 1)$ and various inequalities in the regular case. The main result in this topic is the following characterization theorem of such regularity.

Theorem 2.1. Let $(\Omega, \mathcal{F}, \mu, \{\mathcal{F}_n\}_{n\geq 0})$ be as above. Then following assertions are equivalent. Let $d \geq 1$ be a constant.

(a) The R condition holds

$$\chi_F \leq dE(\chi_F|\mathcal{F}_{n-1}), \quad \forall F \in \mathcal{F}_n, \forall n \geq 1;$$

(b) The condition R_w holds: for all $n \geq 1$, all $F_n \in \mathcal{F}_n$ there exists at least a $G_n \in \mathcal{F}_{n-1}$, such that

$$F_n \subset G_n, \quad |G_n| \leq d|F_n|;$$

(c) For any nonvanishing stopping time T, there exists a stopping time τ such that

$$0 \le \tau < T \quad \text{on the set} \quad \{T < \infty\}, \quad \text{and} \quad |\{\tau < \infty\}| \le d|\{T < \infty\}|;$$

(d) For all nonnegative adapted processes $\gamma = (\gamma_n)_{n\ge 0}$, for all $\lambda \ge \|\gamma_0\|_\infty$, there exists a stopping time τ_λ such that

$$\{M\gamma > \lambda\} \subset \{\tau_\lambda < \infty\}, \quad |\{\tau_\lambda < \infty\}| \le d|\{M\gamma > \lambda\}|,$$

$$\sup_{n\le\tau_\lambda} \gamma_n = M_{\tau_\lambda}\gamma \le \lambda;$$

(e) For all $f \in L^1_+$, for all $\lambda \ge \|f_0\|_\infty, f_0 = E(f|\mathcal{F}_0)$, we have

$$\int_{\{Mf>\lambda\}} f d\mu \le d\lambda|\{M(f) > \lambda\}|;$$

(f) For all nonnegative adapted processes $\gamma = (\gamma_n)_{n\ge 0}$, for all $\lambda \ge \|\gamma_0\|_\infty$, there exist stopping time τ_λ such that besides assertions listed in (d), we have also

$$\lambda_1 \le \lambda_2 \quad \text{implies} \quad \tau_{\lambda_1} \le \tau_{\lambda_2}.$$

The part (d) $\Longrightarrow$ (e) of the theorem is due to Gundy[14], (e) $\Longrightarrow$ (a) is due to B-H-L [2]. All others are due to Long[17,18].

3. Martingale spaces and inequalities

In this topic, we would like only to list an inequality between sharp function operator and maximal function operator, and describe the rearrangement technique. Let $1 \le a \le \infty, f = (f_n)_{n\ge 0}$ be a martingale in L^1_u. Define

$$M_a(f) = \sup_n E(|f|^a|\mathcal{F}_n)^{\frac{1}{a}}, \tag{3.1}$$

$$f^\#_a = \sup_n E(|f - f_{n-1}|^a|\mathcal{F}_n)^{\frac{1}{a}}, \tag{3.2}$$

$$f^\#_{a,\theta} = \sup_n E(|f - \theta_{n-1}|^a|\mathcal{F}_n)^{\frac{1}{a}}, \tag{3.3}$$

where $\theta = (\theta_n)_{n\ge 0}$ is an adapted process, depending on $f = (f_n)_{n\ge 0}$. Let $\Phi(u)$ be a function from $\mathbb{R}^+$ to $\mathbb{R}^+$. $\Phi(u)$ is called "general ", if $\Phi(0) = 0$, and it is increasing, continuous, and of moderate growth in the sense $\Phi(2u) \le C\Phi(u)$, for all $u > 0$. Then we have following result obtained in Long[19].

Theorem 3.1. Let $\Phi(u)$ be general. Then we have

$$E(\Phi(M_a(f)) \leq C_{a,\Phi} E(\Phi(f_a^{\#})), \quad \forall f, \tag{3.4}$$

and more general, we have

$$E(\Phi(M_a(f))) \leq C_{a,\Phi} E(\Phi(f_{a,\theta}^{\#})), \quad \forall f. \tag{3.5}$$

Now we turn to the rearrangement technique in martingale setting. As well known, the concepts of distribution function and of nonincreasing rearrangement functions are very useful in estimating functions themselves. But habitually, people profer to use distribution functions, rather than rearrangement functions. For example, the famous good λ-inequality technique is described in terms of distribution functions. Is it possible to develope a parallel technique in terms of rearrangement function? The answer is positive. C. Herz and Bagby-Kurtz[1] showed this in the classical case first, then Long[20] showed that it was so in the martingale setting. A number of rearrangement function inequalities, of which an example is

$$M_a(f)^*(t) \leq 6 f_a^{\#}(\frac{t}{4}) + M_a(f)^*(\frac{5}{4}t), \quad \forall t > 0, \tag{3.6}$$

where $*$ denotes the nonincreasing rearrangement function operator, were established in Long[20]. As a consequence, there were results like folloiwng in Long[20].

Theorem 3.2. Let $\Phi(u)$ be a convex function from $\mathbb{R}^+$ to $\mathbb{R}^+$ with $p_\Phi = \sup_{u>0} \frac{u\Phi'(u)}{\Phi(u)} < \infty$. Then we have

$$\|M_a(f)\|_\Phi \leq C p_\Phi \|f_a^{\#}\|_\Phi, \quad \forall f, \tag{3.7}$$

$$\|M(f) \vee S(f)\|_\Phi \approx \|M(f) \wedge S(f)\|_\Phi, \quad \forall f, \tag{3.8}$$

with the equivalence constant less that Cp_Φ^2, where $\vee, \wedge$ mean max, min respectively, and $\|\cdot\|_\Phi$ is the norm on the Orlicz space L^Φ.

Remark. The constant like Cp_Φ in (3.7) can not be obtained by making use of good λ-inequality technique.

4. BMO and martingale transforms

BMO is a very important space not only in the classical case, but also in the martingale setting. Here are some related topics we were interested in.

4.1. Decomposition of BMO and factorization of A_p weights

A real martingale $f = (f_n)_{n\geq 0}$ of L_u^1 is called in BLO, if

$$|f_n - f_{n-1}| \leq C, \quad f_n \leq f + C, \quad \forall n \quad (f_n = E(f|\mathcal{F}_n)), \tag{4.1.1}$$

$$\|f\|_{BLO} = \inf\{C : \quad \text{in } (4.1.1.)\}, \tag{4.1.2}$$

A strictly positive process $z = (z_n)_{0\leq n\leq\infty}$ (i.e. $z_n > 0$, a.e. for all n) is called a weight. Let $1 \leq p \leq \infty$. A weight $z = (z_n)$ is called in A_p, if

$$z_n E(z_\infty^{-\frac{1}{p-1}}|\mathcal{F}_n)^{p-1} \leq C, \quad \text{a.e.} \quad \forall n, \quad 1 < p < \infty, \tag{4.1.3}$$

$$z_n \leq C z_\infty, \quad \text{a.e.} \quad \forall n, \quad p = 1, \tag{4.1.4}$$

$$z \in A_q, \quad \text{for some} \quad q, 1 \leq q < \infty, p = \infty. \tag{4.1.5}$$

Obviously, BLO conditions are stronger than BMO's, and the A_1 condition is stronger than A_p' s . In the classical case, Coifman-Rochberg[9] showed that each BMO funcfion can be decomposed as a sum of BLO functions, and Jones[15] showed that each A_p weight can be factoriged in terms of A_1 weights. Such decomposition and factorization are very useful. Long-Peng[23] showed the martingale versions of these results.

Theorem 4.1.1. Let $f = (f_n)_{n\geq 0}$ be a real martingale in BMO, such that the condition $\log A_{\alpha,\beta}$ holds, i.e.

$$\sup_n \|E(e^{\alpha f}|\mathcal{F}_n)^{\frac{1}{\alpha}} E(e^{-\beta f}|\mathcal{F}_n)^{\frac{1}{\beta}}\|_\infty \leq K_{\alpha,\beta} < \infty, \alpha, \beta > 0. \tag{4.1.6}$$

The for any $\varepsilon > 0$, there exists f's decomposition $f = g - h + \varphi$ with $\varphi \in L^\infty$, and $g, h \in BLO$, such that for all $\tau > 0$, g satisfies $\log A_{\alpha-\varepsilon,\tau}$, and h satisfies $\log A_{\beta-\varepsilon,\tau}$, and

$$\|g\|_{BLO} + \|h\|_{BLO} + \|\varphi\|_\infty \leq C\|f\|_{BMO}. \tag{4.1.7}$$

By making use of this decomposition, Long-Peng[23] obtained A_p's factorization as follows.

Theorem 4.1.2. Let $1 \leq p \leq \infty$. Then the special weight $z = (z_n) \in A_p \cap S$, if and only if there exist special weights $z_1, z_2 \in A_1 \cap S$ such that $z = z_1 z_2^{1-p}$, here "special' weight" means $z_n = E(z_\infty|\mathcal{F}_n)(z_\infty > 0$, a.e.) for all n, and S is the weight class

$$S = \{\text{weight} z = (z_n), C z_{n-1} \leq z_n \leq C z_{n-1}, \quad \text{a.e.} \quad \forall n \geq 1\}. \tag{4.1.8}$$

4.2. BMO and Carleson measures

Let ν be a nonnegative measure on $\Omega \times \mathbb{Z}^+$. ν is said to be a Carleson measure if,

$$|||\nu||| = \sup |\{\tau < \infty\}|^{-1} |\{(\omega, k) : k \geq \tau(\omega), \tau(\omega) < \infty\}|_\nu < \infty, \tag{4.2.1}$$

where τ runs throught all stopping times. Long[22] established

Theorem 4.2.1. Let $g = (g_n)_{n\geq 0}$ be a martingale, and $d\nu = |\Delta_k g|^2 \delta_k d\mu$, where δ_k is the Dirac measure centered at k. Then ν is a Carleson measure if and only if $g \in BMO$. And in any case, we have $|||\nu||| = \|g\|^2_{BMO}$.

Theorem 4.2.2. Let $d\nu = \nu_k \delta_k d\mu (\nu_k$ nonnegative random variables) be a Carleson measure, and $0 < p < \infty$. Then for all adapted processes $f = (f_n)_{n\geq 0}$, we have

$$E\left(\sum_{k=0}^{\infty} \nu_k |f_k|^p\right) \leq |||\nu||| E(M(f)^p). \tag{4.2.2}$$

On the contrary, if (4.2.2) holds for some p with $|||\nu|||$ replaced by a constant C_ν, then ν is a Carleson measure and $|||\nu||| \leq C_\nu$.

4.3. Martingale transforms

Let $f = (f_n)_{n\geq 0}$ be a martingale, $v = (v_n)_{n\geq 0}$ be an adapted process, then

$$f \to g = (g_n)_{n\geq 0}, \quad g_n = \sum_1^n v_{k-1} \Delta_k f, \quad n \geq 1, g_0 = 0, \tag{4.3.1}$$

is called a martingale transform (MT). MT were introduced by Burkholder[3]. Garcia[1 discovered that martingale transforms can be used to change h_{p_1} to h_{p_2} for given indices p_1, p_2 , but just only for some couple (p_1, p_2). Chao-Long[7] established such kind of results for general couples of (p_1, p_2).

Theorem 4.3.1.

(a) Let $0 < p_1, p_2 < \infty, f = (f_n)_{n\geq 0} \in h_{p_1}, f_0 = 0$, be given. Then there exist $g = (g_n)_{n\geq 0} \in h_{p_2}$ satisfying

$$C_{p_1,p_2} \|f\|^{p_1}_{h_{p_1}} \leq \|g\|^{p_1}_{h_{p_2}} \leq C_{p_1,p_2} \|f\|^{p_1}_{h_{p_1}}, \tag{4.3.2}$$

and an abapted nonnegative, increasing process $v = (v_n)_{n\geq 0}$ satisfying

$$E(M(v)^\alpha)(= E(v^\alpha_\infty)) < \infty, \quad \alpha = \frac{p_1 p_2}{p_2 - p_1}, \tag{4.3.3}$$

such that f is g's martingale transform. Conversely, in the case $p_1 \leq p_2$, each martingale transform $f = (f_n)_{n\geq 0}, f_n = \sum_1^n v_{k-1} \Delta_k g, f_0 = 0$, is in h_{p_1}, provided

$g \in h_{p_2}$ and (4.3.3) holds. Furthermore, in this case, we have

$$\|f\|_{h_{p_1}} \leq \|M(v)\|_\alpha \|g\|_{h_{p_2}}. \tag{4.3.4}$$

(b) When $p_2 = \infty$, the preceding results remain to be true, with h_{p_2} replaced by

$$bmo_2 = \{\text{martingale} \quad f = (f_n)_{n\geq 0} : E(|f - f_n|^2 | \mathcal{F}_n) \leq C, \quad \text{a.e,} \quad \forall n\}. \tag{4.3.5}$$

When the multipliers $v = (v_n)_{n\geq 0}$ in martingale transforms are in V_∞,

$$V_p = \{ \quad \text{adapted process} \quad v = (v_n)_{n\geq 0} : \|v\|_{v_p} = \|M(v)\|_p < \infty\}, 0 < p \leq \infty, \tag{4.3.6}$$

Burkholder studied the boundedness of martingale transforms in a lot of papers, even he obtained the best possible constants implied in some kind of inequalities concerning martingale transforms, see Burkholder[5]. Chao-Long[6] studied the boundedness of martingale transforms in general case $v \in V_p, 0 < p \leq \infty$, by making use of some extrapolation techniques. Following is a typical extrapolation lemma they used.

Lemma 4.3.2. Let $0 < p_0 \leq r_0 \leq \infty$, T be a linear operator defined on V_∞ and valued in the space of all martingales. Suppose that T is of weak type $(V_{p_0}, H^*_{r_0})$ (means $|\{M(Tv) > \lambda\}| \leq C\lambda^{-r_0}\|v\|^{r_0}_{V_{p_0}}$, for all $\lambda > 0$), with the bound $\|T\|$, and T is commutable with stopping times in the following sense

$$M(T(v - v^{(\tau-1)}))\chi_{\{\tau=\infty\}} = 0, \quad \text{a.e.} \quad \forall\tau, \forall v \in V_\infty, \tag{4.3.7}$$

then for all (p, r) satisfying

$$\frac{1}{p} - \frac{1}{r} = \frac{1}{p_0} - \frac{1}{r_0}, \quad 0 < p \leq p_0, \tag{4.3.8}$$

T is of type (V_p, H^*_r) with the bound $\leq C\|T\|$. And the H^S_r (or h_r) version of the lemma is true too.

By virtue of this kind of extrapolations, the boundedness of martingale transofmrs as operators acting on $\mathsf{V}_p \times \Lambda_\alpha, (v, f) \to g$, where $\Lambda_\alpha(0 \leq \alpha \leq 1)$ are Lip_α spaces of martingales (notice $\Lambda_0 = BMO$) can be answered satisfactorily. For example, we have

Theorem 4.3.3. The operator $(v, f) \to g$ defined by $g_n = \sum_1^n v_{k-1} \Delta_k f$, is of type $(\mathsf{V}_p, \Lambda_\alpha; \Lambda_\beta)$ when $0 < p \leq \infty, 0 \leq \alpha \leq 1, \beta = \alpha - \frac{1}{p} \geq 0$; and of type $(\mathsf{V}_p, \Lambda_\alpha; H^*_r)$ when $0 < p < \frac{1}{\alpha}, 0 \leq \alpha \leq 1, \frac{1}{r} = \frac{1}{p} - \alpha$.

5. Applications of Martingale Theory

5.1. Simplified proof of $T(b)$ theorem

In order to develope a real method to deal with the Analysis on $\mathbb{R}^d$, Calderón-Zygmund studied following operators systematcally in the early 50's

$$Tf(x) = p.v. \int_{\mathbb{R}^d} K(x-y)f(y)dy, \tag{5.1.1}$$

where $K(x)$ is a function defined on $\mathbb{R}^d - \{0\}$, satisfying the size and smoothness conditions as follows: for $r \in (0,1]$,

$$|K(x)| \leq C|x|^{-d}, \quad \forall x \neq 0, \tag{5.1.2}$$

$$|K(x) - K(x')| \leq C|x-x'|^r |x|^{-d-r}, \quad \text{when} \quad |x| \geq 2|x-x'|. \tag{5.1.3}$$

Such T is called a singular integral operator (SIO), and $K(x)$ a Calderón-Zygmund kernel. When some other natural conditions are imposed on the kernel $K(x)$, for example

$$\int_{\alpha<|x|<\beta} K(x)dx = 0, \quad \forall \beta > \alpha > 0, \tag{5.1.4}$$

then by making use of Plancherel theorem, we see that T is L^2-bounded, since

$$\|\widehat{K}(\xi)\|_\infty \leq C, \quad \text{with} \quad \widehat{K}(\xi) = \lim_{\varepsilon \to 0} \int_{\varepsilon \leq |x| \leq \frac{1}{\varepsilon}} K(x) e^{-ix.\xi} dx.$$

Calderón-Zygmund's real method tells us that T is also L^p-bounded for $1 < p < \infty$, and weakly L^1-bounded. In the 80's people were interested in the third generation of singular integral operators, which are not of convolution type, but defined by

$$Tf(x) = p.v. \int_{\mathbb{R}^d} K(x,y)f(y)dy, \tag{5.1.5}$$

where $K(x,y)$ is defined on $\mathbb{R}^d \times \mathbb{R}^d - \{x=y\}$, and satisfying a similar size condition, a similar smoothness condition. A typical example of third generation of SIO is the famous Cauchy integral operator

$$H_p f(x) = p.v. \int_{-\infty}^{\infty} \frac{f(y)}{z(x)-z(y)} dy, \tag{5.1.6}$$

where Γ is a Lipschitz curve in $\mathbb{C}$, and $z(x)$ is its arc-length parameterization. For this kind of SIO, the L^2-boundedness is a very difficult problem. (Even the L^2-boundess of H_Γ, it is a famous conjecture for over a couple of decades). Once the L^2-boundedness is obtained, the Calderón-Zygmund's real method gives other boundedness. Can a general criterion for the L^2-boundedness of such SIO be given? David-Journé [11] gave the pretty $T(1)$ theorem, it says that T is L^2-bounded, if and only if $T(1), T^t(1) \in BMO$, and T has WBP (weak boundedness property). By $T(1)$ theorem, for many SIO T, T's L^2-boundedness is reduced to the simple integration by part. But $T(1)$ theorem is not effective for H_Γ, since 1 is not a good testing function of H_Γ. In order to overcome the shortcoming, David-Journé-Semmes[12] generalized $T(1)$ theorem to $T(b)$ theorem, of which the main ingredient is

Special $T(b)$ theorem. Let $b(x)$ be a pseudoaccretive function in the sense

$$b \in L^\infty, \quad \text{and} \quad |\frac{1}{|I|}\int_I b(x)dx| \geq C > 0, \quad \forall I \in \vartheta, \tag{5.1.7}$$

where I is the set of all quasi-dyadic cubes constructed a little late, such that $T(b) = 0 = T^t(b)$, and T has WBP in the sense

$$| < \chi_I, T(\chi_I) > | \leq C|I|, \forall I \in \vartheta, \tag{5.1.8}$$

where $<,>$ denotes the inner product in L^2, and χ denotes the indicator function. Then T is L^2-bounded.

Now, there are many simplified proof of $T(b)$ theorem. The one introduced by Coifman-Fones-Semmes[8] in the case $d = 1$ is very attractive. According to it, one need only find a nice frame of L^2, whihc is adapted to given $b(x)$, and is convenient for the controll of T's matrix under this frame. By a martingale approach, Long[21] constructed a nice frame of $L^2(\mathbb{R}^d)$, which is very simple, and very convenient for the proof of $T(b)$ theorem. Here is the construction in Long[21].

Construction of the nice frame $\{\alpha_I, \beta_I\}_{I\in\vartheta}$ of $L^2(\mathbb{R}^d)$ adapted to $b(x)$.

Denote $\vartheta_0 = \{$ all dyadic cubes of length $1\}$. Divide each $I \in \vartheta_0$ into two equal parts by hyperplanes perpendicular to the x_1-axis, and let $\vartheta_1 = \{I :$ so produced $\}$. Then continue this way along with the axes $x_2, x_3 \cdots x_d, x_1, \cdots$, and get $\vartheta_2, \vartheta_3 \cdots \vartheta_d, \vartheta_{d+1}, \cdots$ respectively. How about ϑ_k for $k < 0$? They come from the procedure reverse to previous one. Let $\vartheta = \cup\vartheta_k$. Notince that for each $I \in \vartheta_k$, we have $I = I_1 \cup I_2$ with $I_1, I_2 \in \vartheta_{k+1}$ for all $k \in \mathbb{Z}$. this is the main feature of the construction. Now for all k, for all $I \in \vartheta_k$, define

$$\alpha_I = |I|^{-\frac{1}{2}}|I|_b^{-1}(|I_2|_b\chi_{I_1} - |I_1|_b\chi_{I_2}), \tag{5.1.9}$$

$$\beta_I = |I|^{\frac{1}{2}}(|I_1|_b^{-1}\chi_{I_1} - |I_2|_b^{-1}\chi_{I_2}), \tag{5.1.10}$$

where $|I|_b = \int_I b dx$. Then $\{\alpha_I, \beta_I\}_{I\in\vartheta}$ satisfies

$$\int b\alpha_I dx = 0 = \int \beta_I b dx, \quad \forall I \in \vartheta \tag{5.1.11}$$

$$\int |\alpha_I|^2 dx \approx 1 \approx \int |\beta_I|^2 dx, \quad \forall I \in \vartheta, \tag{5.1.12}$$

$$\int \beta_J b \alpha_I dx = \delta_{J,I}, \quad \forall I, J \in \vartheta, \tag{5.1.13}$$

$$f = \sum_I \alpha_I < \beta_I, f >_b, \quad \forall f \in L^2, \tag{5.1.14}$$

$$\|f\|_2 \approx (\sum_I | < \beta_I, f >_b |^2)^{\frac{1}{2}}, \quad \forall f \in L^2, \tag{5.1.15}$$

where the series in (5.1.14) is convergent in L^2. With this nice frame, T's L^2-boundedness is reduced to the l^2-boundedness of T's associated matrix operator $\{< \beta_I, f >_b\}_I \to \{\sum_I < \beta_J, T(b\alpha_I) >_b < \beta_I, f >_b\}_J$. And the later is easy to be checked.

5.2. Applications to the geometry in Banach spaces

In 70's-80's, Banach space-valued martingales have been used by many mathematicians, such as J. Bourgain; D.L. Burkholder; Rubio de Francia, J. L.; G. Pisier; W. A. Woyczynski, etc. to study the geometry in Banach spaces, for example, the smoothness, convexity and UMD property (unconditional martingale difference sequence property) of Banach spaces, see Burkholder's expository paper [4]. In this topic, P.D. Liu made a systematic imestigation.

Let X be a Banach space, the comexity modulus and smootheness modulus of X are defined respecively by

$$\delta_X(\varepsilon) = \inf\{1 - \frac{1}{2}\|x + y\|, \|x\| = \|y\| = 1, \|x - y\| = \varepsilon\}, \quad \varepsilon > 0, \tag{5.2.1}$$

$$\rho_X(\tau) = \sup\{\frac{1}{2}(\|x + y\| + \|x - y\|) - 1, \|x\| = 1, \|y\| = \tau\}, \tau > 0, \tag{5.2.2}$$

X is called q-convexifiable ($2 \le q < \infty$), or p-smoothable ($1 < p \le 2$), if it admits an equivalent norm, such that

$$\delta_X(\varepsilon) \ge c\varepsilon^q, \quad \text{or} \quad \rho_X(\tau) \le C\tau^p, \tag{5.2.3}$$

under this norm. Let $(\Omega, \mathcal{F}, \mu, \{\mathcal{F}_n\}_{n\geq 0})$ be as above, X-valued martingales can be defined similarly. Besides those operators in §1 which can be defined similarly in this case obviously, we define also

$$S_n^{(p)}(f) = (\sum_0^n | \Delta_k f|^p)^{\frac{1}{p}}, S^{(p)}(f) = S_\infty^p(f), \quad 1 \leq p < \infty, \tag{5.2.4}$$

$$\sigma_n^{(p)}(f) = (\sum_0^n E(| \Delta_k f|^p|\mathcal{F}_{k-1}))^{\frac{1}{p}}, \quad \sigma^{(p)}(f) = \sigma_\infty^{(p)}(f), 1 \leq p < \infty. \tag{5.2.5}$$

One of Liu's main results in this topic is following

Theorem 5.2.1. Let X be a Banach space, $2 \leq q < \infty$. Then following assertions are equivalent

(a). X is q-convexifiable,

(b). $S^{(q)}(f) < \infty$, a.e. for every martingale $f = (f_n)_{n\geq 0}$ with $\|f\|_q < \infty$,

(c). $S^{(q)}(f) < \infty$, a.e. for every dyadic martingale $f = (f_n)_{n\geq 0}$ with $\|f\|_q < \infty$.

This result gives a simple criterions for the q-convexity of Banach spaces. Furthermore, Liu gave a lot of criterions for the q-convexity, p-smoothenss and the UMD property of Banach spaces by various fundamental martingale inequalities, of which some typical examples are

Theorem 5.2.2. Let X be a Banach space, $1 < p \leq 2 \leq q < \infty$, and $\Phi(u)$ be any moderate convex function. Then (a) $\Longleftrightarrow$ (b), (c) $\Longleftrightarrow$ (d).

(a). X is q-comexifiable,

(b). $\|S^{(q)}(f)\|_\Phi \leq C\|M(f)\|_\Phi, \forall f$,

(c). X is p-smoothable,

(d). $\|M(f)\|_\Phi \leq C\|S^{(p)}(f)\|_\Phi, \forall f$.

The definition of martingale transforms in X-valued case is similar. If $g = (g_n)_{n\geq 0}$ is f's martingale transform with the multiplier $v = (v_n)_{n\geq 0}$ being real valued and in V_∞, we write $(f, g) \in \mathcal{M}$. Liu's one result concerning martingale transforms and UMD is

Theorem 5.2.3. Let X be a Banach space, then followings are equivalent

(a). $X \in UMD$,

(b). g_n converges in measure for $(f, g) \in \mathcal{M}$, with $\|f\|_1 < \infty$,

(c). $n^{-1}g_n \to 0$, a.e. for $(f, g) \in \mathcal{M}$ with $\sup_n \|\sum_0^n k^{-1} \Delta_k f\|_1 < \infty$,

(d). $n^{-1}g_n \to 0$ in measure for (f, g) as in (c).

The results in §5.2 are refered to Liu's expository paper [16]. Other results mentioned above can be also refered to Long's book [21].

References

[1] Bagby, R. J., Kurtz, D. S. A rearranged good λ-inequality, Trans. A. M. S. 293 (1986), 71–81.

[2] Bru, B., Heinich, H., Lootgieter, J. C., Sur la régularité des filtrations, C. R. Acad. Sc. Paris, 294 (1982), 313–316.

[3] Burkholder, D. L. Martingale transforms, Ann. Math. Sta. 37 (1966), 1494–1504.

[4] — Martingales and Fourier analysis in Banach spaces, Lect. Notes in Math. 1206 (1986), 61–108.

[5] —, Differential subordination of harmonic functions and martingales, Proc. of the Seminar on Harmonic Analysis and PDE (Spain, 1987), Lect. Notes in Math. 1384 (1988), 1–23.

[6] Chao, J. A., Long, R. Martingale transforms with unbounded multipliers, Proc. Amer. M. S. 114 (1992), 831–838.

[7] — Martingale transforms and Hardy spaces, Prob. Theory and Related Fields 91 (1992), 399–404.

[8] Coifman, R. R., Jones, P., Semmes, S., Two elementary proofs of the L^2-boundedness of Cauchy integrals on Lipschitz curves, J. Amer. M. S. 2 (1989), 553–564.

[9] Coifman, R. R., Rochberg, R., Another characterization of BMO, Proc. Amer. M. S. 79 (1980), 249–254.

[10] David, G., Wavelets, Calderón-Zygmund operators and singular integral operators on curves and surfaces, Lect. Notes in Math. 1465(1991).

[11] David, G., Journé, J. L., A boundedness criterion for generalized Calderón-Zygmund operators, Ann. of Math., 120 (1984), 371–397.

[12] David, G., Journé, J. L., Semmes, S., Operateur de Calderón-Zygmund functions paraaccrétifes et interpolation, Rev. Math. Ibe. 1 (1985), 1–55.

[13] Garcia, A., Martingale Inequalities, Sem. Notes on Recent Progrss, Benjamin (1973).

[14] Gundy, R. F., A decomposition for L^1-bounded martingales, Ann. of Math., Statis., 39 (1968), 134–138.

[15] Jones, P. W., Factorization of A_p weight, Ann. of Math. 111 (1980), 511–530.

[16] Liu, P. D., Some new results on martingale inequalities and geometry in Banach spaces, Acta Math. Scientia 12 (1992), 22–32.

[17] Long, R. (=Long, R.L.=Long, J. L.) Martingale régulière et Φ-inégalités avec poids enter $f, S(f), \sigma(f)$, C. R. Acad. Sc. Paris 291 (1980), 31–34.

[18] —. Sur l'espace H_p de martingales régulières $(0 < p \leq 1)$, Ann. Inst. H. Poincare (B), XVII (1981), 123–142.

[19] —. Two classes of martingale spaces, Scientia Sinica A, 26 (1983), 362–375.

[20] —. Rearrangement techniques in martingale setting, Illinois J. of Math., 35 (1991), 506–521.

[21] —. Martingule proof of Clifford valued $T(b)$ theorem on $\mathbb{R}^d$, to appear in Bull. des Sciences Mathématiques.

[22] —. Martingale spaces and inequalities, Peking U. Press and Vieweg Publishing (1993).

[23] Long, R., Peng, L. Z., Decomposition of BMO functions and factorization of A_p weights in martingale setting, Chin Ann. of Math. 4 B(1) (1983), 117–128.

UNIFORM WEAK (1,1) BOUNDS FOR OSCILLATORY SINGULAR INTEGRALS

YIBIAO PAN *
Department of Mathematics and Statistics
University of Pittsburgh
Pittsburgh, PA 15260

1. Introduction. Oscillatory singular integrals arise in many problems in harmonic analysis. Their boundedness properties have been studied quite extensively (see, for example, Stein [13]). In this paper we shall establish a uniform $L^1 \to L^{1,\infty}$ estimate for certain oscillatory singular integral operators with smooth phase functions of finite type. Such estimate has been known previously for operators with real-analytic phases only.

Let $x, y \in \mathbf{R^n}$, $\Phi(x, y)$ be a real-valued smooth function, $K(x, y)$ be a Calderón-Zygmund kernel, $\varphi(x, y) \in C_0^\infty(\mathbf{R^n} \times \mathbf{R^n})$. For $\lambda \in \mathbf{R}$, define T_λ by

$$T_\lambda f(x) = \text{p.v.} \int_{\mathbf{R^n}} e^{i\lambda\Phi(x,y)} K(x, y)\varphi(x, y)f(y)dy. \tag{1}$$

Phong and Stein studied such operators in connection with singular Radon transforms and obtained boundedness results on L^p and Hardy spaces when $\Phi(x, y)$ is a real bilinear form ([8]). Oscillatory integral operators with polynomial phases were studied by Ricci and Stein ([9]), Chanillo and Christ ([1]), among others. To introduce their results we let $P(x, y)$ be a real polynomial and define T by

$$Tf(x) = \text{p.v.} \int_{\mathbf{R^n}} e^{iP(x,y)} K(x, y)f(y)dy, \tag{2}$$

where $K(x, y)$ is a Calderón-Zygmund kernel. The following L^p and weak (1,1) boundedness results are in [9] and [1].

Theorem 1 (Ricci-Stein) *Suppose* $1 < p < \infty$. *Then the operator* T *given in (2) can be extended to be a bounded operator on* $L^p(\mathbf{R^n})$ *to itself. The bound of* $\|T\|$ *can be taken to depend only on the degree of* $P(x, y)$ *and is otherwise independent of the coefficients of* P.

*Supported in part by a grant from the National Science Foundation.

M. Cheng et al. (eds.), Harmonic Analysis in China, 210–219.

Theorem 2 (Chanillo-Christ) *The operator T given in (2) is of weak type (1,1) with a bound depending only on the degree of P.*

Theorem 1 and 2 imply that the operators T_λ's given in (1) are uniformly bounded from L^p to L^p $(1 < p < \infty)$ and from L^1 to $L^{1,\infty}$ if $\Phi(x,y)$ is a polynomial. The focus of our investigation is to determine whether such uniform boundedness holds if Φ is an arbitrary real-valued smooth function.

Earlier we established the uniform $L^p \to L^p$ boundedness for oscillatory singular integral operators with phase functions of finite type. Let $w \in \mathbf{R^n} \times \mathbf{R^n}$. The function $\Phi(x,y)$ is said to be of finite type at w if there are $j, k \in \{1, 2, \ldots, n\}$ such that $\partial^2\Phi/\partial x_j \partial y_k$ does not vanish to infinite order at w. Let $\Delta = \{(x,y) \in \mathbf{R^n} \times \mathbf{R^n} \mid x = y\}$. The following theorem is in [5].

Theorem 3 *Suppose that $\Phi(x,y)$ is of finite type at each point in $\Delta \cap supp(\varphi)$, $1 < p < \infty$. Then there exists a constant $C = C(p)$ which is independent of λ such that*

$$\|T_\lambda f\|_p \leq C\|f\|_p$$

for $f \in L^p(\mathbf{R^n})$.

By a result of Nagel and Wainger ([4]) the uniform L^p boundedness of T_λ may fail if Φ is not of finite type. It is not hard to see that if $\Phi(x,y)$ is a polynomial or real-analytic function and Φ is not of the trivial form $U(x) + V(y)$, then Φ is of finite type.

The question that remains unanswered is whether T_λ's are uniformly bounded from L^1 to $L^{1,\infty}$ if Φ is of finite type. In this paper we provide an answer to this question in the case where the phase function Φ is translation invariant (i.e. $\Phi(x,y) = \phi(x - y)$ for some ϕ). We state our result in the next theorem.

2. Statement of result.

Theorem 4 *Suppose $\Phi(x,y) = \phi(x - y)$, where $\phi \in C^\infty(\mathbf{R^n})$. If there exists an $\alpha \in \mathbf{N^n}$ with $|\alpha| > 1$ such that $\partial^\alpha \phi(0)/\partial x^\alpha \neq 0$, then the operators T_λ's are uniformly bounded from L^1 to $L^{1,\infty}$, i.e. there is a constant $C > 0$ which is independent of λ such that*

$$|\{|T_\lambda f| > \eta\}| \leq C\eta^{-1}\|f\|_1, \tag{3}$$

for $\eta > 0$, $f \in L^1(\mathbf{R^n})$.

Remarks:

(1) The result in Theorem 4 was proved in [6] under the stronger assumption that ϕ is real-analytic. For dimension greater than one, the proof given there uses the real-analyticity in an essential way and one of the key estimates relies on Hardt's slicing and intersection theory of real-analytic varieties. A different approach will be taken here.

(2) When the dimension is one, it is known that T_λ are uniformly bounded from L^1 to $L^{1,\infty}$ if the phase function $\Phi(x,y)$ is real-analytic (Φ is not necessarily translation invariant). See [6].

(3) The condition in Theorem 4 can be replaced by the following weaker condition on $\Phi(x,y)$: for every $w \in \Delta \cap supp(\varphi)$, there exist $\alpha, \beta \in \mathbf{N}^n$ with $|\alpha| \geq 1$, $|\beta| = 1$, such that $\partial_x^\alpha \partial_y^\beta \Phi(w) \neq 0$.

3. Some preliminary estimates. In order to obtain the uniform weak (1,1) estimate for T_λ, we let $f \in L^1$ and make a Calderón-Zygmund decomposition $f = g + \sum b_j$. To estimate $T_\lambda(g)$ one uses the L^2 theory which has been established in Theorem 3. But unlike the usual theory of Calderón-Zygmund singular integrals, one can no longer control the L^1 norm of $T_\lambda(\sum b_j)$. The idea is to use instead a $L^1 \to L^2$ argument which originated in Fefferman [3] and was later used by Chanillo-Kurtz-Sampson [2] and Chanillo-Christ [1]. Our argument follows closely that of Chanillo-Christ. An important ingredient of our proof is a variant of the classical van der Corput lemma (Lemma 2) which may prove useful in other problems.

First let us recall van der Corput's lemma ([11]).

Lemma 1 *Suppose ϕ and ψ are real-valued and smooth in $[a,b]$. If $|\phi(x)| \geq 1$ for $x \in [a,b]$, then*

$$|\int_a^b e^{i\lambda\phi(x)}\psi(x)dx| \leq c_k\lambda^{-1/k}(|\psi(b)| + \int_a^b |\psi'(x)|dx)$$

holds when

(1) $k \geq 2$, or

(2) $k = 1$ if, in addition, it is assumed that $\phi'(x)$ is monotonic.

We now employ a method developed by Sogge and Stein in [10] to establish an estimate for oscillatory integrals. Let $H(x)$ be a fixed C^∞ function such that $H(x) \equiv 0$ for $|x| \leq 1$ and $H(x) \equiv 1$ for $|x| \geq 2$. We have the following lemma:

Lemma 2 *Let $\Psi \in C^\infty(\mathbf{R}^n)$, $\varphi \in C_0^\infty(\mathbf{R}^n)$, $\psi(x) = \dfrac{\partial^\alpha \Psi(x)}{\partial x^\alpha}$, where $\alpha \in \mathbf{N}^n$ and $|\alpha| = k \geq 1$. Let d, D be positive constants such that $|\psi(x)| \leq d \leq D$ for all $x \in supp(\varphi)$. Let*

$V = \{x \in \mathbf{R^n} \mid dist(x, supp(\varphi)) \le d\}$, $A = \sup\{|\frac{\partial^\beta \Psi(x)}{\partial x^\beta}| \mid |\beta| = k+1, x \in V\}$. *Then there exists a constant* C *which depends only on* n, D, A *and* α *such that*

$$|\int_{\mathbf{R^n}} e^{i\lambda\Psi(x)}\varphi(x)H(\rho\psi(x))dx| \le C\lambda^{-1/k}\rho^{1/k}(\rho|V|\|\varphi\|_\infty + \|\nabla\varphi\|_1) \tag{4}$$

for $\lambda, \rho > 0$.

Proof: Let $E = \{\psi \ne 0\} \cap supp(\varphi)$. If $E = \phi$ then we have

$$\int_{\mathbf{R^n}} e^{i\lambda\Psi(x)}\varphi(x)H(\rho\psi(x))dx = 0.$$

We now assume $E \ne \phi$. Let $m = d(n,k)$ be the dimension of the linear space of homogeneous polynomials of degree k in $\mathbf{R^n}$. By the argument in [7] (p. 283), we can find m unit vectors $e_1, e_2, \ldots, e_m$ in $\mathbf{R^n}$, two positive constants $\sigma = \sigma(n,\alpha,A) \le 1$, $\mu = \mu(n,\alpha,A)$, and a sequence $\{x_j\}_{j=1}^\infty \subset E$ such that

(i) $E \subset \bigcup_j B(x_j, r_j)$, where $r_j = \sigma|\psi(x_j)|$;

(ii) $\{B(x_j, r_j/3)\}_{j=1}^\infty$ are disjoint;

(iii) $(1/2)|\psi(x_j)| \le |\psi(y)| \le (3/2)|\psi(x_j)|$ for $y \in B_j^* = B(x_j, 2r_j)$;

(iv) $\sum_j \chi_{B_j^*}(x) \le 36^n$ for $x \in \mathbf{R^n}$;

(v) For every j, there exists a $s_j \in \{1,\ldots,m\}$ such that $|(e_{s_j}\cdot\nabla)^k\Psi(x)| \ge \mu r_j$ for $x \in B_j$.

Let $\{p_j(x)\}_{j=1}^\infty$ be a partition of unity such that

(a) $\sum_j p_j(x) = 1$ for $x \in \bigcup_j B_j$;

(b) $supp(p_j) \subset B_j^*$;

(c) $|\frac{\partial^\gamma p_j(x)}{\partial x^\gamma}| \le C_\gamma r_j^{-|\gamma|}$ for $x \in \mathbf{R^n}$, $\gamma \in \mathbf{N^n}$.

Let j be fixed. If $r_j < \sigma/(2\rho)$, then we have

$$\int_{B_j} e^{i\lambda\Psi(x)}\varphi(x)H(\rho\psi(x))p_j(x)dx = 0. \tag{5}$$

Now assume that $r_j \geq \sigma/(2\rho)$. By using a rotation and a translation if necessary we may assume that $e_{s_j} = (1, 0, \dots, 0)$ and $B_j = B(0, r_j)$. Let $x = (x_1, x')$, where $x' \in \mathbf{R}^{n-1}$. Then by (c), (v) and van der Corput's lemma (or integration by parts if $k = 1$) we have

$$
\begin{aligned}
&|\int_{B_j} e^{i\lambda\Psi(x)}\varphi(x)H(\rho\psi(x))p_j(x)dx| \\
&\leq \; C(\lambda r_j)^{-1/k}\int_{|x'|\leq r_j}(\|\varphi\|_\infty + (M\rho r_j)\|\varphi\|_\infty + \int_{|x_1|\leq\sqrt{r_j^2-|x'|^2}}|\frac{\partial\varphi}{\partial x_1}(x)|dx_1)dx' \\
&\leq \; C\lambda^{-1/k}\rho^{1+1/k}\|\varphi\|_\infty|B(x_j, r_j/3)| + C\lambda^{-1/k}\rho^{1/k}\int_{B_j}|\nabla\varphi(x)|dx.
\end{aligned}
$$

By (a), (ii), and (iv) we have

$$
\begin{aligned}
|\int_{\mathbf{R}^n} e^{i\lambda\Psi(x)}\varphi(x)H(\rho\psi(x))dx| &\leq \sum_{r_j\geq\sigma/(2\rho)}|\int_{B_j} e^{i\lambda\Psi(x)}\varphi(x)H(\rho\psi(x))p_j(x)dx| \\
&\leq C\lambda^{-1/k}\rho^{1/k}[(\rho|V|)\|\varphi\|_\infty + \|\nabla\varphi\|_1]. \qquad (6)
\end{aligned}
$$

The proof of Lemma 2 is complete.

4. Proof of main theorem.

We shall need the following version of Chanillo-Christ theorem ([1]).

Proposition 1 *Let $K(x, y)$ be a fixed Calderón-Zygmund kernel. Then for any $m \in \mathbf{N}$ there exists $C > 0$ such that for all polynomial $P(x, y)$ of degree $\leq m$, all $f \in L^1$, $\eta > 0$ and $\delta > 0$,*

$$|\{x : \; |p.v. \int_{|x-y|<\delta} e^{iP(x,y)}K(x, y)f(y)dy| > \eta\}| \leq C\eta^{-1}\|f\|_1.$$

We are now ready to present the proof of Theorem 4. Let $x, y \in \mathbf{R}^n$, $K(x, y)$ be a Calderón-Zygmund kernel, i.e. K satisfies

(i) $|K(x, y)| \leq A|x - y|^{-n}$;

(ii) $|\nabla K(x, y)| \leq A|x - y|^{-n-1}$;

(iii) $f \to \int_{\mathbf{R}^n} K(x, y)f(y)dy$ extends to a bounded operator on $L^2(\mathbf{R}^n)$.

Let T_λ be given by (1) with $\Phi(x, y) = \phi(x - y)$. Suppose $\partial^\alpha\phi(0)/\partial x^\alpha \neq 0$ for some α with $|\alpha| > 1$. Without loss of generality we may assume that λ is positive and large, $\alpha = (\alpha_1, \alpha_2, \dots, \alpha_n)$ and $\alpha_1 \geq 1$. Let $\beta = (\alpha_1 - 1, \alpha_2, \dots, \alpha_n)$. Then $\beta \in \mathbf{N}^n$ and $|\beta| \geq 1$. To prove (3) we let $\eta > 0$, $f \in L^1(\mathbf{R}^n)$, and make a Calderón-Zygmund decomposition $f = g + \sum_j b_j$ such that

(iv) $\|g\|_\infty < \eta$;

(v) b_j is supported on a cube Q_j with sidelength $2^{k(j)}$;

(vi) The cubes $\{Q_j\}$ are pairwise disjoint except for sets of measure zero;

(vii) $\sum_j |Q_j| \le C\eta^{-1}\|f\|_1, \quad |Q_j|^{-1}\int |b_j| \le C\eta.$

Let $m = 2n(k-1)+k$, $\omega_j = \max\{2^{k(j)}, \lambda^{-1/m}\}$, $h(x)$ be a fixed C^∞ function such that $h(x) = 0$ for $|x| \le 1$ and $h(x) = 1$ for $|x| \ge 2$. Define S_λ^j by

$$S_\lambda^j f(x) = \int_{\mathbf{R^n}} K_\lambda^j(x,y) f(y) dy$$

where

$$K_\lambda^j(x,y) = e^{i\lambda\phi(x-y)} K(x,y)\varphi(x,y) h(\omega_j^{-1}(x-y)).$$

First we shall prove that

$$\|\sum_j S_\lambda^j b_j\|_{L^2(\mathbf{R^n})}^2 \le C\eta\|f\|_{L^1(\mathbf{R^n})}. \tag{7}$$

To this end we let

$$L_{\lambda,i,j}(x,y) = \int_{\mathbf{R^n}} \overline{K_\lambda^i(z,x)} K_\lambda^j(z,y) dz. \tag{8}$$

For fixed i, let $\Lambda_i = \{j \mid k(j) \le k(i)\}$. Then we have

$$\begin{aligned}
\|\sum_j S_\lambda^j b_j\|_{L^2(\mathbf{R^n})}^2 &= < \sum_j S_\lambda^j b_j, \sum_j S_\lambda^j b_j > \\
&\le 2\sum_i \sum_{j\in\Lambda_i} | < (S_\lambda^i)^* S_\lambda^j b_j, b_i > | \\
&= 2\sum_i \sum_{j\in\Lambda_i} | \int_{\mathbf{R^n}}\int_{\mathbf{R^n}} L_{\lambda,i,j}(x,y) b_j(y) dy \overline{b_i(x)} dx| \qquad (9)
\end{aligned}$$

Let $g(x) = \partial^\beta\phi(x)/\partial x^\beta$, $\psi(x,y,z) = g(z-x) - g(z-y)$,

$$M_{\lambda,i,j}(x,y) = \int_{\mathbf{R^n}} \overline{K_\lambda^i(z,x)} K_\lambda^j(z,y)[1 - H(2\lambda^{1/m}\psi(x,y,z))] dz, \tag{10}$$

$$N_{\lambda,i,j}(x,y) = \int_{\mathbf{R^n}} \overline{K_\lambda^i(z,x)} K_\lambda^j(z,y) H(2\lambda^{1/m}\psi(x,y,z)) dz. \tag{11}$$

By the argument in [6], p. 799–801, we have

$$\sum_{j\in\Lambda_i} | \int_{\mathbf{R^n}} M_{\lambda,i,j}(x,y) b_j(y) dy| \le C\eta \tag{12}$$

for $x \in Q_i$. For fixed x, y, let

$$\varphi_{\lambda,i,j}(z) = \overline{K(z,x)}K(z,y)\varphi(z,x)\varphi(z,y)h(\omega_i^{-1}(z-x))h(\omega_j^{-1}(z-y)),$$

and $\Psi(z) = -(\phi(z-x) - \phi(z-y))$. By (i) and (ii) there are constants d, D, and A which are independent of x, y such that

(viii) $|\psi(x,y,z)| \le d \le D$, for $z \in supp(\varphi_{\lambda,i,j})$;

(ix) $|V| \le D$, where $V = \{z \in \mathbf{R^n} \mid dist(z, supp(\varphi_{\lambda,i,j}) \le d\}$;

(x) $|\partial^\gamma \Psi(z)/\partial z^\gamma| \le A$ for $z \in V$, $|\gamma| = k$;

(xi) $\|\varphi_{\lambda,i,j}\|_\infty \le A\lambda^{n/m}\omega_i^{-n}$;

(xii) $\|\nabla\varphi_{\lambda,i,j}\|_1 \le A\lambda^{n/m}\omega_i^{-n}$.

By (viii)–(xii) and Lemma 2 we obtain

$$|N_{\lambda,i,j}(x,y)| \le C\lambda^{-\frac{1}{k-1}+\frac{k}{m(k-1)}+\frac{n}{m}}\omega_i^{-n}. \tag{13}$$

Therefore by (vi) and (vii) we have

$$\begin{aligned}
\sum_{j\in\Lambda_i} \left|\int_{\mathbf{R^n}} N_{\lambda,i,j}(x,y)b_j(y)dy\right| &\le C\lambda^{-\frac{1}{k-1}+\frac{k}{m(k-1)}+\frac{n}{m}}\omega_i^{-n}\sum_{j\in\Lambda_i}\int_{Q_j}|b_j(y)|dy \\
&\le C\eta\lambda^{-\frac{1}{k-1}+\frac{k}{m(k-1)}+\frac{n}{m}}\omega_i^{-n}\sum_{j\in\Lambda_i}|Q_j| \\
&\le C\eta\lambda^{-\frac{1}{k-1}+\frac{k}{m(k-1)}+\frac{n}{m}}\omega_i^{-n}(1+\omega_i^n) \\
&\le C\eta. \qquad (14)
\end{aligned}$$

By (12) and (14) we have, for $x \in Q_i$,

$$\sum_{j\in\Lambda_i}\left|\int_{\mathbf{R^n}} L_{\lambda,i,j}(x,y)b_j(y)dy\right| \le C\eta. \tag{15}$$

By (9), (15) and (vii) we obtain

$$\begin{aligned}
\left\|\sum_j S_\lambda^j b_j\right\|^2_{L^2(\mathbf{R^n})} &\le C\eta\sum_j\int_{\mathbf{R^n}}|b_i(x)|dx \\
&\le C\eta^2\sum_i|Q_i| \\
&\le C\eta\|f\|_{L^1(\mathbf{R^n})},
\end{aligned}$$

which proves (7). Let Q_j^* be the 5-fold of Q_j, $\Omega = (\bigcup_j Q_j^*)^c$, and $\chi(\cdot)$ be the characteristic function of $B(0,2) \subset \mathbf{R}^n$. Define P_λ, Q_λ, and $R_{\lambda,m}$ by

$$\begin{aligned} P_\lambda f(x) &= \int_{\mathbf{R}^n} e^{i\lambda\phi(x-y)} K(x,y)\varphi(x,y)\chi(\lambda^{\frac{1}{m}}(x-y))f(y)dy; \\ Q_\lambda f(x) &= \int_{\mathbf{R}^n} e^{i\lambda\phi(x-y)} K(x,y)\varphi(x,y)(1-\chi)(\lambda^{\frac{1}{m}}(x-y))f(y)dy; \\ R_{\lambda,m} f(x) &= \int_{\mathbf{R}^n} e^{i\lambda\Theta(x-y)} K(x,y)\varphi(x,y)\chi(\lambda^{\frac{1}{m}}(x-y))f(y)dy, \end{aligned}$$

where $\Theta(x)$ is the polynomial given by

$$\Theta(x) = \sum_{|\gamma| \le m-1} \frac{1}{\gamma!} \frac{\partial^\gamma \phi(0)}{\partial x^\gamma} x^\gamma.$$

By (iv), (vii), and Theorem 3, we have $\| \sum_j b_j \|_1 \le C \|f\|_1$ and

$$\begin{aligned} |\{|T_\lambda g| > \eta/4\}| &\le C\eta^{-2} \|T_\lambda g\|^2 \\ &\le C\eta^{-2} \|g\|_2^2 \\ &\le C\eta^{-1} \|f\|_1. \end{aligned} \tag{16}$$

By Proposition 1 we have

$$\begin{aligned} |\{|R_{\lambda,m}(\sum_j b_j)| > \eta/4\}| &\le C\eta^{-1} \|\sum_j b_j\|_1 \\ &\le C\eta^{-1} \|f\|_1, \end{aligned} \tag{17}$$

where C is independent of λ and η. Since $|\phi(x) - \Theta(x)|\chi(\lambda^{\frac{1}{m}}x) \le C|x|^m$, by (i) we have

$$\begin{aligned} &|\{|P_\lambda(\sum_j b_j) - R_{\lambda,m}(\sum_j b_j)| > \eta/4\}| \\ &\quad \le C\eta^{-1} \|P_\lambda(\sum_j b_j) - R_{\lambda,m}(\sum_j b_j)\|_1 \\ &\quad \le C\eta^{-1} \lambda \int_{\mathbf{R}^n} \int_{|x-y| \le 2\lambda^{-\frac{1}{m}}} |x-y|^{m-n} |\sum_j b_j(y)| dy dx \\ &\quad \le C\eta^{-1} \|f\|_1. \end{aligned} \tag{18}$$

For $x \in \Omega$ we have

$$|Q_\lambda b_j(x) - S_\lambda^j b_j(x)| \le C \int_{\omega_j \le |x-y| \le 2\omega_j} |x-y|^{-n} |b_j(y)| dy,$$

which implies that

$$\begin{aligned} \|Q_\lambda b_j - S_\lambda^j b_j\|_{L^1(\Omega)} &\le C \int_{Q_j} |b_j(y)| dy \\ &\le C\eta |Q_j|. \end{aligned} \tag{19}$$

Therefore by (7) and (19) we have

$$\begin{aligned}
&|\{|Q_\lambda(\sum_j b_j)| > \eta/4\}| \\
&\quad \le |\Omega^c| + |\Omega \cap \{|Q_\lambda(\sum_j b_j)| > \eta/4\}| \\
&\quad \le C\eta^{-1}\|f\|_1 + |\Omega \cap \{|\sum_j (Q_\lambda b_j - S_\lambda^j b_j)| > \eta/8\}| + |\Omega \cap \{|\sum_j S_\lambda^j b_j| > \eta/8\}| \\
&\quad \le C\eta^{-1}\|f\|_1 + C\eta^{-1}\|\sum_j (Q_\lambda b_j - S_\lambda^j b_j)\|_{L^1(\Omega)} + C\eta^{-2}\|\sum_j S_\lambda^j b_j\|^2_{L^2(\mathbf{R^n})} \\
&\quad \le C\eta^{-1}\|f\|_1 + C\sum_j |Q_j| \\
&\quad \le C\eta^{-1}\|f\|_1. \qquad (20)
\end{aligned}$$

Since

$$T_\lambda f = T_\lambda g + R_{\lambda,m}(\sum_j b_j) + [P_\lambda(\sum_j b_j) - R_{\lambda,m}(\sum_j b_j)] + Q_\lambda(\sum_j b_j),$$

by (16)–(18) and (20) we have

$$|\{|T_\lambda f| > \eta\}| \le C\eta^{-1}\|f\|_1.$$

The proof of Theorem 4 is now complete.

References

[1] S. Chanillo and M. Christ. *Weak (1,1) bounds for oscillatory singular integrals,* Duke Math. Jour. **55** (1987), 141–155.

[2] S. Chanillo, D. Kurtz and G. Sampson. *Weighted weak (1,1) and weighted L^p estimates for oscillatory kernels,* Tran. Amer. Math. Soc. **295** (1986), 127–145.

[3] C. Fefferman. *Inequalities for strongly singular convolution operators,* Acta Math. **124** (1970), 9–36.

[4] A. Nagel and S. Wainger. *Hilbert transforms associated with plane curves,* Trans. Amer. Math. Soc. **223** (1976), 235–252.

[5] Y. Pan. *Uniform estimates for oscillatory integral operators,* Jour. Func. Anal. **100** (1991), 207–220.

[6] Y. Pan. *Weak (1,1) estimate for oscillatory singular integrals with real-analytic phases,* Proc. Amer. Math. Soc. **120** (1994), 789–802.

[7] Y. Pan. *Boundedness of oscillatory singular integrals on Hardy spaces: II,* Indiana Univ. Math. Jour. **41** (1992), 279–293.

[8] D.H. Phong and E.M. Stein. *Hilbert integrals, singular integrals and Radon transforms, I,* Acta Math. **157** (1986), 99–157.

[9] F. Ricci and E.M. Stein. *Harmonic analysis on nilpotent groups and singular integral, I,* Jour. Func. Anal. **73** (1987), 179–194.

[10] C.D. Sogge and E.M. Stein. *Averages of functions over hypersurfaces in* $\mathbf{R}^n$, Invent. Math. **82** (1985), 543–556.

[11] E.M. Stein. *Oscillatory integrals in Fourier analysis,* Beijing Lectures in Harmonic Analysis, Princeton Univ. Press, Princeton, NJ, 1986.

[12] E.M. Stein. "Singular integrals and differentiablity properties of functions," Princeton Univ. Press, Princeton, NJ, 1970.

[13] E.M. Stein. "Harmonic analysis: real-variable methods, orthogonality, and oscillatory integrals," Princeton Univ. Press, Princeton, NJ, 1993.

PARACOMMUTATORS AND HANKEL OPERATORS

LIZHONG PENG*

Department of mathematics, Peking University

§1 Introduction

This paper is a survey of some works on paracommutators and Hankel operators in China, which are the subjects I have been studying since 1985. The classical Hankel operator is defined on the Hilbert space l^2 by the Hankel matrix $\Gamma = \{b_{n+k}\}_{n,k\geq 0}$, and the Toeplitz operator is defined also on l^2 by the toeplitz matrix $T = \{b_{n-k}\}_{n\geq k\geq 0}$. By the theory of the Fourier series both of them become the operators on the Hardy space $H^2(T) = \{f \in L^2(T) : \hat{f}(n) = 0, n < 0\}$. Let $\overline{H}^2(T)$ denote the conjugate Hardy space and $P, \overline{P}$ denote the ortnogonal projections from $L^2(T)$ into $H^2(T), \overline{H}^2(T)$ respectively. Then the Toeplitz operator is defined as follows

$$T_b(f) = P(\bar{b}f), \text{for} f \in H^2(T), \tag{1.1}$$

and the Hankel operator is defined as follows

$$H_b(f) = \overline{P}(\bar{b}f), \text{for} f \in H^2(T). \tag{1.2}$$

It is clear that we have

$$T_b + H_b = M_b.$$

Toeplitz operators have studied by many authors, and there are many results on them. For the Hankel operator, in 1957 Nehari [25] gave the boundeness of H_b, i.e.H_b is bounded if and only if $b \in BMOA$. In 1958 Hartman [9] gave the compactness of H_b, i.e. H_b is compact if and only if $b \in VMOA$.

In order to describe the Shatten-von Neuman properties of H_b we introduce some notations. The sigular number of an operator A from a Hilbert space to another Hilbert space is defined by

$$s_n = s_n(A) = inf\{\|A - F\| : rank(F) \leq n\}.$$

Schatten-von Neuman class S_p is defined by

$$S_p = \{A : \sum s_n^p < \infty\}, \text{for} 0 < p < \infty.$$

For simplicity we denote the set of all bounded operators by S_∞. For the further properties, see McCarthy [24] and Zhu [51]. Let $B_p^{s,q}$ denote Besov space, simply $B_p^s = B_p^{s,p}$ and $B_p = B_p^{\frac{1}{p}}$. For the further properties see Peetre [27] and Triebel [48].

*Research supported in part by the National Natural Science Foutation of China

M. Cheng et al. (eds.), Harmonic Analysis in China, 220–239.

Theorem A. *$H_b \in S_p$ if and only if $b \in B_p$, for $0 < p \leq \infty$.*

This result is due to Peller [31, 32], Rochverg [45] and Semmes[46]. [31] gave the result for $1 \leq p \leq \infty$, [32] gave the result for $0 < p < 1$. By Caylay transform the unit disk is mapped to the upper half plane U, equivalently one can consider Hankel operator on U. [45] gave the result for $1 \leq p \leq \infty$ on U, [46] gave the result for $0 < p < 1$ also on U.

After the results of Theorem A the Hankel operators and their generalizations have developed into a very active area of Mathematical Analysis with many connections with other areas of Mathematics, both pure and applied. In this development Chinese mathematicians have played an active role.

Paracommutator is a generalization of the classical Hankel operator, but it turns out to be a tool to sdudy the Toeplitz-Hankel operators on Bergman spaces. (see §3.4.5 below.)

§2 Contributions to paracommutators

The paracommutator is an operator of the form:

$$(T_b^{st}(A)f)^\wedge(\xi) = (2\pi)^{-d} \int_{R^d} \widehat{b}(\xi - \eta)A(\xi,\eta)|\xi|^s|\eta|^t\widehat{f}(\eta)\, d\eta, \tag{2.1}$$

which is defined by Janson and Peetre [11]. It synthesizes both paraproduct by Bony and commutator of singular integral by Calderón.

The paracommutators contain many examples, cf. Janson-Peetre [11] and Peng-Qian [40].

Janson and Peetre [11] studied systematically its L^2-boundedness and Schatten-von Neumann properties S_p, for $1 \leq leq\infty$. Peng [35] studied its compactness, [33] studied its S_p-properties for $0 < p < 1$ and gave a complete characterization of the Janson-Wolff phenomena. In the case $0 < p < 1$ S_p is not a Banach space, the method in [11] does not work at all. In fact [33] used wavelet analysis. [39] developed the idea in [33], gave two decompositions of paracommutators by wavelet analysis (standard and non-stahdard), and simplified the proofs of all of the results in [11] and [33, 35]. [18] studied L^p-boundedness of paracommutators. In [12] the theory of paracommutator was applied to the Hankel forms of higher weights, which appeared in tense product representation of the group $SL(2,R)$ as well as in [20] it was applied to compensated compactness.

In [34] Peng studied a multi-fold paracommutator of the form

$$\begin{aligned}&(T_{b_1,\cdots,b_N}^{u_0,u_1,\cdots,u_N}(A)(f))^\wedge(\eta_0)\\ =&(2\pi)^{Nd}\int_{R^{Nd}} \Pi_{j=1}^N \hat{b}_j(\eta_{j-1}-\eta_j)A(\eta_0,\eta_1,\cdots,\eta_N)|\eta_0|^{u_0}|\eta_1|^{u_1}\cdots|\eta_N|^{u_N}d\eta_1\cdots d\eta_N,\end{aligned}$$

and obtained the results of boundedness and S_p-estimates for $1 \leq p \leq \infty$. Applying to a kind of multi-fold commutators $C(b_1,\cdots,b_N) = [b_1,\cdots,[b_N,K]\cdots]$ where K be a Calderón-Zygmund singular integral, Peng obtained a result:

Let X_p denote the space $B_p^{d/p}$ (if$p < \infty$) or BMO (if $p = \infty$). Suppose that $d \geq 2, p \geq 1, d < p_1, \cdots, p_N \leq \infty$ and $\frac{1}{p} = \sum_{j=0}^{N} \frac{1}{p_j}$. Then

$$\|C(b_1, \cdots, b_N)\|_{S_p} \leq C\Pi_{j=1}^{N} \|b_j\|_{X_p}. \tag{2.2}$$

To study big Hankel operator on the upper half plane U, Arazy, Fisher and Peetre [2] found that it is indeed a vector-valued paracommutator. As we will see in §4, the middle hankel operators on U are also vector-valued paracommutators. By a technique one can change them into the above two-fold paracommutators. They constitute an important application of the multi-fold paracommutators.

In [38] Peng considered analytic paracommutators in periodic case, and obtained the simslar results to those in [11] and [33, 35]. This theory was been used to the S_p−estimates of Toeplitz operators and small Hankel operators on the unit disk D. (See §3.)

In [19] Li, Lin and Peng studied the paracommutators on product space. This theory was been used to the middle Hankel operators on $U \times U$ and $D \times D$ in [14] and [50] respectively.

There are also some works on the generalizations of Hankel operators to higher dimensions. In [7] Deng and Peng gave the criteria of Schatten-von neumann ideals for higher dimension Hankel operators in homogeneous self-adjoint cones. In [6] Deng and Fang gave the S_p−estimates for Hankel operators on Hardy space in the ball of C^d. In [49] Xiao gave the results on the boundedness and compactness of Hankel operators with arbitrary symbols on Bergman space. In [37] Peng studied Hankel operators on Paley-Wiener space in the cube, gave a complete answer to Rochberg's problem. In [36] Peng studied Hankel operators on paley-Wiener space in hte disk, gave a partial answer to Rochberg's problem in this case.

§3 Middle Hankel operators on Bergman space

Let $dm(z)$ be the Lebesgue measure on the unit disk D of the complex plane and let, for $-1 < \alpha < \infty$, $d\mu_\alpha(z) = \frac{\alpha+1}{\pi}(1 - |z|^2)^\alpha dm(z)$. Thus $L^{\alpha,2}(D) = L^2(d\mu_\alpha)$ is the space of measurable functions f for which the norm

$$\|f\| = \{\frac{\alpha + 1}{\pi} \int_D |f(z)|^2 (1 - |z|^2)^\alpha dm(z)\}^{\frac{1}{2}}$$

is finite; it is a Hilbert space. The (weighted) Bergman space $A^{\alpha,2}(D) = A_0$ is the subspace of all analytic functions in $L^{\alpha,2}(D)$; this is a closed subspace. It is easy to see that $A_0 = \text{span}\{z^n\}_{n \geq 0}$. The subspace of all anti-analytic functions $f(z)$ with $\int_D f(z)dm(z) = 0$ is denoted by $\overline{A}^{\alpha,2}(D) = \overline{A}_0$. It is obvious that $\overline{A}_0 = \text{span}\{\overline{z}^n\}_{n \geq 1}$.

Let P_0 and $\overline{P}_0$ denote the corresponding orthogonal projection operators from $L^{\alpha,2}(D)$ onto A_0 and $\overline{A}_0$ respectively. For an analytic function $b(z)$ on D, we have the following three classical operators from A_0 to $L^{\alpha,2}(D)$ (see [2]):
(1) the Toeplitz operator:

$$T_b f = P_0 M_{\overline{b}} f \qquad \text{for } f \in A_0, \tag{3.1}$$

(2) the big Hankel operator:

$$H_b f = (I - P_0)M_{\bar{b}} f \qquad \text{for } f \in A_0, \tag{3.2}$$

(3) the small Hankel operator:

$$h_b f = \overline{P}_0 M_{\bar{b}} f \qquad \text{for } f \in A_0. \tag{3.3}$$

For the above three classical operators, there are the following results:

Theorem B. *Let $\alpha > -1$, $0 < p \leq \infty$.*

(1) $T_b \in S_\infty$ if and only if $b \in L^\infty$, and T_b is never compact unless $b \equiv 0$.

(2) If $1 < p \leq \infty, H_b \in S_p$ if and only if $b \in B_p$. H_b is compact if and only if $b \in b_\infty$. If $0 < p \leq 1$, $H_b \in S_p$ only if $b \equiv$ constant (i.e. $H_b \equiv 0$).

(3) $h_b \in S_p$ if and only if $b \in B_p$. h_b is compact if and only if $b \in b_\infty$.

The results (2) of Theorem A are due to Axler [3], Arazy, Fisher and Peetre [2], and the results (3) of Theorem A are due to [2], Peller [31, 32], Rochberg [45] and Semmes [46].

The change in behavior at $p = 1$ of Theorem A (2) is called the cut–off phenomenon (or Janson-Wolff phenomenon). For convenience we say that H_b has cut–off at $p_0 = 1$, that T_b has cut–off at $p_0 = \infty$ and that h_b has cut–off at $p_0 = 0$.

In general, we introduce the partial order $\prec$ between operators from one Hilbert space to another by

$$R \prec S \quad \text{if and only if} \quad R^*R \leq S^*S.$$

It is easy to check that $h_b \prec H_b$.

Professor R. Rochberg proposed a natural problem: does there exist Middle Hankel operator?

In [13] S. Janson and R. Rochberg constructed a middle Hankel operator H_b^{JR}:

$$H_b^{JR} f = \overline{P} M_{\bar{b}} f \qquad \text{for} \quad f \in A_0, \tag{3.4}$$

where $\overline{P}$ is the orthogonal projection operator from $L^{\alpha,2}(D)$ onto span$\{z^n \bar{z}^m : m > n\}$. For an analytic symbol b, they obtained the following results:

Theorem C. *Let $\alpha > -1, 0 < p \leq \infty$.*

(1) $h_b \prec H_b^{JR} \prec H_b$.

(2) $H_b^{JR} \in S_p$ iff $b \in B_p$.

Peng and Zhang [44] constructed a middle Hankel operator H_b^1 with cut–off at $p_0 = \frac{1}{2}$.

Peng, Rochberg and Wu [41] consider the space $L^2((\log \frac{1}{|z|^2})^\alpha dm(z))$ instead of $L^2((1 - |z|^2)^\alpha dm(z))$. They give a complete orthogonal decomposition :

$$L^2((\log \frac{1}{|z|^2})^\alpha dm(z)) = \oplus_{k=0}^\infty (A_k \oplus \overline{A}_k),$$

which turns out to involve Laguerre polynomials:

$$A_k = \text{span}\{L_k^{(\alpha)}((n+1)\log\frac{1}{|z|^2})z^n\}_{n\geq 0},$$
$$\overline{A}_k = \text{span}\{L_k^{(\alpha)}((n+1)\log\frac{1}{|z|^2})\overline{z}^n\}_{n\geq 1}.$$

Denoting the corresponding orthogonal projection operator from $L^2((\log\frac{1}{|z|^2})^\alpha dm(z))$ onto A_k by P_k, they define a series of middle Hankel operators H_b^k by

$$H_b^k = (I - \sum_{\nu=0}^{k} P_\nu)M_{\overline{b}}P_0$$

and prove that H_b^k has cut−off at $\frac{1}{k+1}$.

In [42] Peng and Xu give a complete decomposition to $L^{\alpha,2}(D)$, which turns out to involve Jacobi polynomials.

Theorem 3.1.

$$L^{\alpha,2}(D) = \oplus_{k=0}^{\infty}(A_k \oplus \overline{A}_k),$$

where

$$A_k = span\{P_k^{(\alpha,n)}(2|z|^2-1)z^n\}_{n\geq 0},$$
$$\overline{A}_k = span\{P_k^{(\alpha,n)}(2|z|^2-1)\overline{z}^n\}_{n\geq 1}.$$

Denote the corresponding orthogonal projection operators from $L^{\alpha,2}(D)$ onto A_k and $\overline{A}_k$ again by P_k and $\overline{P}_k$ respectively. For an analytic function $b(z) = \sum_{n=0}^{\infty}\hat{b}(n)z^n$, define formally $P_l b(z)$ by

$$P_l b(z) = \sum_{n=0}^{\infty}\hat{b}(n)P_l^{(\alpha,n)}(2|z|^2-1)z^n$$

Now one can define three kinds of Toeplitz-Hankel operators:
(1) the Toeplitz type operator $T_b^{(k,l,k')}$,

$$T_b^{(k,l,k')}f = P_k M_{\overline{P_l b}}f \quad \text{for } f \in A_{k'}, \tag{3.5}$$

(2) the small Hankel type operator $h_b^{(k,l,k')}$,

$$h_b^{(k,l,k')}f = \overline{P}_k M_{\overline{P_l b}}f \quad \text{for } f \in A_{k'}\ , \tag{3.6}$$

(3) and the middle Hankel type operator $H_b^{(k,l,k')}$,

$$H_b^{(k,l,k')}f = (I - \sum_{\nu=0}^{k} P_\nu)M_{\overline{P_l b}}f \quad \text{for } f \in A_{k'}. \tag{3.7}$$

It is obvious that $T_b^{(0,0,0)}$ is the classical Toeplitz operator, $h_b^{(0,0,0)}$ is the classical small Hankel operator, $H_b^{0,0,0}$ is the classical big Hankel operator, and that $H_b^{1,0,0}$ is just the middle Hankel operator in [44]. For simplicity denote $H_b^k = H_b^{k,0,0}$.

Now the sequence of middle Hankel operators H_b^k, $k = 0, 1, 2, \cdots$, links H_b, h_b and H_b^{JR} in the following sense

$$H_b = H_b^0 \succ H_b^1 \succ \cdots \succ H_b^k \succ H_b^{k+1} \succ \cdots \succ \lim_{k\to\infty} H_b^k = H_b^{JR} \succ h_b.$$

For the three types of operators, [42] proved the following three theorems.

Theorem 3.2. *Let $\alpha > -1$ and $0 < p \leq \infty$, then*
(a) $h_b^{(k,0,k')} \in S_p$ iff $b \in B_p$, and $h_b^{(k,0,k')}$ is compact iff $b \in b_\infty$.
(b) (i) if $k < l$, then $h_b^{(k,l,0)} \equiv 0$,
(ii) if $k \geq l$ then $h_b^{(k,l,0)} \in S_p$ iff $b \in B_p$, and $h_b^{(k,l,0)}$ is compact iff $b \in b_\infty$.
(c) (i) if $k' < l$, then $h_b^{(0,l,k')} \equiv 0$,
(ii) if $k' > l$, then $h_b^{(0,l,k')} \in S_p$ iff $b \in B_p$, and $h_b^{(0,l,k')}$ is compact iff $b \in b_\infty$.

Theorem 3.3. *Let $\alpha > -1$ and $0 < p \leq \infty$.*
(a) (1) If $k > k'$, then
(i) for $\frac{1}{k-k'} < p \leq \infty$, $T_b^{(k,0,k')} \in S_p$ iff $b \in B_p$,
(ii) $T_b^{(k,0,k')}$ is compact iff $b \in b_\infty$,
(iii) for $0 < p \leq \frac{1}{k-k'}$, $T_b^{(k,0,k')} \in S_p$ only if $T_b \equiv 0$ i.e. $b = \sum_{j=0}^{s} \hat{b}(j) z^j$ and $s < k - k'$.
(2) If $k = k'$, then
(i) $T_b^{(k,0,k)} \in S_\infty$ iff $b \in L^\infty$,
(ii)$T_b^{(k,0,k)}$ is never compact unless $b \equiv 0$.
(3) If $k < k'$, then
$T_b^{(k,0,k')} \equiv 0$.
(b) (1) If $k > 0$, then
(i) for $\frac{1}{k} < p \leq \infty$, $T_b^{(k,l,0)} \in S_p$ iff $b \in B_p$,
(ii) $T_b^{(k,l,0)}$ is compact iff $b \in b_\infty$
(iii) for $0 < p \leq \frac{1}{k}$, $T_b^{(k,l,0)} \in S_p$ only if $T_b^{(k,l,0)} \equiv 0$ i.e. $b = \sum_{j=0}^{S} \hat{b}(j) z^j$ and $s < k$.
(2) If $k = 0$, then
(i) $T_b^{(0,l,0)} \in S_\infty$ iff $b \in L^\infty$,
(ii)$T_b^{(0,l,0)}$ is never compact unless $b \equiv 0$.
(c) (1) If $l \geq k' > 0$, then
(i) for $\frac{1}{k'} < p \leq \infty$, $T_b^{(0,l,k')} \in S_p$ iff $b \in B_p$,
(ii) $T_b^{(0,l,k')}$ is compact iff $b \in b_\infty$,
(iii) for $0 < p \leq \frac{1}{k'}$, $T_b^{(0,l,k')} \in S_p$ only if $T_b^{(0,l,k')} \equiv 0$ i.e. $b \equiv 0$.
(2) If $l \geq k' = 0$, then
(i) $T_b^{(0,l,0)} \in S_\infty$ iff $b \in L^\infty$,

(ii) $T_b^{(0,l,0)}$ *is never compact unless* $b \equiv 0$.
(3)If $l < k'$, *then*

$$T_b^{(0,l,k')} \equiv 0.$$

Theorem 3.4. *For* $\alpha > -1, \quad 0 < p \leq \infty$,
(i) if $\frac{1}{k+1} < p \leq \infty$, *then* $H_b^k \in S_p$ *iff* $b \in B_p$,
(ii) H_b^k *is compact iff* $b \in b_\infty$,
(iii) if $0 < p \leq \frac{1}{k+1}$, *then* $H_b^k \in S_p$ *iff* $b = \sum_{j=0}^{l} \hat{b}(j) z^j$ and $l \leq k$.

By using the orthonormal basis e_k^n of A_k, straighforward calculation shows that the Toeplitz type operator and small Hankel operator become the analytic paracommutators in periodic case, which are studued in [38]. Then theorems in [38] give the proofs of Theorem 3.1 and Theorem 3.2. For example, $T_b^{(k,0,k')}$ is determined by the matrix

$$\{\langle T_b^{(k,0,k')}(e_{k'}^n), e_k^m \rangle\}_{n \geq 0, m \geq 0},$$

a straightforward calculation involving the integral formulas in the book [8] shows that

$$\langle T_b^{(k,0,k')}(e_{k'}^n), e_k^m \rangle$$
$$= \begin{cases} \bar{\hat{b}}(n-m) A(\alpha,k,k') B_1(n;\alpha,k,k') B_2(m;\alpha,k,k') C(n-m;\alpha,k,k'), \\ \qquad\qquad \text{if } k \geq k' \text{ and } n-m \geq k-k', \\ 0, \qquad\qquad \text{if } k < k' \text{ or } n-m < k-k', \end{cases}$$

where

$$A(\alpha,k,k') = \sqrt{\frac{\Gamma(\alpha+k+1)k!}{\Gamma(\alpha+k'+1)k'!} \frac{1}{(k-k')!}},$$

$$B_1(n;\alpha,k,k') = \frac{\sqrt{(\alpha+n+2k'+1)\Gamma(\alpha+n+k'+1)\Gamma(n+k'+1)}}{\Gamma(\alpha+n+k+k'+2)},$$

$$B_2(m;\alpha,k,k') = \sqrt{\frac{\alpha+m+2k+1}{\Gamma(\alpha+m+k+1)\Gamma(m+k+1)}} \Gamma(\alpha+m+k+k'+1),$$

$$C(n-m;\alpha,k,k') = \frac{\Gamma(n-m+1)}{\Gamma(n-m+k'-k+1)}.$$

By a property of the Gamma function, we can get the following estimates:

$$B_1(n;\alpha,k,k') \asymp (n+1)^{-k-\frac{\alpha+1}{2}},$$
$$B_2(m;\alpha,k,k') \asymp (m+1)^{k'+\frac{\alpha+1}{2}},$$
$$C(n-m;\alpha,k,k') \asymp (n-m+1)^{k-k'}.$$

So if $k < k'$ or $k > k'$ but $b(z) = \sum_{j=0}^{s} \hat{b}(j)z^j$ where $s < k - k'$, then $T_b^{k,0,k'} \equiv 0$. If $k = k'$ or $k > k'$ and $b(z)$ is not a polynomial with degree not more than $k - k' - 1$, then $T_b^{(k,0,k')} \in S_p$ (or compact) iff $I^{(k'+\frac{\alpha+1}{2})}T_{I^{k'-k}b}I^{k+\frac{\alpha+1}{2}} \in S_p$, (or compact), this gives the proof of the part (a) of Theorem 3.3.

The proof of Theorem 3.4 is more complicated, we omit it here.

As is well known that Jacobi polynomial can be expressed by hypergeometric fynction:

$$P_n^{(\alpha,\beta)}(x) = \frac{(\alpha+1)_n}{n!}{}_2F_1(-n, n+\alpha+\beta+1; \alpha+1; \frac{1-x}{2}).$$

The relation between hypergeometric function and operator theory was established. The theory of hypergeometric functions gave the properties of the operators. On the other hand theory of the Toeplitz-Hankel operators can give also the properties of hypergeometric functions. For example, in [42] it follows that

Theorem 3.5. *Let $n, j_1, j_2 \in Z_+$, $\alpha > -1$ then*

(i)

$$\begin{aligned}
&{}_4F_3(-j_1, -j_2, \alpha+1, n+\alpha+1; n+1, n+j_1+\alpha+2, n+j_2+\alpha+2; 1)\\
&+\frac{2(\alpha+1)(n+\alpha+1)j_1j_2}{(n+1)(n+j_1+\alpha+2)(n+j_2+\alpha+2)}\times\\
&{}_4F_3(-j_1+1, -j_2+1, \alpha+2, n+\alpha+2; n+2, n+j_1+\alpha+3, n+j_2+\alpha+3; 1)\\
&=\frac{(n+j_1+1)_{j_2}(n+\alpha+2)_{j_2}}{(n+1)_{j_2}(n+j_1+\alpha+2)_{j_2}},
\end{aligned}$$

(ii)

$$\begin{aligned}
&{}_3F_2(-j_1, -j_2, 1; n+j_1+2, n+j_2+2; 1)+\\
&\frac{2j_1j_2}{(n+j_1+2)(n+j_2+2)}{}_3F_2(-j_1+1, -j_2+1, 2; n+j_1+3, n+j_2+3; 1)\\
&=\frac{(n+j_1+1)(n+j_2+1)}{(n+1)(n+j_1+j_2+1)}.
\end{aligned}$$

Now another natural problem is raised: for any $0 < p < \infty$, does there exist an operator T such that it has a cut-off at p?

The paper [38] give an affirmative answer.

Let $f_n^s = \frac{r^{n^s}e^{in\theta}}{\gamma_{n^s,\alpha}}$, and $B^{\alpha,2}(D) = span\{f_n^s\}_{n\geq 0}$, P_α^s denote the corresponding projection. We define $H_b^s = P_\alpha^s M_b P_\alpha$. If $s = 1$, $H_b^s = T_b$ is the Toeplitz operator on the Bergman space. If $s \neq 1$, it is a Hankel-like operator. This kind of operators is also analytic paracommutator T_b^{st} in periodic case. Their cut-off phenomena were characterised in the following

Theorem 3.6. *(a) $\alpha > -1, s > 1$ and $0 < p \leq \infty$.*

(i) If $\frac{1}{p} < (s-1)\frac{\alpha+1}{2}$, then $H_b^s \in S_p$ iff $b \in B_p^{(1-s)(\alpha+1)/2+\frac{1}{p}}$.

(ii) If $0 < p < \infty$ *and* $\frac{1}{p} \geq (s-1)\frac{\alpha+1}{2}$, *then* $H_b^s \in S_p$ *only if* $b \equiv 0$.

(iii) H_b^s *is compact iff* $b \in b_\infty^{(1-s)(\alpha+1)/2}$.

(b) $s = 1, H_b^1$ *is the Toeplitz operator, so* H_b^1 *is bounded iff* $b \in L^\infty$, *and if* H_b^1 *is compact only if* $b \equiv 0$.

(c) Let $\alpha > -1, 0 < s < 1$ *and* $0 < p \leq \infty$.

(i) If $\frac{1}{p} < (1-s)(\alpha+1)/2$, *then* $H_b^s \in S_p$ *iff* $b \in B_p^{(s-1)(\alpha+1)/2+\frac{1}{p}}$.

(ii) If $0 < p < \infty$ *and* $\frac{1}{p} \geq (1-s)(\alpha+1)/2$, *then* $H_b^s \in S_p$ *only if* $b \equiv 0$.

(iii) H_b^s *is compact iff* $b \in b_\infty^{(s-1)(\alpha+1)/2}$.

(d) Let $\alpha > -1, -\infty < s < 0$ *and* $0 < p \leq \infty$.

(i) If $\frac{1}{p} < \frac{\alpha+1}{2}$, *then* $H_b^s \in S_p$ *iff* $b \in B_p^{-(\alpha+1)/2+\frac{1}{p}}$.

(ii) If $\frac{1}{p} \geq \frac{\alpha+1}{2}$, *then* $H_b^s \in S_p$ *only if* $b \equiv 0$.

(iii) H_b^s *is compact iff* $b \in b_\infty^{-\frac{\alpha+1}{2}}$.

[14] and [50] constructed a series of middle Hankel operators on $U \times U$ and $D \times D$ and showed that they can be compact and can belong to S_P- ideals, respectively. These results are important, because such domains have rank 2. [4] has shown that the big Hankel operator on the domain with rank≥ 2 can not be compact unless it is trivial. The proofs need the theory of paracommutators on product space in [20].

§4 Wavelet transforms and Hankel type operators

The Cayley transform maps the unit disk D onto the upper half plane U. In [15, 16] Jiang and Peng find that the Toeplitz-Hankel operator theory connects with wavelet transform, which is anather active area.

Let $U = \{x + iy, y > 0\}$ be the upper half-plane. For $-1 < \alpha < \infty$, the space $L^{\alpha,2}(U)$ consists of all functions on U for which the integral $\|f\|_2^2 = \int_U |f(x,y)|^2 y^\alpha dxdy$ is finite. Let $A^{\alpha,2}(U)$ denote the (weighted) Bergman space (or Dzhrbashyan space) on U, i.e., the subspace of all analytic functions in $L^{\alpha,2}(U)$ and let $\overline{A}^{\alpha,2}(U)$ denote the subspace of all anti-analytic functions in $L^{\alpha,2}(U)$.

Let G denote the affine group. It consists of all couples $\{(x,y) : y > 0, x \in R\}$ with the law $(x_1, y_1)(x,y) = (y_1 x + x_1, y_1 y)$. It is a locally compact nonunimodular group with right Haar measure $d\mu_R(x,y) = dxdy/y$ and left Haar measure $d\mu_L(x,y) = dxdy/y^2$. It can be identified as the quotient group of $SL(2,R)$ by $SO(2,R)$. The identification is made by

$$g = (x,y) \Leftrightarrow \begin{pmatrix} \sqrt{y} & x/\sqrt{y} \\ 0 & 1/\sqrt{y} \end{pmatrix}.$$

Define a representation U_g of G on $L^{\alpha,2}(U)$ by

$$U_g f(z_1) = (1/y^{\frac{\alpha+2}{2}}) f(\frac{z_1 - x}{y}).$$

A function ψ on R is called as an (weighted) admissible wavelet if it satisfies the admissibility condition:

$$\int_G |(\psi, U_g\psi)|^2 d\mu_L(g) < \infty, \tag{4.1}$$

where $(\cdot,\cdot)$ is the usual scalar product on $L^{\alpha,2}(U)$.

Let AW denote the space consisting of the admissible wavelets whose Fourier transforms are supported in $[0. + \infty)$ and let $\overline{AW} = \{f : \overline{f} \in AW\}$.

For $s \in R$, let $H_s^2(R)$ denote the subspace of the Sobolev space consisting of all functions such that $supp\hat{f} \subset [0, +\infty)$ and

$$\|f\|_{H_s^2(R)}^2 = \int_0^\infty |\hat{f}(\xi)|^2 \xi^{-s} d\xi < \infty,$$

i.e., $H_s^2 = I_{-\frac{s}{2}}(H^2)$, here $I_{-\frac{s}{2}}$ is the fractional integral operator and H^2 is the Hardy space on R. And let $\overline{H}_s^2(R) = \{f : \overline{f} \in H_s^2(R)\}$.

For $\psi \in AW$, the admissibility condition becomes

$$\begin{aligned}&\iint |(\psi, U_g\psi)|^2 d\mu_L(g)\\ &= \frac{1}{(2\pi)^2 4^{\alpha+1}} \int_0^\infty \frac{|\hat{\psi}(\xi)|^2}{\xi^{\alpha+1}} d\xi \int_0^\infty |\hat{\psi}(y)|^2 \frac{dy}{y^{2+\alpha}}.\end{aligned}$$

Thus

$$AW = \{\psi : 0 < \|\psi\|_{H^2_{\alpha+2}}, \|\psi\|_{H^2_{\alpha+1}} < \infty \quad \text{and} \quad supp\hat{\psi} \subset [0, +\infty)\}$$

Let $L_k^{(\alpha)}(x) = \sum_{\nu=0}^k \binom{k+\alpha}{k-\nu}(-x)^\nu/\nu!$ be the Laguerre polynomials(see [47]).

For $k \in Z^+$, let ψ^k and $\bar{\psi}^k$ be functions defined by

$$\hat{\psi}^k(\xi) = \begin{cases} L_k^{(\alpha)}(2\xi)e^{-\xi}, & \text{for } \xi \geq 0 \\ 0, & \text{for } \xi < 0 \end{cases}$$

and $(\bar{\psi}^k)^\wedge(\xi) = \hat{\psi}^k(-\xi)$, define A_k and $\overline{A}_k$ by

$$\begin{aligned} A_k &= \{\psi_y^k * f(x) : f \in H_{\alpha+1}^2\}, \\ \overline{A}_k &= \{\bar{\psi}_y^k * f(x) : f \in \overline{H}_{\alpha+1}^2\}, \end{aligned} \tag{4.2}$$

where $\phi_y(x) = \frac{1}{y}\phi(\frac{x}{y})$. Then the orthogonal decomposition of AW indused a complete orthogonal decomposition of $L^{\alpha,2}(U)$.

Theorem 4.1.

$$L^{\alpha,2}(U) = \oplus_{k=0}^{\infty}(A_k \oplus \overline{A}_k).$$

The orthonormal basis of A_k is as follows

$$e_{nk}(x,y) = \frac{2^{\alpha+1}\Gamma(n+\alpha+2)}{2\pi n!}\sum_{j=0}^{n}\sum_{\nu=0}^{k}\binom{n}{j}\binom{\alpha+k}{k-\nu}\cdot \tag{4.3}$$

$$\cdot\binom{\alpha+\nu+j+1}{\nu}\left(\frac{-2}{y+1-ix}\right)^{j+\nu}\frac{y^{\nu}}{(y+1-ix)^{\alpha+2}},$$

and in particular if $k=0$, then

$$e_{n0}(z) = \frac{2^{\alpha+1}\Gamma(n+\alpha+2)}{2\pi n!}\left(\frac{z-i}{z+i}\right)^{n}\frac{1}{(-i(z+i))^{\alpha+2}},$$

where $z = x+iy$. When $\alpha = 0$, they are just the images of z^n in $A^{0,2}(D)$ under the Caylay transform.

The reproducing kernel of A_k is as follows

$$K_z^{(k)}(w) = \frac{1}{2\pi c_k}\sum_{\nu=0}^{k}\sum_{j=0}^{k}\binom{\alpha+k}{k-\nu}\binom{\alpha+k}{k-j}\cdot \tag{4.4}$$

$$\cdot\frac{\Gamma(\nu+j+\alpha+2)}{\nu!j!}\left(\frac{2iy}{\bar{z}-w}\right)^{\nu}\left(\frac{2i\nu}{\bar{z}-w}\right)^{j}\frac{1}{[i(\bar{z}-w)]^{\alpha+2}}.$$

Let P_k (resp.$\overline{P}_k$) be the orthogonal projection from $L^{\alpha,2}(U)$ onto A_k (resp. $\overline{A}_k$). Let $\overline{P}_l b = \overline{P}_l b(x,y)$ be functions defined by

$$(\overline{P}_l b)^{\wedge}(\xi,y) = \begin{cases} e^{\xi y}L_l^{(\alpha)}(-2\xi y)\hat{b}(\xi), & \text{for } \xi < 0 \\ 0, & \text{for } \xi \geq 0. \end{cases} \tag{2.13}$$

Define three kinds of Toeplitz-Hankel operators as same as those in §3.

The main results about the above three kinds of operators on U are the following

Theorem 4.2. *Let $h_b^{k,l,k'}$ be operators defined by (2.15), then*
(1) If $k+k' < l$,then $h_b^{k,l,k'}$ is the zero operator;
(2) If $k+k' \geq l$, then $h_b^{k,l,k'} \in S_p$ iff $b \in \overline{B}_p$.

Theorem 4.3. *Let $T_b^{k,l,k'}$ be operators defined by (2.14), then*
(1) If $k+l < k'$, then $T_b^{k,l,k'}$ is the zero operator;
(2) If $k = k'$, then $T_b^{k,l,k'} \in S_\infty$ iff $b \in L^\infty$; and $T_b^{k,l,k'}$ is never compact unless $b \equiv 0$;
(3) If $k+l \geq k', k \neq k'$ and $\frac{1}{|k-k'|} < p \leq \infty$, then $T_b^{k,l,k'} \in S_p$ iff $b \in \overline{B}_p$;
(4) If $k+l \geq k', k \neq k', 0 < p \leq \frac{1}{|k-k'|}$, and $T_b^{k,l,k'} \in S_p$, then $b \equiv 0$

Theorem 4.4. *Let $H_b^{k,l,k'}$ be operators defined by (2.16), then*
(1) If $k < k'$, then $H_b^{k,l,k'} \in S_\infty$ iff $b \in L^\infty$;
(2) If $k < k'$ and $H_b^{k,l,k'}$ is compact, then b is a constant;
(3) If $k \geq k'$ and $\frac{1}{k+1-k'} < p \leq \infty$, then $H_b^{k,l,k'} \in S_p$ iff $b \in \overline{B}_p$;
(4) If $k \geq k'$ and $0 < p \leq \frac{1}{k+1-k'}$ and $H_b^{k,l,k'} \in S_p$, then b is a constant.

From the Theorem 4.3 and Theorem 4.4, we know that if $k + l \geq k', k \neq k'$ then $T_b^{k,l,k'}$ have cut-off phenomena at points $\frac{1}{|k-k'|}$ and that if $k \geq k'$, then $H_b^{k,l,k'}$ have cut-off phenomena at points $\frac{1}{k+1-k'}$.

Here $-1 < \alpha < \infty$ in the definition of $L^{\alpha,2}(U)$. In [JP1] Jiang and Peng studied the case $\alpha = -2$. In that case the space $L^{-2,2}(U)$ (with the similar definition) contains no analytic function, so it is the limiting case. There are also an orthogonal decomposition of $L^{-2,2}(U)$ and three kinds of Ha-plitz operators $h_b^{k,k'}$, $T_b^{k,k'}$ and $H_b^{k,k'}$. In fact for $\alpha = -2$, one also can define Ha-plitz operators of the above more general types: $h_b^{k,l,k'}$, $T_b^{k,l,k'}$, $H_b^{k,l,k'}$ for $l \in Z^+$ and get the similar results to those in this paper.

The proofs of Theorems 4.2, 4.3, 4.4 need again paracommutator theory. By Fourier transform, they become vector-valued paracommutators (see [2]). By a technique, [15] changes $T_b^{k,l,k'}, h_b^{k,l,k'}$ into usual paracommutators (studied by Janson and Peetre [5], Peng [10] and others) and change $H_b^{k,l,k'}$ into a two-fold paracommutator (studied by Peng [11]), then by the theory of paracommutator and multi-paracommutator, obtains the desired results for them.

For example, $T_b^{k,l,k'}$ become a vector-valued paracommutator

$$(T_b^{k,l,k'}F)^\wedge(\xi,y) = \frac{1}{2\pi}\int_0^\infty \hat{b}(\xi-\eta)\hat{f}(\eta)A_y^{k,l,k'}(\xi,\eta)d\eta, \tag{4.5}$$

where

$$A_y^{k,l,k'}(\xi,\eta) \tag{4.6}$$
$$= \begin{cases} 1/(c_k 2^{\alpha+1})(\frac{\xi}{\eta}^{\alpha+1} e^{-y\xi} L_k^{(\alpha)}(2y\xi)c^{k,l,k'}(\frac{\xi}{\eta}), & \text{for } 0 \leq \xi \leq \eta \\ 0, & \text{elsewhere.} \end{cases}$$

If $k + l < k', c^{k,l,k'}(t) \equiv 0$, thus $T_b^{k,l,k'} \equiv 0$ and (1) of Theorem 4.3 is true. In the following, one can assume $k + l \geq k'$. By a technique it can changed into an usual paracommutator $t_b^{k,l,k'}$.

$$(t_b^{k,l,k'}f)^\wedge(\xi) = \frac{1}{2\pi}\int_0^\infty \hat{b}(\xi-\eta)\hat{f}(\eta)A^{k,l,k'}(\xi,\eta)d\eta, \tag{4.7}$$

where

$$A^{k,l,k'}(\xi,\eta) = \begin{cases} 1/(c_k 2^{\alpha+1})(\frac{\xi}{\eta})^{\frac{\alpha+1}{2}} c^{k,l,k'}(\frac{\xi}{\eta}), & \text{for } 0 \leq \xi \leq \eta \\ 0, & \text{elsewhere.} \end{cases}$$

If $k = k'$, $A^{k,l,k'}(\xi,\xi) = \frac{1}{c_k 2^{\alpha+1}} c^{k,l,k'}(1) = 1$, then $T_b^{k,l,k'} \in S_\infty$ iff $b \in L^\infty$ and that $T_b^{k,l,k'}$ is not compact unless $b \equiv 0$ by Theorem 12.1, 12.2 of [JP1]. Thus (2) of Theorem 4.3 is true.

About (3) of Theorem 4.3 , one has to estimate the degree of $\xi - \eta$ as $\eta - \xi \to 0^+$. If $k < k' \leq k + l$,

$$c^{k,l,k'}(t) = c_{k,l,k'}(1-t)^{k'-k}(1 + c_1(1-t) + \cdots). \tag{4.8}$$

If $k' < k$,

$$c^{k,l,k'}(t) = P_1(t)(1-t)^{k-k'} + P_2(t)(1-t)^{k-k'+1} + \cdots ,$$

where $P_1(t)$ is a polynomial and $P_1(1) \neq 0$, $P_2(1) \neq 0$.

So $A^{k,l,k'}(\xi,\eta)$ satisfies the conditions $A_0, A_1, A_3(|k-k'|), A_4$, and $A_{4\frac{1}{2}}$. Thus by the theory of paracommutator, (3) of Theorem 4.3 is true for $p < \infty$ (cf. [11], [33]). For $p = \infty$, the proof is much easy.

§5 Weighted Plancherel formula and invariant Hankel operators

The Möbius group $G = SU(1,1)$ consists of all 2×2 complex matrices

$$g = \begin{pmatrix} \alpha & \beta \\ \overline{\beta} & \overline{\alpha} \end{pmatrix}$$

such that $|\alpha|^2 - |\beta|^2 = 1$. It acts on D by means of the maps

$$z \longmapsto gz = g(z) = \frac{\alpha z + \beta}{\overline{\beta} z + \overline{\alpha}}, \qquad (z \in D).$$

All holomorphic automorphisms are so obtained. Set

$$Z = \begin{pmatrix} i & 0 \\ 0 & -i \end{pmatrix}, \; A = \begin{pmatrix} 0 & 1 \\ 1 & 0 \end{pmatrix}, \; B = \begin{pmatrix} 0 & i \\ -i & 0 \end{pmatrix}.$$

Then $\{Z, A, B\}$ is a basis of the Lie algebra $su(1,1)$ of $SU(1,1)$. The Casimir element is

$$\Box = Z^2 - A^2 - B^2.$$

For $g \in SU(1,1)$, define

$$T^\nu(g): \qquad f(z) \longmapsto f(gz)\{g'(z)\}^{\frac{\nu}{2}} = f\left(\frac{\alpha z + \beta}{\overline{\beta} z + \overline{\alpha}}\right)\left(\overline{\beta} z + \overline{\alpha}\right)^{-\nu}. \tag{5.1}$$

Then T^ν gives a genuine representation of the universal covering group of $SU(1,1)$. If $\nu \in \mathbb{Z}$, then T^ν gives a continuous unitary representation of the group $SU(1,1)$. T^ν induces a representation of the Lie algebra $su(1,1)$ and its universal enveloping algebra on the space of C^∞- vectors for T^ν, which will be denoted by T^ν also. It is easy to get

$$T^\nu(Z) = 2iz\frac{\partial}{\partial z} - 2i\overline{z}\frac{\partial}{\partial \overline{z}} + i\nu,$$
$$T^\nu(A) = (1-z^2)\frac{\partial}{\partial z} + (1-\overline{z}^2)\frac{\partial}{\partial \overline{z}} - \nu z,$$
$$T^\nu(B) = i(1+z^2)\frac{\partial}{\partial z} - i(1+\overline{z}^2)\frac{\partial}{\partial \overline{z}} + i\nu\overline{z}.$$

Therefore,

$$\Box_\nu = T^\nu(\Box) = -4(1-|z|^2)^2\frac{\partial^2}{\partial z\partial \overline{z}} + 4\nu(1-|z|^2)\overline{z}\frac{\partial}{\partial \overline{z}} - \nu^2 + 2\nu. \tag{5.2}$$

(see [29]). $\Box_\nu$ is called again as the Casimir operator or invariant Laplacian. When $\nu \neq 0$, Helgason establishes Plancherel formula, which is equivalent to the irreducible decomposition of the unitary representation T^ν, and gives the results of eigenspace representations as well. Peetre, Peng and Zhang [29] studies the case of $\nu > 0$ and ν non odd integer, and establishes a corresponding weighted Plancherel formula. It is different from the case of $\nu = 0$, for the case of $\nu > 1$ the Casimir operator $\Box_\nu$ has not only continuous spectrum, but also finite discrete spectra.

As is well known D is holomorphically equivalent to the upper half plane $U = \{z : Imz > 0\}$, and $SU(1,1)$ is isomorphic to the holomorphic automorphic group $SL(2,\mathbb{R})$ of U. The classification of the irreducible unitary representations of $SU(1,1)$ is due to Bargmann. The non-trivial irreducible unitary representations of $SU(1,1)$ have three classes: principal series, discrete series and complementary series:

First(even) principal series	$\pi^e_{is},\ s \in \mathbb{R}, \pi^e_{is} \sim \pi^e_{-is}$
Second(odd) principal series	$\pi^o_{is},\ s \in \mathbb{R}\setminus\{0\}, \pi^o_{is} \sim \pi^o_{-is}$
Holomorphic discrete series	$\pi^+_n,\ n \in \mathbb{N}^+$, n is the lowest weight,
Conjugate holomorphic discrete series	$\pi^-_{-n}, n \in \mathbb{N}^+$, $-n$ is the highest weight,
Complementary series	$\pi^e_s,\ s \in (-1,1)\setminus\{0\}$,

where π^+_1 and π^-_{-1} are limits of discrete series representations. They and the complementary series do not appear in the irreducible decomposition of the left regular representation.

The main results are the following

Theorem 5.1. *Assume that $\nu \in \mathbb{R}, k = \max\{j \in \mathbb{Z} : j < \frac{|\nu|-1}{2}\}$. Then for $f \in \mathcal{D}(D)$, we have*

(i) the inversion formula

$$f(z) = \int_B \int_{\mathbb{R}} \widehat{f}(\lambda, b) e^{\nu}_{\lambda,b}(z) \rho_\nu(\lambda)\, d\lambda db$$
$$+ \sum_{l=0}^{k} \frac{(|\nu| - 1 - 2l)}{\pi} \int_B \widehat{f}(-i(|\nu| - 1 - 2l), b) e^{\nu}_{-i(|\nu|-1-2l),b}(z)\, db,$$

(ii) the Plancherel formula

$$\int_D |f(z)|^2\, d\mu_\nu(z) = \int_B \int_{\mathbb{R}} |\widehat{f}(\lambda, b)|^2 \rho_\nu(\lambda)\, d\lambda db + \sum_{l=0}^{k} \frac{(|\nu| - 1 - 2l)}{\pi} \cdot$$
$$\int_B \widehat{f}(-i(|\nu| - 1 - 2l), b) \overline{\widehat{f}(i(|\nu| - 1 - 2l), b)}\, db,$$

and the integral

$$\int_B \widehat{f}(-i(|\nu| - 1 - 2l), b) \overline{\widehat{f}(i(|\nu| - 1 - 2l), b)}\, db$$

is nonnegative,
(iii) the operators P_l defined by

$$P_l f(z) = \frac{(|\nu| - 1 - 2l)}{\pi} \int_B \widehat{f}(-i(|\nu| - 1 - 2l), b) e^{\nu}_{-i(|\nu|-1-2l),b}(z)\, db$$

are jointly orthogonal projections,
and (iv) denote $A^\nu_l(D) = P_l L^2(D, d\mu_\nu)$, then the map $f(z) \longmapsto \widehat{f}(\lambda, b)$ extends to an unitary isometry from

$$A^\nu_\omega(D) = L^2(D, d\mu_\nu) \ominus \sum_{l=0}^{k} {}^{\oplus} A^\nu_l(D)$$

onto

$$L^2(\mathbb{R}^+ \times B, 2\rho_\nu(\lambda) d\lambda db).$$

Theorem 5.2. *For $\nu \in \mathbb{Z}$, the unitary representation T^ν of $SU(1,1)$ is decomposed uniquely into the sum of the irreducible representations as follows*

$$\text{(i)} \quad \text{if } \nu = 0, \qquad T^\nu \sim \int_{-\infty}^{\infty} \pi^e_{i\lambda} \rho_\nu(\lambda)\, d\lambda,$$

$$\text{(ii)} \quad \text{if } \nu = 2, 4, \cdots, \qquad T^\nu \sim \int_{-\infty}^{\infty} \pi^e_{i\lambda} \rho_\nu(\lambda)\, d\lambda \oplus \pi^+_2 \oplus \pi^+_4 \cdots \oplus \pi^+_\nu,$$

$$\text{(iii)} \quad \text{if } \nu = -2, -4, \cdots, \qquad T^\nu \sim \int_{-\infty}^{\infty} \pi^e_{i\lambda} \rho_\nu(\lambda)\, d\lambda \oplus \pi^-_{-2} \oplus \pi^-_{-4} \cdots \oplus \pi^-_\nu,$$

$$\text{(iv)} \quad \text{if } \nu = 1, 3, \cdots, \qquad T^\nu \sim \int_{-\infty}^{\infty} \pi^o_{i\lambda} \rho_\nu(\lambda)\, d\lambda \oplus \pi^+_3 \oplus \pi^+_5 \cdots \oplus \pi^+_\nu,$$

$$\text{(v)} \quad \text{if } \nu = -1, -3, \cdots, \qquad T^\nu \sim \int_{-\infty}^{\infty} \pi^o_{i\lambda} \rho_\nu(\lambda)\, d\lambda \oplus \pi^-_{-3} \oplus \pi^-_{-5} \cdots \oplus \pi^-_\nu,$$

where π_0^o should be replaced by π_1^+ in (iv) and π_{-1}^- in (v) respectively.

The result of the case $\nu = 0$ is due to Helgason, the result of the case $\nu = 2, 4, \cdots$ is due to Peetre-Peng-Zhang [29], the others are due to Liu and Peng [22]. It is clear that they are very different. So it is necessary to study the cases of different ν. Moreover, when ν is odd, a limit of discrete series representation appears in the decomposition. The different ν determines the irreducibility of eigenspace representations.

By Cayley transform: $z \to \frac{1+z}{1-z}i$, then the unit disc D becomes the upper half plane $U := \{x + iy : x \in R, y > 0\}$. The automorphism group of U is $SL(2, R)$. In [17] Jiang and Peng define the representation T^ν on $L^{\alpha,2}(U)$ similarly. The Casimir operator Δ_ν in this case becomes

$$\Delta_\nu := -y^2(\frac{\partial^2}{\partial x^2} + \frac{\partial^2}{\partial y^2}) + i\nu y(\frac{\partial}{\partial x} + i\frac{\partial}{\partial y}) - \frac{\nu(\nu-2)}{4}. \tag{5.3}$$

Let $L_+^{\alpha 2}$, $L_-^{\alpha 2}$ be the subspaces defined as follows:

$$\begin{aligned} L_+^{\alpha 2} &:= \{f(x,y) : f \in L^{\alpha 2}(U), \hat{f}(\xi, y) = 0, \xi \le 0\}, \\ L_-^{\alpha 2} &:= \{f(x,y) : f \in L^{\alpha 2}(U), \hat{f}(\xi, y) = 0, \xi \ge 0\}. \end{aligned} \tag{5.4}$$

An analyzing admissible wavelet ψ is a function on R satisfies:

$$C_\psi := \int_0^{+\infty} \frac{|\hat{\psi}(\xi)|^2}{\xi} d\xi < \infty, \quad \mathrm{supp}\hat{\psi} \subset [0, +\infty), \quad \hat{\psi} \quad \text{is real.}$$

The continuous wavelet transform of f with wavelet ψ is:

$$\begin{aligned} W_\psi f(b,a) :&= \frac{1}{\sqrt{a}} \int \overline{\psi}(\frac{x-b}{a}) f(x) dx \\ &= \frac{\sqrt{a}}{2\pi} \int_0^{+\infty} \hat{\psi}(a\xi) e^{i\xi b} \hat{f}(\xi) d\xi. \end{aligned}$$

The transform $f \to W_\psi f$ is an isometry from H^2 (Hardy space) into $L^2(U, \frac{da db}{a^2})$. Thus for $f \in H^2$, $a^{-\frac{\alpha+2}{2}} W_\psi f(b,a) \in L_+^{\alpha 2}$ and we denote

$$T_\psi f(b,a) := a^{-\frac{\alpha+2}{2}} W_\psi f(b,a) = \frac{1}{2\pi a^{\frac{\alpha+1}{2}}} \int_0^{+\infty} \hat{\psi}(a\xi) e^{i\xi b} \hat{f}(\xi) d\xi. \tag{5.5}$$

In the following, let Δ_ν $(\nu = \alpha + 2)$ denote

$$\Delta_\nu := -y^2(\frac{\partial^2}{\partial x^2} + \frac{\partial^2}{\partial y^2}) + i\nu y(\frac{\partial}{\partial x} + i\frac{\partial}{\partial y}).$$

Let Δ_ν^+ and Δ_ν^- be the restrict of Δ_ν on $L_+^{\alpha 2}$ and $L_-^{\alpha 2}$ respectively, i. e.

$$\Delta_\nu^+ := \Delta_\nu|_{L_+^{\alpha 2}}, \quad \Delta_\nu^- := \Delta_\nu|_{L_-^{\alpha 2}}.$$

By (5.5) and a direct calculation, [17] gets

$$\begin{aligned}\triangle_\nu T_\psi f(b,a) &= \triangle_\nu^+ T_\psi f(b,a)\\ &= \frac{1}{2\pi a^{\frac{\alpha+1}{2}}}\int_0^{+\infty}(D_+\hat{\psi})(a\xi)e^{i\xi b}\hat{f}(\xi)d\xi,\end{aligned}$$

where

$$D_+ := -x^2\frac{d^2}{dx^2} - x\frac{d}{dx} + x^2 - \nu x + \frac{(\nu-1)^2}{4}. \tag{5.6}$$

The differential operator D_+ was studied in [5], it is just the Schrodinger operator with the Morse potential. It has spectra:

$$\sigma(D_+) = \{(\frac{\nu-1}{2})^2 - (\frac{\nu-1}{2} - k)^2, k \in N, k < \frac{\nu-1}{2}\} \cup \{[(\frac{\nu-1}{2})^2, +\infty)\}.$$

Its eigenvector with respect to discrete spectrum λ_k is the function $\hat{\psi}_k$ on $R_+ = [0,\infty)$ satisfying

$$D_+\hat{\psi}_k = \lambda_k\hat{\psi}_k, \tag{5.7}$$

where $\lambda_k = (\frac{\nu-1}{2})^2 - (\frac{\nu-1}{2} - k)^2, k < \frac{\nu-1}{2}$. It turns out to be expressed by the Whittaker's function:

$$\hat{\psi}^k(x) = h_k(x) := (2x)^{\frac{\nu-1}{2}-k}e^{-x}{}_1F_1(-k;\nu-2k;2x), \tag{5.8}$$

where

$$M_{\mathcal{N},\mu_k}(z) = e^{-\frac{z}{2}}z^{\mu-k+\frac{1}{2}}{}_1F_1(\mu_k + \frac{1}{2} - \mathcal{N}; 1+2\mu_k; z).$$

Since $\frac{\nu-1}{2} - k > 0$, $\psi^k(x)$ is an analyzing admissiable wavelet. Let H^2 denote the Hardy space, then

$$A_k = \{T_{\psi^k}f(b,a) : \hat{\psi}^k(t) = h_k(t), f \in H^2\} \tag{5.9}$$

is the eigenspace of $\triangle_\nu$ with the eigenvalue $\lambda_k = (\frac{\nu-1}{2})^2 - (\frac{\nu-1}{2} - k)^2$ and it is the range of wavelet transform of functions in H^2 with wavelet ψ^k.

By (5.9), it is easy to get the reproducing kernel $K^{(k)}(z,w)$ of A_k :

$$K^{(k)}(z,w) = c_k y^{-\frac{\alpha+2}{2}}v^{-\frac{\alpha+2}{2}}\psi_y^k * \psi_v^k(x-u) \tag{5.10}$$

where $z = x+iy, w = u+iv$, $c_k = (C_{\psi^k})^{-1}$.

For $k < \frac{\nu-1}{2} = \frac{\alpha+1}{2}$, let P_k be the projection from $L^{\alpha 2}$ onto A_k, define the Toeplitz type operators:

$$T_b^{kk'} := P_k M_{\bar{b}} P_{k'} \tag{5.11}$$

where b is analytic on U. For their S_p−estimates, there is the following

Theorem 5.3. *For $T_b^{kk'}$ defined as above, then*

(1) If $k < k'$, then $T_b^{kk'} = 0$;

(2) If $k = k'$, then $T_b^{kk'} \in L^\infty$ iff $b \in L^\infty$, and $T_b^{kk'}$ never compact unless it is zero;

(3) If $k > k'$, $1 < p \le \infty$, then $T_b^{kk'} \in S_p$ iff $b \in B_p^{\frac{1}{p}}$;

(4) If $k > k'$, $0 < p \le 1$, $T_b^{kk'} \in S_p$, then $T_b^{kk'} = 0$.

From Theorem 5.4, for all $k, k', k > k'$, $T_b^{kk'}$ are operators having cut-off phenomenon at 1.

A parallel result to Theorem 5.4 on the case D was obtained by Peng and Zhang [43].

§6 Open Problems

1. As we seen in §4, the results of three kinds of Toeplitz-Hankel operators on U in Theirem 4.2, 4.3 and 4.4 are quite complete. But the corresponding results on D in Theirem 3.2, 3.3 and 3.4 are not yet complete. How to prove the complete results?

In the case U, the proofs used the integral formula involving three Laguerre polynomials, wuich can be calcuted not too hard. But in the case D, we have to calculate the integral formula involving three Jacobi polynomials, which are too hard. If this problem can be solved, the complete results can be done.

2. The Toeplitz type operators $T_b^{k,l,k'}$, when $k = k'$, are never compact unless $b = 0$. It is interesting to develop their spectral theory and index theory.

3. The generalizations of Hankel operators to higher dimension have been done by many authors. The most important case is the case of bounded symmetric domains or Siegel upper half planes. Both the invariant function theory and the invariant operator (i.e Toeplitz-Hankel operator) theory in this case are very active areas. [1], [28] and [29] are some of pioneer works.

References

1. J. Arazy, *Realization of the invariant inner products on the highest quotient of the composition series*, Ark. Mat. **30** (1992), 1–24.
2. J. Arazy, S. Fisher and J. Peetre, *Hankel operators in Bergman spaces*, Amer. J. Math. **110** (1988), pp.989–504.
3. S. Axler, *The Bergman kernel, the Bloch space, and commutators of multiptlication operators*, Duke Math. J. **53** (1986), pp. 315–332.
4. C. Berger, L. Coburn and K. Zhu, *BMO on the Bergman spaces of bounded symmetric domains*, Bull. Amer. Math. Soc. **17** (1987), 133–136.
5. I. Daubechies, J. Klauder and T. Paul, *Wiener measures for path integrals with affine kinematic variables*, J. Math. Phys. **28** (1987).
6. D. Deng and Q. Fan, *Trace ideal criteria for Hankel operators on Hardy space in the ball of C^n*, Acta Sci. Nat. Uni. Pekinensis **24:3** (1988), 257–268.
7. D. Deng and L. Peng, *High dimension Hankel operators in homogeneous self-adjoint cones and Schatten-von Neumann class, (in Chinese)*, Acta Math. Sinica **31:5** (1988), 623–633.
8. A. Erdélyi et al, "Tables of integral transforms", McGraw–Hill, New York–Toronto– London **Vol. 2** (1954).
9. P. Hartman, *On completely continuous Hankel matrices*, Proc. Amer. Math. Soc. **9** (1958), 852–866.
10. S. Janson, *Hankel operators between weighted Bergman spaces*, Ark. Mat. **26** (1988), pp205–219.

11. S. Janson and J. Peetre, *Paracommutators-boundedness and Schatten-von Neumann properties*, Trans. Amer. Math. Soc. **305** (1988), 467–504.
12. S. Janson and J. Peetre, *A new generalization of Hankel operators (the case of higher weights)*, Math. Nachr. **132** (1987), 313–328.
13. S. Janson and R. Rochberg, *Intermediate Hankel operators on the Bergman space*, J. Oper. Theory (to appear).
14. Q. Jiang, *Hankel type operators on $U \times U$*, Aproxamation Theory and Its Applications (1993) (to appear).
15. Q. Jiang and L. Peng, *Wavelet transform and Toeplitz- Hankel type operators*, Math. scand. **70** (1992), 247–264.
16. Q. Jiang and L. Peng, *Toeplitz and Hankel type operators on the upper half-plane*, Integral Equations and operator theory **15** (1992), 744–767.
17. Q. Jiang and L. Peng, *Casimir operator and wavelet transforms*, preprint.
18. C. Li, *Boundedness of paracommutators on L^p-spaces*, Acta Math. Sinica, New Series **6** (1990), 131–147.
19. C. Li, P. Lin and L. Peng, *Paracommutators in product space*, Acta Mathematica Scientia **13:1** (1993), 39–55.
20. C. Li, A. Mcintosh, Z. Wu and K. Zhang, *Compesated copactness, paracommutators, and Hardy spaces*, Macquarie Mathematics Reports **92–120** (1992).
21. P. Lin and L. Peng, *Besov spaces of Paley-Wiener type*, Lecture Notes in Math., Springer-Verlag **1494** (1991), 95- 112.
22. P. Lin and L. Peng, *Two-dimension Hankel operator of Schatten-von Neumann class*, Acta Sci. Nat. Uni. Pekinensis **26:1** (1990), 38–4.
23. H. Liu and L. Peng, *Weighted Plancheral formula. Irreducible unitary representations and eigenspace representations*, Math. Scand. (1993) (to appear).
24. C. A. McCarthy, *Cp*, Israel J.Math. **5** (1967), 249–271.
25. Z. Nehari, *On bounded bilinear forms*, Ann. Math. **65** (1957), 153–62.
26. T. Paul, *Functions analytic on the half-plane as quantum mechanical states*, J. Math. Phys. **25** (1985), pp3252–3263.
27. J. Peetre, "New thoughts on Besov spaces", Duke University, Durham, N. C. (1976).
28. J.Peetre, *Hankel forms of arbitrary weitht over a symmetric domain via the transvectant*, Technical Reports, Univ. Stochholm **16** (1992).
29. J. Peetre, L. Peng and G. Zhang, *A weighted Plancherel formula I, case of the disk. Applications to Hankel operators.*, Report Univ. Stockholm, Sweden **11** (1990).
30. J. Peetre and G. Zhang, *A weighted Plancherel formula III. The case of the hyperbolic matrix ball*, Technical Reports, Odense Universitet **27** (1992).
31. V.V. Peller, *Hankel operators of class γ_p and applications*, Math. USSR. Sbornik **41** (1982), pp. 443–479.
32. V.V. Peller, *A description of Hankel operators of class γ_p for $p > 0$ and investigation of the rate of rational approximation, and other application*, Math. USSR. Sbornik **50** (1985), pp. 465–494.
33. L. Peng, *Paracommutator of Schatten-von Neumann class $S_p, 0 < p < 1$*, Math. Scand. **61** (1987), pp68–92.
34. L. Peng, *Multilinear singular integrals of Schatten-von Nemann class S_p*, Approx. Theory & its appl. **4:1** (1988), 103–137.
35. L. Peng, *On the compactness of paracommutators*, Ark. Mat. **26:2** (1988), 315–325.
36. L. Peng, *Hankel operators on the Paley-Wiener space in disk*, Proc. CMA. ANU. **16** (1988), 173–183.
37. L. Peng, *Hankel operators on the Paley-Wiener space in R^d*, Integral Equations and Operator Theory **12** (1989), 567–591.
38. L. Peng, *Toeplitz and Hankel operators on Bergman space*, Mathematika **40** (1993), 345–356.
39. L. Peng, *Wavelets and paracommutators*, Ark. Mat. **31:1** (1993), 83–99.
40. L. Peng and T. Qian, *A kind if multilinear operator and the Schatten-von Neumann classes*, Ark. Mat. **27:1** (1989), 145–154.

41. L. Peng, R. Rochberg and Z. Wu, *Orthogonal polynomials and middle Hankel operators on Bergman space*, Studia Math. **102** (1992), 57–75.
42. L. Peng and C. Xu, *Jacobi polynomials and Toeplitz-Hankel type operators on Bergman space*, Complex Variables **23** (1993), 47–amstex hnkl 71.
43. L. Peng and C. Zhang, *Invariant Hankel operators on Bergman space (preprint).*
44. L. Peng and G. Zhang, *Middle Hankel operators on Bergman space*, In "Function Spaces", edited by K.Jarosz, Lecture Notes in Pure and Applied Math. Series 136, Marcel Dekker Dec. (1991), 225–236.
45. R. Rochberg, *Trace ideal criteria for Hankel operators and commutators*, Indiana Univ. Math. J. **31** (1982).
46. S. Semmes, *Trace ideal criterion for Hankel operators,* $0 < p < 1$, Integral Equations Oper. Theory **7** (1984), pp. 241–281.
47. G. Szegö, "Orthogonal polynomials", Amer. Math. Soc. Colloq. Publications **Vol. 23** (1939).
48. H. Triebel, "Theory of Function Spaces", Birkhäuser Verlag , Basel-Boston-Stuttgart (1983).
49. J. Xiao, *Boundedness and compactness of Hankel operators on Bergman space*, Acta Mathematica Scientia **13:1** (1993), 56–65.
50. C. Xu and C. Zhang, *Middle Hankel operators on* $D \times D$, Acta Math. Sinica (1993) (to appear).
51. K. Zhu, "Operator theory on function spaces", Marcel Dekker (1990).

OPERATORS-DERIVATIVES-SPACES-DIFFERENTIAL EQUATIONS ON LOCALLY COMPACT VILENKIN GROUPS

Su Weiyi*

Nanjing University, China

Abstract

Our main purpose is to establish the theroy of Gibbs type differential operators and equations over locally compact Vilenkin groups. In this paper, as the first part, we study the following topics : pseudo–differential operators and their boundedness on Sobolev spaces; the concept of para–differential operators, Gibbs type differential operators and equations. Then [21] as the second part is to study the boundedness of para–differential operators and the para–linearization of non–linear Gibbs type differential equations.

§ 1. *Introduction*

It is a very interesting topic to establish the theory of Gibbs differential operators and equations on locally compact Vilenkin groups (simply, Vilenkin groups), and certainly, it is difficult to study this topic because the structure of a Vilenkin group G is quite different from that of $\mathbb{R}^n$ so that classical derivatives and other powerful tools do not work on G. Thus, we have to introduce new concept of derivatives on G. Fortunately, Gibbs derivative, simply, G–derivative may play this role .

Gibbs, J.E. introduced the new concept of derivative in 1967 [5], named G–derivative. Then lots of mathematicians have improved the definition, studied properties and given applications during the last 25 years (see [2–4], [6–11], [13–15], [18,19], [23,24]). The author of [14] pointed out that there are three important marks of development of G–derivatives : the paper [5] published in 1967, paper [3] in 1972 and paper [10] in 1977. Some special Gibbs differential equations also introduced and investigated (see [4], [7,8], [11], [15]). However, there is no systemetic discussion for G–differential operators and equations yet. In the series papers, including this one and [21], we will devote to these topics and try to establish basic theory of G–differential equations partially on the Vilenkin groups. This pa-

* Supported by the National Natural Science Foundation of China

M. Cheng et al. (eds.), Harmonic Analysis in China, 240–255.

per is arranged as follows : in § 2, we list some basic notation and concepts; then in § 3, prove the decomposition theorem of symbols of pseudo–differential operators, give some examples of evaluating G–derivatives and applying to solve a simple G–differential equation; in § 4, we define so called para–differential operators, introduce general nonlinear G–differential equations as a preparing for paper [21].

§ 2. *Notation and Definitions*

Let G be a locally compact Abelian topological group containing a strictly decreasing sequence of open and compact subgroups $\{ G_n \}_{n=-\infty}^{\infty}$ which satisfy

(i) $\bigcup_{n=-\infty}^{\infty} G_n = G$, and $\bigcap_{n=-\infty}^{\infty} G_n = \{0\}$,

(ii) $2 \leqslant M = \sup\{ \text{order}\,(G_n / G_{n+1}) : n \in \mathbb{Z} \} < +\infty$.

This G is called a Vilenkin group. The most important difference between the Vilenkin group G and the real line $\mathbb{R}$ (regarded as a group under the operation "+") is that G is totally disconnected but $\mathbb{R}$ is connected.

If G is a Vilenkin group, then it can be equipped with a non–Archimedean norm $|\cdot|$ which has discrete values $|\cdot| \in \{ m_k : k \in \mathbb{Z} \}$, where

$$0 \leqslant \cdots < m_{k-1} < m_k < m_{k+1} < \cdots, \qquad k \in \mathbb{Z}, \qquad \text{and} \qquad m_0 = 1 .$$

This norm satisfies (i) $|x| \geqslant 0$, $|x| = 0 \Leftrightarrow x = 0$, (ii) $|x+y| \leqslant \max\{ |x|, |y| \}$.

Moreover, the zero element $0 \in G$ has an open and compact topological base, hence G is 0 dimensional. The local fields, including p–series fields, p–adic number fields and their finite algebraic extensions are the examples of such groups (see [10], [22]).

Let μ be the Haar measure on a Vilenkin group G with $\mu(G_0) = 1$ and $\mu(G_n) = m_n^{-1}$, $n \in \mathbb{Z}$. Set $\kappa_n = \text{order}\,(G_n / G_{n+1}) \leqslant M$, then $m_{n+1} = \kappa_n m_n$. Each $x \in G$ can be written as

$$x = x_n + x_{n+1} + \cdots, \qquad n \in \mathbb{Z},$$

where $x_j \in G_j \setminus G_{j+1}$, $j = n, n+1, \cdots$, and if $x_n \neq 0$, then $|x| = m_n^{-1}$.

If Γ is the character group of G, i.e., Γ is the collection of all functions which are bounded and multiplicative on G, then by the duality theory, for $\chi \in \Gamma$, $\chi : G \to \mathbb{C}$, we have $\langle \chi, x \rangle \in \{ z \in \mathbb{C} : |z| = 1 \}$, and for $\chi, \eta \in \Gamma$,

$$\langle \chi, x+y \rangle = \langle \chi, x \rangle \langle \chi, y \rangle, \qquad x, y \in G,$$

$$\langle \chi + \eta, x \rangle = \langle \chi, x \rangle \langle \eta, x \rangle, \qquad x \in G,$$

Denote by Γ_n the annihilator of G_n, i.e.,

$$\Gamma_n = \{\xi \in \Gamma : \langle \xi, x \rangle = 1 \quad \text{for all } x \in G_n\}.$$

Then $\{\Gamma_n\}_{n=-\infty}^{\infty}$ forms a strictly increasing sequence of open and compact subgroups in Γ, and

(i)′ $\bigcup_{n=-\infty}^{\infty} \Gamma_n = \Gamma$, and $\bigcap_{n=-\infty}^{\infty} \Gamma_n = \{I\}$,

(ii)′ $\text{order}(\Gamma_{n+1} / \Gamma_n) = \text{order}(G_n / G_{n+1})$, $n \in \mathbf{Z}$.

The Haar measure λ on Γ can be chosen such that $\lambda(\Gamma_0) = \mu(G_0) = 1$ and $\lambda(\Gamma_n) = (\mu(G_n))^{-1} = m_n$, $n \in \mathbf{Z}$. Each $\xi \in \Gamma$ has the expression

$$\xi = \xi_n + \xi_{n-1} + \cdots, \qquad n \in \mathbf{Z},$$

where $\xi_j \in \Gamma_j \setminus \Gamma_{j-1}$, $j = n, n-1, \cdots$, and if $\xi_n \neq 0$, then $|\xi| = m_n$.

Hence, we have the correspondence $\xi \rightarrow y$ from Γ onto G, and it is one–to–one, such that $\langle \xi, x \rangle \equiv \chi_\xi(x) \rightarrow \chi_y(x)$. And if G is a local field, then $\chi_y(x)$ is determined by $\chi(xy)$. Moreover, $\chi \in \Gamma$ can be chosen such that it is trivial on G_0 but non–trivial on G_{-1} (for details we refer to [22] and [24]).

Dyadic Walsh system, the well–known function system, $\{w_y(x) : x, y \in G\}$, on which Gibbs, J.E. [5] and Butzer, P.L. et al [3] introduced a new concept of derivatives and differential equations firstly, turns out the character group Γ of the following group G:

$$G = \left\{ x = \sum_{j=s}^{\infty} x_j 2^{-j} : x \in \{0, 1\},\ x_s \neq 0,\ s \in \mathbf{Z} \right\}.$$

Obviously, G is an Abelian group with the addition "+" : $x + y = (x_j + y_j \mod 2)$, moreover, $0 \in G$ has a basic neighborhood system $\mathscr{B} = \{\mathscr{B}_k : k \in \mathbf{Z}\}$, here $\mathscr{B}_k = \{y \in G : y = (y_k, y_{k+1}, \cdots),\ y_j \in \{0, 1\},\ j = k, k+1, \cdots \in \mathbf{Z},\ y_k \neq 0\}$. Thus G is a Vilenkin group underlying on the set $(-\infty, \infty)$ equipped with another topological structure which is quite different from the usual $\varepsilon-$ neighborhood topology on $\mathbb{R}$.

§ 3. *Pseudo–differential operators and Gibbs type derivatives*

Saloff–Coste, L. introduced the pseudo–differential operators on Vilenkin groups in 1986 [12]. He defined firstly the basis function class $S(G)$ and $S(\Gamma)$, symbol class $S^m_{\rho,\delta}$, $m \in \mathbb{R}$, $\delta \geqslant 0$, $\rho \geqslant 0$; gave the corresponding symbolic calculus : composition, transpose, adjoint; then defined pseudo–differential operators and studied its L^p–continuity, $1 < p$

$<\infty$, i.e., for pseudo-differential operator T_σ with the symbol class $\sigma \in S^m_{\rho,\delta}$, $m \in \mathbb{R}$, $0 < \delta \leqslant \rho < 1$, or $0 \leqslant \delta < \rho \leqslant 1$, if $(1-\rho)\left|\frac{1}{2} - \frac{1}{p}\right| \leqslant -m$, he proved that T_σ preserves the L^p-boundedness. Su improved the definition of the symbol class in [16] and proved the $B(\alpha, r, s)$-continuity for those operators with the symbol in $S^m_{\rho,\delta}$, $m < 0, m + ([\alpha]+1) + 2(1-\rho) < 0, 1 \leqslant r, s < \infty, \alpha > 0$. Furthermore, in [18], we defined the Gibbs type derivatives and integrals on G by means of the pseudo-differential operators and studied their certain important properties.

Here we give some definitions and properties, which have more or less revisions of those in [12], [16] and [18]. In the end of this section, we show four examples for evaluating Gibbs type derivatives and solving for a simple Gibbs type differential equation.

Definition 1. (basis function class $S(G)$) If function $\varphi : G \to \mathbb{C}$ satisfies

(i) for any $N \in \mathbb{P}$, $\mathbb{P} = \{0, 1, \cdots\}$, there is a constant $c_N > 0$ such that

$$|\varphi(x)| \leqslant c_N <x>^{-N}, \qquad x \neq 0, \tag{1}$$

where $<x> = \begin{cases} \max\{1, |x|\}, & x \neq 0, \\ 0, & x = 0; \end{cases}$

(ii) for any $(\mu, N) \in \mathbb{P} \times \mathbb{P}$, there is a constant $c_{\mu,N} > 0$ such that for $h \in G$,

$$|\Delta_h \varphi(x)| \leqslant c_{\mu,N} |h|^\mu <x>^{-N}, \qquad x \neq 0, \tag{2}$$

where $\Delta_h \varphi(x) = \varphi(x+h) - \varphi(x)$. Then we say that φ is a basic function on G, and denote by $S(G)$ the set of all basic functions. $S(G)$ is a Fréchet space with certain semi-norms [12].

Definiton 2. (symbol class $S^m_{\rho,\delta}$) For $m \in \mathbb{R}$, $\rho, \delta \geqslant 0$, if function $\sigma(x, \xi)$: $G \times \Gamma \to \mathbb{C}$ satisfies :

(i) there exists a constant $c > 0$, such that

$$|\sigma(x, \xi)| \leqslant c <\xi>^m, \qquad \xi \neq 0, \tag{3}$$

where $<\xi> = \begin{cases} \max\{1, |\xi|\}, & \xi \neq 0, \\ 0, & \xi = 0; \end{cases}$

(ii) for any $(\mu, \nu) \in \mathbb{P} \times \mathbb{P}$, there exist constants $c_{\mu,\nu}, c_\mu, c_\nu > 0$, such that

$$\left|\Delta^x_h \Delta^\xi_\zeta \sigma(x, \xi)\right| \leqslant c_{\mu,\nu} |h|^\mu |\zeta|^\nu <\xi>^{m+\delta\mu-\rho\nu}, \qquad \xi \neq 0, \tag{4}$$

$$\left|\Delta^x_h \sigma(x, \xi)\right| \leqslant c_\mu |h|^\mu <\xi>^{m+\delta\mu}, \qquad \xi \neq 0, \tag{5}$$

$$\left|\Delta^\xi_\zeta \sigma(x, \xi)\right| \leqslant c_\nu |\zeta|^\nu <\xi>^{m-\rho\nu}, \qquad |\zeta| < <\xi>, \qquad \xi \neq 0, \tag{6}$$

where $h \in G$, $\zeta \in \Gamma$. Then we say that $\sigma(x,\xi) \in S^m_{\rho,\delta}$, and call $S^m_{\rho,\delta}$ the symbol class with order m and type (ρ,δ).

With certain semi–norms, $S^m_{\rho,\delta}$ is also a Fréchet space.

Definition 3. (pseudo–differential operator) The pseudo–differential operator T_σ on $S(G)$ with symbol $\sigma \in S^m_{\rho,\delta}$ is defined by

$$T_\sigma f(x) = \int_\Gamma \left\{ \int_G \sigma(x,\xi) f(t) \bar{\chi}_\xi(t-x)\, dt \right\} d\xi, \tag{7}$$

for $f \in S(G)$.

We have the decomposition theorem of $S^m_{\rho,\delta}$ which will be used in § 3.

Theorem 1. *Let* $\sigma \in S^m_{\rho,\delta}$, *and* Φ_{Γ_0}, $\Phi_{\Gamma_j \setminus \Gamma_{j-1}}$ *be the characteristic functions of* Γ_0, $\Gamma_j \setminus \Gamma_{j-1}$, $j > 0$, *respectively. Then*

$$\sigma(x,\xi) = \sum_{k,j=0}^{\infty} \omega_{kj}(x)\, \varphi_{kj}(\xi), \tag{8}$$

where

$$\omega_{kj}(x) = \begin{cases} \displaystyle\int_\Gamma \sigma(x,\xi)\, \Phi_{\Gamma_0}(\xi)\, \bar{\chi}_{V(k)}(\xi)\, d\xi, & j = 0, \\ \displaystyle\int_\Gamma \sigma(x,\eta)\, \Phi_{\Gamma_0 \setminus \Gamma_{-1}}(\xi)\, \bar{\chi}_{V(k)}(\xi)\, d\xi, & j > 0, \end{cases} \tag{9}$$

with $|\eta| = m_j |\xi|$, *and*

$$\varphi_{kj}(\xi) = \begin{cases} \Phi_{\Gamma_0}(\xi)\, \chi_{V(k)}(\xi), & j = 0, \\ \Phi_{\Gamma_j \setminus \Gamma_{j-1}}(\xi)\, \chi_{V(k)}(\theta) = \Phi_{\Gamma_0 \setminus \Gamma_{-1}}(\theta)\, \chi_{V(k)}(\theta), & j > 0, \end{cases} \tag{10}$$

with $|\theta| = m_j^{-1} |\xi|$. *Moreover, when* $m \in \mathbb{R}$, $\rho > 1$, $\delta \geqslant 0$, *or* $m + 2(1-\rho) < 0$, the series (8) *is absolutely uniformly convergent.* $\{V(k)\}_{k=0}^{\infty}$ *in* (9) *and* (10) *is the set of complete list of distinct coset representatives of* G_0 *in* G *(we refer to* [22] *for determinating* $\{V(k)\}_{k=0}^{\infty}$ *);* $\{ \chi_{V(k)}|_{\Gamma_0} \}_{k=0}^{\infty}$ *is the orthonormal complete system on* G_0.

Remark 1. This theorem is the generalization of the Lemma 2 in [16] .

Proof. By the Fourier series theory on groups [1], it is easy to expand $\sigma(x,\xi)$ as

$$\sigma(x,\xi) \sim \sum_{k,j=0}^{\infty} \omega_{kj}(x)\, \varphi_{kj}(\xi) \tag{11}$$

with (9) and (10), (see [16] also). If $m \in \mathbb{R}$, $\rho > 1$, $\delta \geqslant 0$, using the following formula for $j > 0$

$$|\chi_{V(k)}(\zeta)-1||\omega_{kj}(x)| = \left|\int_\Gamma \Delta_\zeta^\xi \sigma(x,\eta)\Phi_{\Gamma_j\setminus\Gamma_{j-1}}(\eta)\,\bar{\chi}_{V(k)}(\xi)\,d\xi\right| \tag{12}$$

with $|\eta| = m_j|\xi|$ and $|\zeta|<1$, by the inequality (6) of $\sigma(x,\xi)$ with $\nu=\beta+1$, and in virtue of the Lemma 2 in [18], we get

$$\begin{aligned}|V(k)|^\beta\,|\omega_{kj}(x)| &\leqslant \left(\int_{|\zeta|<1}\frac{|\chi_{V(k)}(\zeta)-1|}{|\zeta|^{\beta+1}}\,d\zeta\right)|\omega_{kj}(x)|\\ &\leqslant \left(\int_{|\zeta|<1}\frac{|\chi_{V(k)}(\zeta)-1||\omega_{kj}(x)|}{|\zeta|^{\beta+1}}\,d\zeta\right)\\ &\leqslant \int_{|\zeta|<1} c_\beta m_j^{\,m+(1-\rho)\beta}\,d\zeta \leqslant c_\beta m_j^{\,m+(1-\rho)\beta}\,.\end{aligned}$$

This implies

$$|\omega_{kj}(x)| \leqslant c_\beta m_j^{\,m+(1-\rho)\beta}\,|V(k)|^{-\beta}\,. \tag{13}$$

Noting that $|V(k)|\geqslant m_k$ for large k (see [22]), and choosing β large enough, we know that the series (11) converges absolutely uniformly for $m\in\mathbb{R}$, $\delta\geqslant 0$, $\rho>1$. Similar argument holds for the case $m+2(1-\rho)<0$. The theorem is proved.

Having verified that $\sigma(x,\xi) = <\xi>^m \in S^m_{\rho,\delta}$ (see [18]), we may define the Gibbs type derivatives.

Definition 4. (Gibbs type derivative) For a Haar measurable function f on G, if the integral

$$T_{<\cdot>^m} f(x) = \int_\Gamma\left\{\int_G <\xi>^m f(t)\,\bar{\chi}_\xi(t-x)\,dt\right\}d\xi\,,\quad m\geqslant 0\,, \tag{14}$$

exists at $x\in G$, then it is called (pointwise) Gibbs type derivative of order m of f at x, denoted by $f^{<m>}(x)$. Moreover, let

$$f_n(x) = \begin{cases} f(x)\,, & |x|\leqslant m_n\,,\\ 0\,, & |x|>m_n\,,\end{cases}\qquad n\in\mathbb{Z}\,.$$

If for $m\geqslant 0$, there exists $g\in L^r(G)$, $1\leqslant r<\infty$, such that

$$\|g - T_{<\cdot>^m} f_n\|_{L^r}\to 0\,,\qquad n\to+\infty\,,$$

then g is called L^p–strong Gibbs type derivative, denoted by $D^{<m>}f$.

Remark 2. Onneweer, C.W. introduced the Gibbs derivatives on p–series fields and p–adic number fields firstly [10,11]. Zheng, W.X. defined the other form of G–derivatives independently [23]. They both gave some interesting properties and applications. On the other hand, It is clear that the above definition 4 contains the cases of integer and fractional order derivatives. Moreover, this definition, as we have pointed out in [18], has many advantages in theory, calculation and applications. We now just list two special useful theorems as follows (see [18]).

Theorem 2. *The operator* $T_{<\cdot>^m}$ *is a homeomorphism from* $S(G)$ *onto* $S(G)$, $m \in \mathbb{R}$. *Furthermore, it is also a homeomorphism from the test function class* $\mathcal{S}(G)$ *of* G *onto itself. And for any* $m \geqslant 0$ *and* $\varphi \in S(G)$ *or* $\mathcal{S}(G)$, *the following formula holds*

$$\varphi^{<m>}(x) = D^{<m>}\varphi(x) . \tag{15}$$

By this theorem, we know that the Gibbs type derivative as an operation is closed on $S(G)$ and $\mathcal{S}(G)$, thus it makes us to generalize definition 4 to distributions in the distribution space $S'(G)$ or $\mathcal{S}'(G)$.

Definition 5. (G–derivative of distribution) For $f \in S'(G)$, (resp. $\mathcal{S}'(G)$) , and $m \geqslant 0$, if there exists a distribution $g \in S'(G)$, (resp. $g \in \mathcal{S}'(G)$), which satisfies

$$< g, \varphi > \; = \; < f, \varphi^{<m>} > , \qquad \text{for all } \varphi \in S(G) ,$$

$$(\text{resp.} \quad < g, \varphi > \; = \; < f, \varphi^{<m>} > , \qquad \text{for all } \varphi \in \mathcal{S}(G)) ,$$

then we say that $g \equiv f^{<m>}$ is the Gibbs type derivative of f in the distribution sense.

Noting that the Fourier transform is a homeomorphism from $S'(G)$ onto $S'(\Gamma)$, (resp. $\mathcal{S}'(G)$ onto $\mathcal{S}'(\Gamma)$), we can prove

Theorem 3. *If* $f \in S(G)$ *or* $f \in \mathcal{S}(G)$, *then*

$$(f^{<m>}(\cdot))^{\wedge}(\xi) = <\xi>^m f^{\wedge}(\xi) , \qquad a.e. \ \xi \in \Gamma . \tag{16}$$

And if $f \in S'(G)$ *or* $f \in \mathcal{S}'(G)$, *then*

$$(f^{<m>})^{\wedge} = <\xi>^m f^{\wedge} , \qquad a.e. \ \xi \in \Gamma , \tag{17}$$

in the distribution sense. Where " $\wedge$ " is the operation of Fourier transform.

Example 1. Evaluate the Gibbs type derivative of $f(x) = 1$, $x \in G$.

Since $f(x) = 1 \in L_{loc}(G)$, and $f^{\wedge} = 1^{\wedge} = \delta$, so we have

$$(f^{<m>})^{\wedge} = <\xi>^m f^{\wedge} = <\xi>^m \delta , \tag{18}$$

in the distribution sense. If "$\vee$" is the operation of inverse Fourier transform, then (18) implies

$$f^{<m>} = (<\xi>^m \delta)^{\vee} = <\xi>^m |_{\xi=0} \delta^{\vee} = 0 ,$$

the second equality holds due to the following evaluation : for $\varphi \in \mathcal{S}(G)$,

$$< (<\xi>^m \delta)^{\vee}, \varphi > = < <\xi>^m \delta, \varphi^{\vee} > = < \delta, <\xi>^m \varphi^{\vee} >$$

$$= \{ <\xi>^m \varphi^{\vee} \}|_{\xi=0} = <\xi>^m |_{\xi=0} \varphi^{\vee}(0) = <\xi>^m |_{\xi=0} < \delta, \varphi^{\vee} >$$

$$= <\xi>^m\big|_{\xi=0} < \delta^\vee, \varphi > = < <\xi>^m\big|_{\xi=0} \delta^\vee, \varphi > .$$

Thus

$$f^{<m>} = <\xi>^m\big|_{\xi=0} \delta^\vee = <\xi>^m\big|_{\xi=0} \times 1 = 0 .$$

Example 2. Let $f(x) = \begin{cases} |x|, & |x| \leqslant 1, \\ 0, & |x| > 1, \end{cases}$ evaluate $f^{<m>}$.

By definition

$$f^{<m>}(x) = \int_\Gamma <\xi>^m \int_{G_0} |t| \, \bar{\chi}_\xi(t) \, dt \, \chi_x(\xi) \, d\xi ,$$

and applying for the formula for $j \geqslant 0$

$$\int_{|\xi| = m_j} \chi_x(\xi) \, d\xi = \begin{cases} m_j \left(1 - \dfrac{m_{j-1}}{m_j} \right), & |x| \leqslant m_j^{-1}, \\ -m_{j-1}, & |x| = m_{j-1}^{-1}, \\ 0, & |x| > m_{j-1}^{-1}, \end{cases}$$

we get for $l \geqslant 0$ (omit the detials)

$$f^{<m>}(x) = \begin{cases} 0, & |x| > 1, \\ -r_{l+1} m_l + \sum\limits_{j=1}^{l} r_j (m_j - m_{j-1}) + \sum\limits_{j=-\infty}^{0} m_j (m_j - m_{j-1}), \\ & |x| = m_l^{-1}, \end{cases}$$

with

$$r_j = -m_{j+1} m_{-j} + \sum_{k=-\infty}^{j} m_k (m_k - m_{k-1}), \qquad j \in \mathbb{P} .$$

Example 3. Let $f(x) = \begin{cases} 1, & |x| \leqslant 1 \\ 0, & |x| > 1 \end{cases} = \Phi_{G_0}$ evaluate $f^{<1>}$.

It follows that

$$f^{<1>}(x) = \int_\Gamma <\xi> \int_G \Phi_{G_0}(t) \, \bar{\chi}_\xi(t) \, dt \, \chi_x(\xi) \, d\xi$$

$$= \int_\Gamma <\xi> \Phi_{\Gamma_0}(\xi) \, \chi_x(\xi) \, d\xi = \int_\Gamma \Phi_{\Gamma_0}(\xi) \, \chi_x(\xi) \, d\xi$$

$$= \Phi_{G_0}(x) = f(x) .$$

It is esay to check that (16) holds : $[f^{<1>}(\cdot)]^\wedge(\xi) = <\xi> f^\wedge(\xi)$, a.e.

Example 4. Solve for the following Gibbs type differential equation (simply, G-differential equation) :

$$f^{<2>}(x) + f(x) = \delta , \qquad x \in G .$$

Using the Fourier Transform Method [1] and the formula (16),

$$<\xi>^{2} f^{\wedge}(\xi) + f^{\wedge}(\xi) = \delta^{\wedge} ,$$

it turns out

$$f^{\wedge}(\xi)\,(1 + <\xi>^{2}) = 1 ,$$

thus

$$f(x) = \int_{\Gamma} \frac{1}{1 + <\xi>^{2}} \chi_x(\xi)\, d\xi = \frac{1}{2}\int_{\Gamma_0} \chi_x(\xi)\, d\xi +$$

$$+ \int_{\Gamma\setminus\Gamma_0} \frac{1}{1 + |\xi|^2} \chi_x(\xi)\, d\xi \equiv J_1 + J_2 .$$

For J_2 , it follows that

$$J_2 = \int_{\Gamma\setminus\Gamma_0} \frac{1}{1 + |\xi|^2} \chi_x(\xi)\, d\xi = \sum_{j=1}^{\infty} \int_{|\xi| = m_j} \frac{1}{1 + |\xi|^2} \chi_x(\xi)\, d\xi$$

$$= \sum_{j=1}^{\infty} \frac{1}{1 + m_j^{\,2}} \int_{|\xi| = m_j} \chi_x(\xi)\, d\xi$$

$$= \sum_{j=1}^{\infty} \frac{1}{1 + m_j^{\,2}} \begin{cases} m_j\left(1 - \dfrac{m_{j-1}}{m_j}\right), & |x| \leqslant m_j^{-1} , \\ -m_{j-1} , & |x| = m_{j-1}^{-1} , \\ 0 , & |x| > m_{j-1}^{-1} , \end{cases}$$

$$= \begin{cases} -\dfrac{m_k}{1 + m_{k+1}^{\,2}} + \displaystyle\sum_{j=1}^{k} \frac{m_j - m_{j-1}}{1 + m_j^{\,2}} , & |x| = m_k^{-1}, \quad k \geqslant 1 , \\ -\dfrac{m_0}{1 + m_1^{\,2}} , & |x| = m_0^{-1} , \\ 0 , & |x| > 1 , \end{cases}$$

For J_1 , it is easy to know

$$J_1 = \frac{1}{2}\int_{\Gamma_0} \chi_x(\xi)\, d\xi = \begin{cases} \dfrac{1}{2} , & |x| \leqslant 1 , \\ 0 , & |x| > 1 . \end{cases}$$

Thus we get

$$f(x) = \begin{cases} \dfrac{1}{2} - \dfrac{|x|}{|x|^2 + \kappa_{k+1}^{\ 2}} + \sum\limits_{j=1}^{k} \dfrac{m_j - m_{j-1}}{1 + m_j^{\ 2}}, & |x| = m_k^{\ -1}, \quad k \geq 1, \\ \dfrac{1}{2} - \dfrac{|x|}{|x|^2 + \kappa_1^{\ 2}}, & |x| = m_0^{\ -1}, \\ 0, & |x| > 1, \end{cases}$$

There are we other examples for evaluating the G–derivatives of fractals in [19].

§ 4. *The boundedness of* T_σ *on the space* W^s

As we know, in classical case the Sobolev space is a very important function class that plays essential role in partial differential equations, harmonic analysis and approximation theory. It is nature to ask if we can define Sobolev type space on Vilenkin groups.

We define the Lebesgue type space $L(s, r)$ firstly.

$$L(s,r) = \left\{ f \in \mathcal{S}'(G) : \| f \|_{L(s,r)} = \| (< \cdot >^s f^\wedge(\cdot))^\vee \|_{L^r} < +\infty \right\}$$

with $s \in \mathbb{R}$, $1 \leq r \leq \infty$. And the Sobolev type space $W^s = L(s, 2)$:

$$W^s = \left\{ f \in \mathcal{S}'(G) : \| f \|_{W^s} = \| (< \cdot >^s f^\wedge(\cdot))^\vee \|_{L^2} < +\infty \right\}. \tag{19}$$

Obviously, W^s has the equivalent norm by the Parseval theorem :

$$\| f \|_{W^s} = \left\{ \int_\Gamma < \xi >^{2s} | f^\wedge(\xi) |^2 \, d\xi \right\}^{\frac{1}{2}}.$$

There are the equivalence theorems for $L(s, 2)$ and W^s, Theorem 4, which is the improvement of Theorem 1 in [20].

Theorem 4. *If* $s \in \mathbb{R}$, *then the following statements are equivalent* :

(i) $f \in W^s$,

(ii) f *has a decomposition* $f = \sum\limits_{j=0}^{\infty} f_j$, *and there exist constants* c_j, $j \in \mathbb{P}$, *such that*

① $\operatorname{supp} f_0^\wedge \subset \Gamma_0$, $\operatorname{supp} f_j^\wedge \subset \Gamma_j \setminus \Gamma_{j-1}$, $j > 0$,

② $\| f_j \|_{L^2} \leq c_j m_j^{-s}$, $j = 0, 1, 2, \cdots$, with $\left\{ \sum\limits_{j=0}^{\infty} c_j^{\ 2} \right\}^{\frac{1}{2}} < \infty$.

(iii) f *has a decomposition* $f = \sum\limits_{j=0}^{\infty} f_j$, *and there exist constants* c_j, $j \in \mathbb{P}$, *such that*

① $\operatorname{supp} f_0^\wedge \subset \Gamma_0$, $\operatorname{supp} f_j^\wedge \subset \Gamma_j$, $\quad j>0$,

② $\|f_j\|_{L^2} \leqslant c_j m_j^{-s}$, $\quad j=0,1,2,\cdots$, $\quad$ with $\quad \left\{\sum_{j=0}^{\infty} c_j^2\right\}^{\frac{1}{2}} < \infty$.

This theorem also holds for $L(s,r)$, $1<r<\infty$ if ② is replaced by ②′

②′ $\|f_j\|_{L^2} \leqslant c_j m_j^{-s}$, $\quad j=0,1,2,\cdots$, $\quad$ with $\quad \left\{\sum_{j=0}^{\infty} c_j^r\right\}^{\frac{1}{r}} < \infty$.

We now prove the W^s boundedness of pseudo-differential operators.

Theorem 5. *If* $\sigma \in S^m_{\rho,\delta}$, $m,\delta \geqslant 0$, $\rho>0$, *or* $m+2(1-\rho)<0$, *then the operator* $T_\sigma : W^s \to W^{s-m}$ *is bounded, i.e., there exists a constant c, such that*

$$\|T_\sigma u\|_{W^{s-m}} \leqslant c\,\|u\|_{W^s}. \tag{20}$$

Proof. Take the Littlewood–Paley decomposition [20] of $u \in W^s$

$$u = u*\Delta_0 + \sum_{j=1}^{\infty} u*(\Delta_j - \Delta_{j-1}) = \sum_{j=0}^{\infty} u_j,$$

where $\Delta_n(x) = m_n \Phi_{G_n}(x)$, and Φ_{G_n} is the characteristic function of G_n. Thus, the statement (ii) in Theorem 4 is satisfied. Using the decomposition formula (8) of σ with (9) and (10), we reduce by $\operatorname{supp}\varphi_{kj} \cap \operatorname{supp} u_l^\wedge = \emptyset$, $j \neq l$

$$T_\sigma u(x) = \int_\Gamma \sigma(x,\xi)\, u^\wedge(\xi)\,\chi_x(\xi)\,d\xi = \int_\Gamma \sigma(x,\xi) \sum_{l=0}^{\infty} u_l^\wedge(\xi)\,\chi_x(\xi)\,d\xi$$

$$= \sum_{l=0}^{\infty} T_\sigma u_l(x) = \sum_{l=0}^{\infty} \int_\Gamma \sum_{k,j=0}^{\infty} \omega_{kj}(x)\,\varphi_{kj}(\xi)\,u_l^\wedge(\xi)\,\chi_x(\xi)\,d\xi$$

$$= \sum_{l=0}^{\infty}\sum_{k=0}^{\infty} \int_\Gamma \omega_{kl}(x)\,\varphi_{kl}(\xi)\,u_l^\wedge(\xi)\,\chi_x(\xi)\,d\xi \equiv \sum_{l=0}^{\infty}\sum_{k=0}^{\infty} I_{kl}.$$

Then by (10) and (13) for $l \geqslant 1$, it turns out (recall that $|\theta| = m_l^{-1}|\xi|$)

$$\|T_\sigma u_l\|_{L^2} = \left\{ \int_G |T_\sigma u_l(x)|^2\,dx \right\}^{\frac{1}{2}}$$

$$= \left\{ \int_G \left| \sum_{k=0}^{\infty} \int_\Gamma \omega_{kl}(x)\,\Phi_{\Gamma_l \setminus \Gamma_{l-1}}(\xi)\,\chi_{V(k)}(\theta)\,u_l^\wedge(\xi)\,\chi_x(\xi)\,d\xi \right|^2 dx \right\}^{\frac{1}{2}}$$

$$\leqslant \sum_{k=0}^{\infty} \left\{ \int_G \left| \omega_{kl}(x) \int_\Gamma u_l^\wedge(\xi)\,\chi_{V(k)}(\theta)\,\chi_x(\xi)\,d\xi \right|^2 dx \right\}^{\frac{1}{2}}$$

$$\leqslant \sum_{k=0}^{\infty} \| \omega_{kl} \|_{L^\infty} \| u_l \|_{L^2} \leqslant \sum_{k=0}^{\infty} c_\beta m_l^{m+(1-\rho)\beta} | V(k) |^{-\beta} \| u_l \|_{L^2}$$

$$\leqslant c_\beta m_l^{m+(1-\rho)\beta} c_l' m_l^{-s} \| u \|_{W^s} \leqslant c_l m_l^{-(s-m)} \| u \|_{W^s} .$$

Moreover, we have

$$(T_\sigma u_l(\cdot))^\wedge(\xi) = \sum_{k=0}^{\infty} I_{kl}{}^\wedge(\xi) = \sum_{k=0}^{\infty} \int_\Gamma \omega_{kl}{}^\wedge(\xi - \eta) u_l{}^\wedge(\eta) \chi_{V(k)}(\theta) \, d\eta$$

$$= \sum_{k=0}^{\infty} \int_\Gamma \omega_{kl}{}^\wedge(\eta) u_l{}^\wedge(\xi - \eta) \Phi_{\Gamma_l \setminus \Gamma_{l-1}}(\xi - \eta) \chi_{V(k)}(\theta') \, d\eta$$

with $| \theta' | = m_l^{-1} | \xi - \eta |$. It is clear that $\operatorname{supp} (I_{kl})^\wedge \subset \Gamma_l$, thus, by (iii) in Theorem 4, $T_\sigma : W^s \to W^{s-m}$ is bounded. The proof is complete.

Remark 3. We have seen that the above result in the Theorem 5 is quite different from that of in $\mathbb{R}$ case. On the other hand, Saloff–Coste, L. has proved the boundedness of W^s under the condition $0 \leqslant \delta < \rho \leqslant 1$, or $0 < \delta \leqslant \rho < 1$ [12]. Therefore, the result of Theorem 5 is that of generalization of [12].

§ 5. *Para–differential operators*

To establish the theory of Gibbs differential operators and equations, it is necessary to introduce the concept of para–differential operators on Vilenkin groups.

We now introduce the symbol class $\mathcal{F}_\rho^m$ for para–differential operators firstly. (Note that it is different from the symbol class $S_{\rho,\delta}^m$ for pseudo–differential operators, compare with the Definition 2).

Definition 6. (symbol class $\mathcal{F}_\rho^m$) If function $l(x, \xi) : G \times \Gamma \to \mathbb{C}$ satisfies

(i) $l(\cdot, \xi) \in C^\rho$ [20], $\rho \in \mathbb{R}^+ \setminus \mathbb{N}$,

$l(x, \cdot) \in C^\infty \equiv \{ g \in \mathcal{S}' : g$ has any r order $G-$derivative, $r \geqslant 0 \}$,

(ii) for each $v \in \mathbb{P}$, there exists a constant c_v, such that for $m \in \mathbb{R}$ and $\eta \in \Gamma$

$$\left\| \Delta_\eta^\xi l(x, \xi) \right\|_{C^\rho} \leqslant c_v | \eta |^v < \xi >^{m-v},$$

then it is called a symbol function, or simply, symbol. The set of all symbols is said to be the symbol class $\mathcal{F}_\rho^m$.

To define the para–differential operator, let $\omega(\xi, \eta) : \Gamma \times \Gamma \to \mathbb{C}$ be a para–truncated function defined by [20]

$$\omega(\xi,\eta) = \psi(\xi,\eta)\, s(\eta) \tag{21}$$

with

$$\psi(\xi,\eta)=\begin{cases} 1, & |\xi| \leqslant m_{N_1}|\eta|, \\ 0, & |\xi| > m_{N_2}|\eta|, \end{cases}$$

$$s(\eta) = \begin{cases} 0, & |\eta| \leqslant m_{N_3}, \\ 1, & |\eta| > m_{N_4}. \end{cases} \tag{22}$$

where $N_1, N_2 \in \mathbf{Z}$, $N_1 < N_2$, and $N_3, N_4 \in \mathbf{Z}$, $N_3 < N_4$ being fixed integers.

Definition 7. (para–differential operator) Let $l \in \mathcal{F}_p^m$, and $\omega(\xi,\eta) = \psi(\xi,\eta)\, s(\eta)$ be a para–truncated function defined in (21) and (22). Then the para–differential operator P_l is defined by

$$(P_l u)^{\wedge}(\xi) = \int_\Gamma \omega(\xi-\eta,\eta)\, l^{\wedge}(\xi-\eta,\eta)\, u^{\wedge}(\eta)\, d\eta \tag{23}$$

for $u \in S(G)$, P_l is called a para–differential operator with symbol $l \in \mathcal{F}_p^m$.

Remark 4. When $l(x,\xi) = a(x)$, (21) reduces to the para–product (see [20]). And when the para–truncated function disappears, (21) becomes the pseudo–differential operator (7) in § 3 formally. Thus, the concept of para–differential operators is certain generalization of that of pseudo–differential operators and para–product operators.

The decompostion theorem for symbols will prove in [21].

Consider the partial G–derivatives and G–differential equations as follows.

Suppose that $x = (x_1, \cdots, x_n) \in G^n$. If $u(x) : G^n \to \mathbb{C}$ is a function on G^n. Its partial G–derivatives are defined as follows : for $r_1, r_2, r_3, \cdots \in \mathbb{R}^+$,

$$\partial_{x_1}^{<r_1>} u(x_1, \cdots, x_n) =$$

$$= \int_\Gamma <\xi_1>^{r_1} \int_G u(t_1, x_2, \cdots, x_n)\, \bar{\chi}_{\xi_1}(t_1 - x_1)\, dt_1\, d\xi_1,$$

$$\partial_{x_1}^{<r_1>} \partial_{x_2}^{<r_2>} u(x_1, \cdots, x_n) = \int_\Gamma <\xi_1>^{r_1} \int_\Gamma <\xi_2>^{r_2} \cdot$$

$$\cdot \int_G \int_G u(t_1, t_2, x_3, \cdots, x_n)\, \bar{\chi}_{\xi_1}(t_1 - x_1)\, \bar{\chi}_{\xi_2}(t_2 - x_2)\, dt_1\, dt_2\, d\xi_1\, d\xi_2,$$

....

And we write $\partial^{<r>} \equiv \partial_{x_1}^{<r_1>} \cdots \partial_{x_n}^{<r_n>}$, $|r| = r_1 + \cdots + r_n$.

Let $y = (y^1, \cdots, y^N) \in \mathbb{C}^N$, and $F(x, y) : G^n \times \mathbb{C}^N \to \mathbb{C}$, we call

$$F(x, u, \cdots, \partial^{<r>} u, \cdots)_{|r| \leq m} = 0 \tag{24}$$

G–differential equation, where $u : G^n \to \mathbb{C}$ is a function of x on G^n.

For a nonlinear G–differential equation, we will prove the para–linearization theorem in [21]

Theorem 6. *If $u \in W^{s+m}$, $s - \frac{n}{2} > 0$, is a real solution of non–linear G–differential equation* (23), *and $F : G^n \times \mathbb{R}^N$ is $C^\infty(G^n \times \mathbb{R}^N)$ function whose partial derivatives about y are bounded on $G^n \times K$ for any compact set $K \subset \mathbb{R}^N$. Then* (24) *deduces to the following para–differential equation*:

$$P_l u = R \tag{25}$$

with $R \in W^{2(s - \frac{n}{2})}$, where P_1 *is a para–differential operator*:

$$P_l = \sum_{0 \leq |r| \leq m} T_{\frac{\partial F}{\partial y^r}} \partial^{<r>} \tag{26}$$

with symbol

$$l(x, \xi) = \sum_{0 \leq |r| \leq m} \frac{\partial F}{\partial y^r} (x, u(x), \cdots, \partial^{<m>} u(x)) <\xi>^r, \tag{27}$$

here $l \in \mathcal{P}^m_{s - \frac{n}{2}}$, and the principal symbol[21] *is*

$$l_m(x, \xi) = \sum_{|r| = m} \frac{\partial F}{\partial y^r} <\xi>^r.$$

We may consider quasi–linear G–differential equations, semi–linear G–differential equations, and so on, by the above method. Sometimes, the R may have more better regularities (see [21]).

References

1 P.L. BUTZER & J.L. NESSEL, 'Fourier Analysis and Approximation', *I, Birkhäuser Basel and Academic Press, New York*, 1971.

2 P.L. BUTZER & P. STANKOVIC Eds., 'Theory and Applications of Gibbs Derivatives', *Proc. of the First International Workshop on Gibbs Derivatives, held Sep.* 26–28, 1989, *Kupari–Dubrovnik, Yugoslavia, Matematicki Institut, Beograd,* 1990.

3 P.L. BUTZER & H.J. WAGNER, 'Approximation by Walsh Polynomials and the Concept of a Derivative', *Proc. Symp. Applic., Walsh Functions, Washington, D.C.*, 1972, 388–392.

4 P.L. BUTZER & H.J. WAGNER, 'Walsh–Fourier Series and the Concept of a Derivative', *Applicable Anal.*, 3, 1973, 29–46.

5 J.E. GIBBS, 'Walsh Spectrometry, a Form of Spectral Analysis Well Suited to Binary Digital Computation', *Nat. Phys. Lab., Teddington, Middlesex, UK*, 1967, 24pp.

6 J.E. GIBBS, 'Some Properties of Functions on the Non–negative integers less than 2^n, ' *NPL DES Rept.* No.3, 1969, ii+23pp.

7 J.E. GIBBS & M.J. MILLARD, 'Walsh Functions as Solutions of a Logical Differential', *NPL DES Rept.* No.1, 1969, 9pp.

8 J.E. GIBBS & MILLARD, M.J., 'Some Methods of Solution of Linear Ordinaty Logical Differential Equations', *NPL DES Rept.* No.2, 1969, ii+33pp.

9 J.E. GIBBS & R.S. STANKOVIC, 'Why IWGD–89? A Look at the Bibliography of Gibbs Derivatives' , *Proc. of the First International Workshop on Gibbs Derivatives, held Sep.* 26–28, 1989, *Kupari–Dubrovnik, Yugoslavia, Matematicki Institut, Beograd,* 1990, XI – XXIV.

10 C.W. ONNEWEER, 'Differentiability for Rademacher Series on Groups', *Acta. Sci. Math.* (*Szeged*), 39, 1977, 121–128.

11 C.W. ONNEWEER, 'Differentiation on a p–adic or p–series Fields', *Linear Spaces and Approx., Birkhauser Verlag Basel*, 1978, 187–198.

12 L. SAOFF–COSTE, 'Opérateurs Pseudo–Différentiels sur Certains Groupes Totalement Discontinus', *Studia Math., T.LXXXIII*, 1986, 205–228.

13 R.S. STANKOVIC, 'A Note on Differential Operators on Finite Non–Ableian Groups', *Applicable Anal.*, 21, 1986, 31–41.

14 R.S. STANKOVIC, 'Gibbs Derivatives', *to appear.*

15 SU WEIYI, 'On an Extremum Problem for *n*–Variable Walsh Transform', *J. of Nanjing University* (*Natural Science issue*), No.2, 1980, 6–14.

16 SU WEIYI, 'Pseudo–Differential Operators in Besov Spaces over Local Fields', *Approx. Theory & its Appl.* 4:2, 1988, 119–129.

17 SU WEIYI, 'Para–Product Operators over Locally Compact Vilenkin Groups', *A Friendly Collection of Mathematical Papers I, Jilin University Press, Changchun, China, Sept.* 1990, 1–5.

18 SU WEIYI, 'Pseudo–Differential Operators and Derivatives on Locally Compact Vilenkin Groups', *Science in China* (*A*), 35:7, 1992, 826–836.

19 SU WEIYI, 'Gibbs Derivatives and Their Applications', *to appear in* 《 *Recent Chinese Works on Spectral Techniques : Theory and Applications* 》, *Germany.*

20 SU WEIYI, 'Para–Product Operators and Para–Linearization on Locally Compact Vilenkin Groups', *to appear.*

21 SU WEIYI, 'Derivatives, Operators and Gibbs type Differential Equations on Locally Compact Vilenkin Groups (II), *to appear.*

22 M.H. TAIBLESON, 'Fourier Analysis on Local Fields', *Princeton University Press, Princeton,* 1975.

23 ZHENG WEIXING, 'Derivatives and Approximation Theorems on Local Fields', *Rocky Mountain J. of Math.*, 15:4, 1985, 803–817.

24 ZHENG WEIXING, SU WEIYI & JIANG HUIQUN, 'A Note to the Concept of Derivatives on Local Fields', *Approx. Theory & its Appl.*, 6:3, 1990, 48–58.

ON SELFSIMILARITY OF FUNCTIONS

Zheng Weixing

Nanjing University, China

Abatract We discuss the self-similarity of functions in the setting of the p-series field and p-adic field. A characterization of self-similar functions is given by means of a convolution operator that is of product type. Some local properties are established. Their Fourier expansions and derivatives have the advantage is deduce useful expressions of some typical interesting functions such as the p-adic Cantor functions.

1. Introduction

Let K be the p—adic field or p—series field[3], $|\cdot|$ the nonarchimedean norm satisfying the following axioms:

(i) $|x|\geqslant 0$; $|x|=0$ iff $x=0$,

(ii) $|xy|=|x|\,|y|$,

(iii) $|x+y|\leqslant\max\{|x|,|y|\}$,

for $x, y\in K$, and ξ its prime element, $|\xi|=p^{-1}$. For $k\in Z$, the ball of radius p^{-k} is $P^k=\{x\in K: |x|\leqslant p^{-k}\}$ and write $P=P^1$, $O=P^0$ for the prime ideal, the ring of integers in K, respectively. We have $O/P\cong\{\varepsilon_0=0, \varepsilon_1,\cdots,\varepsilon_{p-1}\}\equiv Z_p$. Every $x\in K$ is expressed as the formal series with $x_\nu\in Z_p$:

$$x=\sum_{\nu=-n}^{\infty} x_\nu\xi^\nu.$$

As usual, a character $\chi\in K^+$ is fixed such that it is trivial on O but nontrivial on P^{-1}. Then a completely orthonormal system $\{\chi_j\}_{j\in Z^+}$ can be obtained, where $\chi_j(t)=\chi(\lambda^{-1}(j)t)$ and λ is the mapping $Z^+\to K$ defined in [5]. For example, in the case of p-adic field as χ we may take

$$\chi(x)=exp(2\pi i\sum_{\nu=-n}^{\infty} x_\nu p^\nu)$$

and in the case of p-series field as χ we may take

$$\chi(x)=exp(2\pi i x_{-1}p^{-1}).$$

The following formulae are valid:

$$\chi_j(x)=\chi_j(x_0)\chi_{[j/p]}(x_1)\cdots\chi_{[j/p^r]}(x_r),\ x\in O,$$

$$\chi_j(x+y)=\chi_j(x)\chi_j(y),\ x,\ y\in K$$

and $j\in Z^+$, $j=j_{-r}p^r+\cdots+j_0$ with $j_{-r}\neq 0$ for $j\neq 0$.

Let a function $f: K\to C$ be given on K. For $N\in \mathcal{N}$, set

M. Cheng et al. (eds.), Harmonic Analysis in China, 256–265.

$$\triangle_N f(x) = \sum_{k=-N}^{N} p^{-N-k+1} \sum_{l=0}^{p^N-1} \sum_{\nu=1}^{p-1} exp(\frac{-2\pi i}{p}\nu l) f(x + l\zeta^{-k}). \qquad (1)$$

If the limit $\lim_{N\to\infty} \triangle_N f(x)$ exists and is finite, it is called a (p)type derivative[3] of $f(x)$ at x and is denoted by $f^{\langle 1\rangle}(x)$, the definition applies for the p-adic field case. As for the p-series field case, all the above remain unchanged but using

$$\triangle'_N f(x) = \sum_{k=-N}^{N} p^k \{ \sum_{\nu=0}^{p-1} A_\nu f(x + \zeta^k \varepsilon_\nu)\} \qquad (2)$$

instead of $\triangle_N f(x)$, where $A_0=(p-1)/2$, $A_\nu=\omega^\nu(1-\omega^\nu)^{-1}$, $\omega=\exp(2\pi i/p)$, $\nu=1, \cdots, p-1$. The (p) type derivative of f in L^q-sense $(1\leqslant q<\infty)$ may be defined analogously and is denoted by $D^{\langle 1\rangle}f$. The same thing is for the weak and higher order (p) type derivatives.

The following theorem was proved in [4].

Theorem 1. 1 (i) For any $\xi\in K$, the character $\chi_\xi(x)\equiv\chi(\xi x)$ has a (p) type derivative $\chi_\xi^{\langle 1\rangle}(x)$ at each $x\in K$ and

$$\chi_\xi^{\langle 1\rangle}(x) = |\xi|\chi_\xi(x).$$

(ii) If a function f has a (p) type derivative at x, then so does $f(\zeta^s x)$ and

$$\{f(\zeta^s x)\}^{\langle 1\rangle} = |\zeta^s| f^{\langle 1\rangle}(\zeta^s x), \; s \in Z.$$

2. Expansion of self-similar functions

Let $p=2s+1$, a prime number. The ring O of integers can be decomposed into the union of disjoint cosets of P:

$$O = \bigcup_{\nu=0}^{p-1} (\varepsilon_\nu + P).$$

Let $f_s(x)=h(x)\kappa_{\varepsilon_s+P}(x)$ be given, where $h(x)$ is an initial-valued function, where κ_E is the characteristic function of E. In the following we always make the arsumption that $f_s(x)$ has an absolutely eonvergent series $f_s(x)=\sum_{k=0}^{\infty} h_k\chi_k(x)$, h_k: the Fourier coefficients of f_s with respect to the system $\{\chi_k(x)\}_{k\in Z^+}$. A function f on O is said to be self-similar with the initial valued function h [7],[8], if for each $\nu=0, 1, \cdots, p-1$, $\nu\neq s$, $x\in\varepsilon_\nu+P$, we have

$$f(\varepsilon_\nu + x') = a_{\varepsilon\nu} + t_{\varepsilon\nu} f(x'\zeta^{-1}), \; x' \in P, \qquad (3)$$

and for $\nu=s$, $f(\varepsilon_s+x')=h(\varepsilon_s+x')$, where $a_{\varepsilon\nu}$ are constants and $t_{\varepsilon\nu}\in(0, 1)$.

A function f defined on K is said to have the period ζ^{-1}, if for every $m\in N$, we have $f(x+\zeta^{-m})=f(x)$, $x\in K$. Throughout the paper it is assumed that all functions on K will have the period ζ^{-1}.

Lemma 2. 1 Let f be selfsimilar with the initial h and $f\in L^1(O)$, then the Fourier coefficients c_k of f with respect to the system $\{\chi_k\}_{k\in Z^+}$ satisfy the relation:

$$c_k = h_k + \frac{1}{p}A(k)\delta_{0,[k/p]} + \frac{1}{p}A(k)c_{[k/p]}, \; k \in Z^+, \qquad (4)$$

where

$$A(k) = \sum_{v\neq s} a_{\varepsilon v}\bar{\chi}_k(\varepsilon_v), \tag{5}$$

$$\Lambda(k) = \sum_{v\neq s} t_{\varepsilon v}\bar{\chi}_k(\varepsilon_v). \tag{6}$$

Proof In view of the selfsimilarity of f, we have

$$\begin{aligned} c_k &= \int_O f(x)\bar{\chi}_k(x)dx \\ &= \int_O f_s(x)\bar{\chi}_k(x)dx + \sum_{v\neq s}\int_{\varepsilon v+P}\{a_{\varepsilon v}+t_{\varepsilon v}f(\zeta^{-1}x')\}\bar{\chi}_k(x)\,dx, \end{aligned}$$

where $x=\varepsilon_v+x'$, $x'\in P$.

Note that $\chi_k(\varepsilon_v+x')=\chi_k(\varepsilon_v)\chi_k(x')$ and $\chi_k(x')=\chi_{[k/p]}(\zeta^{-1}x')$, we obtain immediately

$c_k=h_k+\frac{1}{p}\sum_{v\neq s} a_{\varepsilon v}\bar{\chi}_k(\varepsilon_v)\int_O\bar{\chi}_{[k/p]}(x)dx+\frac{1}{p}\sum_{v\neq s} t_{\varepsilon v}\bar{\chi}_k(\varepsilon_v)\int_O f(x)\bar{\chi}_{[k/p]}(x)dx$, which reduces to the formula (4) by the notation (5), (6).

We point out when $k=o$, from (4) we obtain the relation

$$c_0 = h_0 + \frac{1}{p}\sum_{v\neq s} a_{\varepsilon v} + \frac{1}{p}\sum_{v\neq s} t_{\varepsilon v}c_0,$$

hence the expression of c_0:

$$c_0 = (h_0 + \frac{1}{p}\sum_{v\neq s} a_{\varepsilon v})/(1-\frac{1}{p}\sum_{v\neq s} t_{\varepsilon v}). \tag{7}$$

Lemma 2.2 The explicit expression of c_k is

$$\begin{aligned} c_k &= h_k+\frac{1}{p}\Lambda(k)h_{[k/p]}+\cdots+p^{-r}\Lambda(k)\Lambda([k/p])\cdots\Lambda([k/p^{r-1}])h_{[k/p^r]} \\ &\quad +p^{-r-1}\Lambda(k)\cdots\Lambda([k/p^{r-1}])(A([k/p^r])+c_0\Lambda([k/p^r])), \end{aligned} \tag{8}$$

where $k=k_{-r}p^r+\cdots+k_0$, the p—adic expansion of k, $k_{-r}\neq 0$ for $k\neq 0$. In which, c_0 is read as (7) and $c_1, \cdots, c_{p-1}$ are read as

$$c_k = h_k + p^{-1}(A(k) + c_0\Lambda(k)),\ k = 1, \cdots, p-1. \tag{9}$$

Proof We see the formula (8) is true for $k\in\{0, 1, \cdots, p-1\}$ under the obvious notation. Assume that (8) is true for all $k<p^{r+1}$, $r\in Z^+$, then for $p^{r+1}\leqslant k<p^{r+2}$, we have from (4)

$$c_k = h_k = \frac{1}{p}A(k)\delta_{0,[k/p]} + \frac{1}{p}\Lambda(k)c_{[k/p]}. \tag{10}$$

In this case, $p^r\leqslant[k/p]<p^{r+1}$, by the induction assumption, we have

$$\begin{aligned} c_{[k/p]} &= h_{[k/p]}+p^{-1}\Lambda([k/p])h_{[k/p2]}+\cdots+p^{-r}\Lambda([k/p]\cdots\Lambda([k/p^r])h_{[k/p^{r+1}]} \\ &\quad +p^{-r-1}\Lambda([k/p]\cdots\Lambda([k/p^r])(A([k/p^{r+1}])+c_0\Lambda([k/p^{r+1}])). \end{aligned}$$

Put this expression into (10) we find the formula (8) is true for $p^{r+1}\leqslant k<p^{r+2}$, $r\in Z^+$.

Corollary 2. 3 Let $h(x)=\alpha_s$, $x\in\varepsilon_s+P$, the coefficients c_k reduce to

$$c_k=p^{-r-1}\Lambda(k)\Lambda([k/p])\cdots\Lambda([k/p^{r-1}])(A([k/p^r])+c_0\Lambda[k/p^r]),$$
$$k=k_{-r}p^r+\cdots+k_0. \tag{11}$$

Proof In fact, we have

$$h_k=\alpha_s\int_{\varepsilon_s+P}\overline{\chi}_k(x)dx=\alpha_s\overline{\chi}_k(\varepsilon_s)\int_P\overline{\chi}_k(x')dx'=\alpha_s\overline{\chi}_k(\varepsilon_s)\delta_{0[k/p]},$$

So that $h_k=\cdots=h_{[k/p^r]}=0$ for $p^r\leqslant k<p^{r+1}$. Thus (11) follows from (8).

Lemma 2. 4 Let $h_k=\int_{\varepsilon_s+P}h(x)\overline{\chi}_k(x)dx$, $k\in Z^+$. If $\nu\neq s$, then for any $j\in Z^+$,

$$\sum_{k=jp}^{jp+p-1}h_k\chi_k(\varepsilon_\nu)=0; \tag{12}$$

if $\nu\neq s$, then

$$\sum_{k=jp}^{jp+p-1}h_k\chi_k(\varepsilon_\nu)=ph_o. \tag{13}$$

Proof For $\nu\neq s$, we have

$$\sum_{k=jp}^{jp+p-1}h_k\chi_k(\varepsilon_\nu)=\sum_{k=jp}^{jp+p-1}\int_{\varepsilon_s+P}h(x)\chi_k(\varepsilon_\nu-x)dx=\int_{\varepsilon_s+P}h(x)\sum_{k=jp}^{jp+p-1}\chi_k(\varepsilon_\nu-\varepsilon_s)dx=0;$$

and for $\nu=s$, the result is clearly ph_o.

Lemma 2. 5 For any $j\in Z^+$, we have

$$\sum_{k=jp}^{jp+p-1}\Lambda(k)\chi_k(\varepsilon_\nu)=pt_{\varepsilon\nu},\ for\ \nu\neq s; \tag{14}$$

$$\sum_{k=jp}^{jp+p-1}A(k)\chi_k(\varepsilon_\nu)=pt_{\varepsilon\nu},\ for\ \nu\neq s. \tag{15}$$

Proof For $\nu\neq s$, we have

$$\sum_{k=jp}^{jp+p-1}\Lambda(k)\chi_k(\varepsilon_\nu)=\sum_{k=jp}^{jp+p-1}\sum_{\mu\neq s}t_{\varepsilon\mu}\chi_k(\varepsilon_\nu-\varepsilon_\mu)$$
$$=\sum_{\mu\neq s}t_{\varepsilon\mu}\sum_{l=0}^{p-1}\chi_l(\varepsilon_\nu-\varepsilon_\mu)\chi_{jp}(\varepsilon_\nu-\varepsilon_\mu)$$
$$=pt_{\varepsilon\nu};$$

and similarly,

$$\sum_{k=jp}^{jp+p-1}A(k)\chi_k(\varepsilon_\nu)=\sum_{k=jp}^{jp+p-1}\sum_{\mu\neq s}\alpha_{\varepsilon_\mu}\chi_k(\varepsilon_\nu-\varepsilon_\mu)$$
$$=\sum_{\mu\neq s}\alpha_{\varepsilon_\mu}\sum_{l=0}^{p-1}\chi_l(\varepsilon_\nu-\varepsilon_\mu)\chi_{jp}(\varepsilon_\nu-\varepsilon_\mu)$$
$$=p\alpha_{\varepsilon\nu}.$$

Theorem 2. 6 Let $f\in L^1(O)$. Then f is selfsimilar with the initial h iff f has the following expansion

$$f(x)=\sum_{k=0}^{\infty}c_k\chi_k(x),\ x\in O, \tag{16}$$

where the coefficients c_k are defined by (8).

Proof In view of Lemmas 2. 1 and 2. 2, we need only to show that if the equality

(16) holds, where the c_k are defined by (8), then f is selfsimilar with the initial h. In this case the function term series in (16) is absolutely and uniformly convergent and represents the function f itself. We may indeed varify f is selfsimilar.

For this end, let $x \in \varepsilon_v + P$. Write $x = \varepsilon_v + x'$, $x' \in P$. Then we have

$$f(x) = \sum_{k=0}^{\infty} c_k \chi_k(\varepsilon_v) \chi_{[k/p]}, \ y = \zeta^{-1} x^1 \in O. \tag{17}$$

For $k \in \{0, 1, \cdots, p-1\}$, the corresponding sum in (17) is

$$\begin{aligned} \sum_{k=0}^{p-1} c_k \chi_k(\varepsilon_v) &= \sum_{k=0}^{p-1} h_k \chi_k(\varepsilon_v) + \sum_{k=0}^{p-1} p_{-1}(A(k) + c_0 \Lambda(k)) \chi_k(\varepsilon_v) \\ &= 0 + p^{-1}(p a_{e_v} + c_0 p t_{e v}) = a_{e_v} + t_{e_v} c_0, \ for\ v \neq s, \end{aligned} \tag{18}$$

by (9), Lemma 2. 4 and Lemma 2. 5.

In general, let $J = \{k;\ [k/p] = j\}$, $j \geqslant p$. It is easy to see that $J = \{jp, jp+1, \cdots, jp+p-1\}$. Thus

$$\begin{aligned} \sum_{k \in J} c_k \chi_k(\varepsilon_v) = & \sum_{\mu=0}^{p-1} \{h_{jp+\mu} + p^{-1} \Lambda(jp+\mu) h_j + \cdots + p^{-r-1} \Lambda(jp+\mu) \Lambda(j) \cdots \Lambda([j/p^{r-1}]) h_{[j/p^r]} \\ & + p^{-r-2} \Lambda(jp+\mu) \cdots \Lambda([j/p^{r-1}]) (A([j/p^r] + c_0 \Lambda([j/p^r])) \chi_k(\varepsilon_v) \\ = & h_j t_{ev} + \cdots + p^{-r} \Lambda(j) \cdots \Lambda([j/p^{r-1}] h_{[j/p^r]} t_{ev} \\ & + p^{-r-1} \Lambda(j) \cdots \Lambda([j/p^{r-1}]) (A([j/p^r])) + c_0 \Lambda([j/p^r])) t_{ev}, \end{aligned} \tag{19}$$

by Lemmas 2. 4 and 2. 5.

On the other hand, we have

$$a_{e_\iota} + t_{e_\iota} f(y) = (a_{e_\iota} + t_{e_\iota} c_0) + \sum_{k=1}^{\infty} t_{e_\iota} c_k \chi_k(y), \ y \in O, \ for\ \iota \neq s, \tag{20}$$

hence the two expressions (17) and (20) are in agreement for $\iota \neq s$.

If $\iota = s$, $x = \varepsilon_s + x'$, $x' \in P$, then

$$f(\varepsilon_s + x') = \sum_{k=0}^{\infty} h_k \chi_k(x),$$

and by our assumption, the series is absolutely convergent. Now we have

$$\begin{aligned} & \sum_{k=jp}^{jp+p-1} \Lambda(k) \cdots \Lambda([k/p^{r-1}]) (A([k/p^r] + c_0 \Lambda([k/p^r])) \chi_k(\varepsilon_s + x') \\ = & \sum_{\mu=0}^{p-1} \Lambda(jp+\mu) \Lambda(j) \cdots \Lambda([j/p^{r-2}]) (A([j/p^{r-1}]) + c_0 \Lambda([j/p^{r-1}])) \chi_{jp+\mu}(\varepsilon_s + x') \\ = & \Lambda(j) \cdots \Lambda([j/p^{r-2}]) (A([j/p^{r-1}]) + c_0 \Lambda([j/p^{r-1}])) \sum_{\mu=0}^{p-1} \Lambda(jp+\mu) \chi_\mu(\varepsilon_s) \chi_{jp}(x) \\ = & 0, \end{aligned}$$

by Lemma 2. 5, where $j \geqslant 1$. And for $j = 0$, we also have

$$\begin{aligned} \sum_{\mu=0}^{p-1} c_\mu \chi_\mu(x) &= \sum_{\mu=0}^{p-1} h_\mu \chi_\mu(x) + p^{-1} \sum_{\mu=0}^{p-1} (A(\mu) + c_0 \Lambda(\mu)) \chi_\mu(x) \\ &= \sum_{\mu=0}^{p-1} h_\mu \chi_\mu(x). \end{aligned}$$

Hence for $x \in \varepsilon_s + P$,

$$f(x) = \sum_{k=0}^{\infty} c_k \chi_k(x) = \sum_{k=0}^{\infty} h_k \chi_k(x) = h(x).$$

Therefore the proof of the selfsimilarity of f is complete.

3. A characterization by the operator of product type

Introduce the operator $L_{n,r}(f; \circ) = f * k_{n,r}$ with the so-called kernel of product type[6],[7],[8]

$$k_{n,r}(x) = \prod_{j=0}^{n-1} (1 + a_{j1}^{(n)}(r)\chi_{pj}(x) + \cdots + a_{j,p-1}^{(n)}(r)\chi_{pj}^{p-1}(x)), \tag{21}$$

where we make the following assumptions:

(i) $\lim_{\substack{n\to\infty \\ r\to 1}} a_{j,l}^{(n)}(r) = 1$, for any fixed $j \in Z^+$ and $l \in \{1, \cdots, p-1\}$. (22)

(ii) there exist $r' \in (0, 1]$ tending to 1 together with r such that

$$a_{j+1,l}^{(n)}(r) = a_{j,l}^{(n-1)}(r'), \text{ for } j \in \{0, 1, \cdots, n-2\};\ l \in \{1, \cdots, p-1\}. \tag{23}$$

In this section the discusion is in the setting of p-series field.

Lemma 3. 1 Assume that f is selfsimilar with the initial valued function h and that the kernel $k_{n,r}(x)$ is of product type satisfying the conditions (i), (ii), then we have the following relation:

$$L_{n,r}(f;\, x) = \frac{1}{p} \sum_{l=0}^{p-1}{}' \eta_{0,r}^{(n)}(\omega^l)(a_{x_0-\varepsilon_l} + t_{x_0-\varepsilon_l} L_{n-1,r'}(f; \zeta^{-1}x')) + L_{n,r}(f_s; x), \tag{24}$$

where $\eta_{0,r}^{(n)}(z) = 1 + a_{01}^{n}(r)z + \cdots + a_{0,p-1}^{(n)}(r)z^{p-1}$, $x = x_0\zeta^0 + x'$, $x' \in P$, $\omega = \exp(2\pi i/p)$ and Σ' means the sum without the term l such that $x_0 - \varepsilon_l = \varepsilon_s$.

Proof By the definition of selfsimilarity,

$$L_{n,r}(f;x) = L_{n,r}(f_s;x) + \sum_{l=0}^{p-1}{}' \int_P \{a_{x_0-\varepsilon_l} + t_{x_0-\varepsilon_l} f(\zeta^{-1}(x' - t'))\} k_{n,r}(\varepsilon_l + t')dt'$$

$$= L_{n,r}(f_s;x) + \frac{1}{p} \sum_l{}' a_{x_0-\varepsilon_l} \int_o k_{n,r}(\varepsilon_l + \zeta t)dt$$

$$+ \frac{1}{p} \sum_l{}' t_{x_0-\varepsilon_l} \int_o f(\zeta^{-1}x' - t) k_{n,r}(\varepsilon_l + \zeta t)dt.$$

Note that by (i) and (ii), we can deduce

$$k_{n,r}(\varepsilon_l + \zeta t) = \eta_{0,r}^{(n)}(\omega^l) k_{n-1,r'}(t),$$

hence

$$L_{n,r}(f;x) = L_{n,r}(f_s;x) + p^{-1} \sum_l{}' a_{x_0-\varepsilon_l} \eta_{0,r}^{(n)}(\omega^l) + p^{-1} \sum_l{}' t_{x_0-\varepsilon_l} \eta_{0,r}^{(n)}(\omega^l) L_{n-1,r'}(f; \zeta^{-1}x'),$$

which is (24).

Theorem 3. 2 Assume that $L_{n,r}(f;x)$ is the operator of psoduct type which is of approximation identity. Then a function f on O is selfsimilar iff the relation (24) holds.

Proof In virtue of Lemma 3. 1, we need only to show the 'only if' part of the theorem. Then, assume that (24) holds. At first we see, as $n\to\infty$, $r\to 1$,

$$\eta_{\delta,r}^{(n)}(1) = 1 + a_{\delta 1}^{(n)}(r) + \cdots + a_{\delta,p-1}^{(n)}(r) \to p,$$
$$\eta_{\delta,r}^{(n)}(\omega^l) = 1 + a_{\delta 1}^{(n)}(r)\omega^l + \cdots + a_{\delta,p-1}^{(n)}(r)\omega^{(p-1)l} \to 0 \text{ for } l = 1,\cdots,p-1.$$

So from (24) we have for $x_0 \neq \varepsilon_s$,

$$f(x) = \lim_{\substack{n\to\infty\\ r\to 1}} L_{n,r}(f;x) = \frac{1}{p}\lim_{\substack{n\to\infty\\ r\to 1}} \eta_{\delta,r}^{(n)}(1)(a_{x_0} + t_{x_0} L_{n-1,r'}(f;\zeta^{-1}x'))$$
$$= a_{x_0} + t_{x_0} f(\zeta^{-1}x');$$

and for $x_0 = \varepsilon_s$,

$$f(x) = \lim_{\substack{n\to\infty\\ r\to 1}} L_{n,r}(f_s;x) = h(x).$$

This means f is selfsimilar on O with the initial valied function h.

Lemma 3. 3 If the operator $L_{n,r}(f;x)$ converges to $f(x)$ a. e. on the coset $\varepsilon_s + P$, then $L_{n,r}(f;x)$ converges to $f(x)$ a. e. on the whole set O.

Proof We have for $x = \varepsilon_l + x'$, $x' \in P$,

$$L_{n,r}(f;\varepsilon_l + x') = \frac{1}{p}\sum_l \eta_{\delta,r}^{(n)}(\omega^l)(a_{\varepsilon_0-\varepsilon_l} + t_{\varepsilon_0-\varepsilon_l} L_{n-1,r'}(f;\zeta^{-1}x)) + L_{n,r}(f_s;x),$$

hence if we put $x' = \varepsilon_s\zeta + x''$, $x'' \in P^2$, then

$$L_{n,r}(f;\varepsilon_l + \varepsilon_s\zeta + x'') = \frac{1}{p}\sum_l \eta_{\delta,r}^{(n)}(\omega^l)(a_{\varepsilon_0-\varepsilon_l} + t_{\varepsilon_0-\varepsilon_l} L_{n-1,r'}(f;\varepsilon_l\zeta^{-1} + \varepsilon_s + x''\zeta^{-1})) + L_{n,r}(f_s;x)$$
$$= \frac{1}{p}\sum_l \eta_{\delta,r}^{(n)}(\omega^l)(a_{\varepsilon_0-\varepsilon_l} + t_{\varepsilon_0-\varepsilon_l} L_{n-1,r'}(f;\varepsilon_s + x''\zeta^{-1})) + L_{n,r}(f_s;x).$$

Note that $x''\zeta^{-1} \in P$ when $x''\zeta^{-1} \in P^2$. Thus by the hypothesis of the lemma,

$$\lim_{\substack{n\to\infty\\ r\to 1}} L_{n,r}(f;\varepsilon_l + \varepsilon_s\zeta + x'') = \frac{1}{p}\lim \eta_{\delta,r}^{(n)}(1)(a_{\varepsilon_l-\varepsilon_0} + t_{\varepsilon_l-\varepsilon_0} f(\varepsilon_s + x'))$$
$$= a_{\varepsilon_l} + t_{\varepsilon_l} f(\varepsilon_s + x')$$
$$= f(\varepsilon_l + \varepsilon_s\zeta + x''), \text{ for each } l \neq s.$$

This means the operator $L_{n,r}(f;x)$ converges to $f(x)$ a. e. on the union of $\{\varepsilon_l + \varepsilon_s\zeta + P^2;\ l \neq s\}$. Similarly, we see $L_{n,r}f(f;x)$ converges to $f(x)$ a. e. on the sets $\varepsilon_{l_0} + \varepsilon_{l_1}\zeta + \varepsilon_s\zeta^2 + P^3$, l_0, $l_1 \neq s$. Finally by induction, we see the convergence holds on each of the cosets

$$\varepsilon_{l_0} + \varepsilon_{l_1}\zeta + \cdots + \varepsilon_{l_v}\zeta^v + P^{v+1},\ l_0, \cdots, l_v \neq s;\ v \in Z^+.$$

But the measure of the union of the sets $\{\varepsilon_{l_0} + \cdots + \varepsilon_{l_v}\zeta^v + P^{v+1};\ l_0, \cdots, l_v \neq s,\ v \in Z^+\}$ is 1, so $L_{n,r}(f;x)$ is convergent to $f(x)$ a. e. on O as $n\to\infty$, $r\to 1$.

A neighbourhood U is called normal, if U contains either a ball $u_0\zeta^0 + \cdots + u_{k-1}\zeta^{k-1} + P^k$ with all $u_0, \cdots, u_{k-1} \neq \varepsilon_s$ or a ball $u_0\zeta^0 + \cdots + u_{k-2}\zeta^{k-2} + \varepsilon_s\zeta^{k-1} + P^k$ with all $u_0, \cdots, u_{k-2} \neq \varepsilon_s$ for some $k \in Z^+$.

Theorem 3. 4 Assume that the operator $L_{n,r}(f;x)=f * k_{n,r}(x)$ is of product type, uniformly bounded and that f is selfsimilar with the initial valued function h, then the operator $L_{n,r}(f;x)$ converges to f a. e. on O as $n\to\infty$, $r\to 1$, provided the convergence holds on any fixed normal neighbourhood $U\subset O$ and $L_{n,r}(f_s;x)\to f_s(x)$ a. e., no matter how small it is.

Proof We may select a ball $u+P_k\subset U$, where k is some positive integer and $u\in O\backslash P^k$. Write

$$u = u_0\zeta^0 + u_1\zeta + \cdots + u_{k-1}\zeta^{k-1} + u' , \; u' \in P^k.$$

There are two cases to be considered.

Case 1 All $u_0, \cdots, u_{k-1}$ are different from ε_s.

By the relation (24), we see the convergence of $L_{n,r}(f;x)$ to $f(x)$ on $u+P^k$ implies $L_{n-1,r'}(f;x)$ converges to $f(x)$ on $\zeta^{-1}u+P^{k-1}$ as $n\to\infty$, $r\to 1$ $(r'\to 1)$. By induction we conclude that $L_{n,r}(f;x)$ converges to $f(x)$ on $u_{k-1}+P$ as $n\to\infty$, $r\to 1$. The selfsimilarity of f then implies the convergence holds on the union

$$O=\bigcup_{l=0}^{p-1}(\varepsilon_l+P).$$

Case 2 For $u=u_0\zeta^0+u_1\zeta+\cdots+u_{k-1}\zeta^{k-1}+u'$, $u'\in P^k$, $u_{k-1}=\varepsilon_s$ but $u_0,\cdots,u_{k-2}\neq\varepsilon_s$.

Then applying Lemma 3. 3 by using the interval

$$u_0\zeta^0 + \cdots + u_{k-2}\zeta^{k-2} + P^{k-1}$$

instead of O, we conclude that $L_{n,r}(f;x)$ converges to $f(x)$ a. e. on the interval $u_0\zeta^0+\cdots+u_{k-2}\zeta^{k-2}+P^{k-1}$. Note that all $u_0,\cdots,u_{k-2}\neq\varepsilon_s$, the proof is completed by reducing it to Case 1.

Theorem 3. 5 Under the same hypothesis as in the previous theorem, the operator $L_{n,r}(f;x)$ converges to f on O in L^q-sense provided the convergence holds on any fixed normal neighbourhood $U\subset O$ and $L_{n,r}(f_s;\circ)\to f_s$ in L^q-sense, no matter how small it is.

Proof I. At first we prove that if $U\supset x_0+P$, $x_0\neq s$, then we have

$$\| L_{n,r}(f;\circ) - f(\circ) \|_q \to 0 \text{ as } n\to\infty,\; r\to 1. \tag{25}$$

To this end, we write for $x_0\neq\varepsilon_s$,

$$L_{n,r}(f;x) - f(x)\kappa_{x_0+P}(x)$$

$$= (\frac{1}{p}\eta\delta_{0,r}^{(n)}(1) - 1)a_{x_0} + t_{x_0}\{\frac{1}{p}\eta\delta_{0,r}^{(n)}(1)(L_{n-1,r'}(f;\zeta^{-1}x) - f(\zeta^{-1}x)) + (\frac{1}{p}\eta\delta_{0,r}^{(n)}(1) - 1)f(\zeta^{-1}x)\}$$

$$+ \frac{1}{p}\sum_{l=1}^{p-1}{}' \eta\delta_{0,r}^{(n)}(\omega^l)\{a_{x_0-\varepsilon_l} + t_{x_0-\varepsilon_l}L_{n-1,r'}(f;\zeta^{-1}x)\} + L_{n,r}(f_s;x).$$

Let the constants A, B, C be such that

$$|a_{\varepsilon_l}| \leqslant A,\; 0 < B \leqslant t_{\varepsilon_l} < 1,\; \| L_{n,r}(f;x) \|_q \leqslant C \| f \|_q,$$

and for any given $\varepsilon>0$ ($\varepsilon<1/2$), take n_0, r_0 so that as $n>n_0$, $r>1-r_0$,

$$1-\varepsilon\leqslant\frac{1}{p}\eta\delta_{n,r}^{(n)}(1)\leqslant 1+\varepsilon,\quad -\varepsilon\leqslant\frac{1}{p}\eta\delta_{n,r}^{(n)}(\omega^l)\leqslant\varepsilon,\ l\in\{1,\cdots,p-1\}.$$

Then we have

$$\| L_{n,r}(f;\circ)-f(\circ)\|_{L^q(x_0+P)}\geqslant\frac{1}{2}B\| L_{n-1,r'}(f;\circ)-f(\circ)\|_q-\varepsilon(1+A+(1+C)\|f\|_q)$$
$$-\| L_{n,r}(f_s;\circ)\|_{L^q(x_0+P)},\ as\ n>n_0,\ r>1-r_0.$$

It is easy to see $\| L_{n,r}(f_s;\circ)\|_{L_q(x^0+P)}\to 0$ as $n\to\infty$, $r\to 1$. So the above inequality shows (25) is valid.

Ⅱ. Assume that there is a ball $u+P^k\subset U$, k some positive integer,

$$u=u_0\zeta^0+\cdots+u_{k-1}\zeta^{k-1}+u',\ u'\in P_k,$$

and all $u_0,\cdots,u_{k-1}$ are different from ε_s. Then applying the part I with $u_0\zeta^0+\cdots+u_{k-1}\zeta^{k-1}+P^k$ instead of x_0+P, we get that $\| L_{n,r}(f;\circ)-f(\circ)\|\to 0$ in $L^q(u_0+\cdots+u_{k-2}\zeta^{k-2}+P^{k-1})$- sense as $n\to\infty$, $r\to 1$. By induction we conclude that $\| L_{n,r}(f;\circ)-f(\circ)\|_q\to 0$ as $n\to\infty$, $r\to 1$.

Ⅲ. Assume that $u=u_0+\cdots+u_{k-1}\zeta^{k-1}+u'$, $u'\in P^k$, $u_{k-1}=\varepsilon_s$ but $u_0,\cdots,u_{k-2}\neq\varepsilon_s$.

In this case the proof of the theorem is somewhat similar to the previous theorem that we omit it.

For special selfsimilar functions, i.e. for an initial constant function $h(x)=\alpha_{\varepsilon_s}$, we have established the following result[9].

Theorem 3.6 Let $f\in L^1(O)$ be a selfsimilar function on O with the initial constant function $h(x)=\alpha_{\varepsilon_s}$ ($x\in\varepsilon_s+P$), then:

(i) if $tp<1$, we have $f^{<1>}\in L^1(O)$, where $t=\max\{t_0,\cdots,t_{p-1}\}$.

(ii) if $tp<1$, then the fundamental formula in calculus

$$f(x)-\hat{f}(0)=(f^{<1>}*W_1)(x)\tag{26}$$

holds a.e. for $x\in O$, where $W_1(x)=\sum_{k=1}^{\infty}k^{-1}\chi_k(x)$.

One should take notice that f is a singular function in the theorem, so that (26) is impossible in the classical case.

References

[1] Su, W. Y., The kernels of product type on local fields, Approx. Theory & its Appl., 1:2 (1985), 93—109.

[2] ———, ibid, 2:2 (1986), 95—111.

[3] Taibleson, M. H., Fourier Analysis on Local Fields, Princeton Univ. Press, 1975.

[4] Zheng, W. X., Derivative and approximation theorems over local fields, Rocky Mount. Jour. Math., 15:4 (1985), 803—817.

[5] ———, Further on a class of approximation identity operators over local fields, Scientia Sinica, Ser. A. 30 (1987), 908—920.

[6] ———, Remarks on the kernel of product type, A Friendly Collective Papers, Dalian (1990), 43—46.

[7] ———, On p-adic Cantor functions, Lecture Notes in Math., No. 1494 (1990), 219—226.

[8] ———, Expansion of self-similar functions, Proc. Asian Math. Confer. '90 (1991), 564—569.

[9] ———, On self-similar functions on local fields, Chin. Ann. of Math., 14A: 1 (1993), 93—98.

HARMORNIC ANALYSIS ON COMPACT LIE GROUPS AND COMPACT HOMOGENEOUS SPACES IN CHINA

XUE-AN ZHENG

Anhui University, Hefei, China

ABSTRACT

Late Professor Loo-keng Hua (L. K. Hua), having completed his famous work "*Harmonic Analysis on Classial Domains in the Theory of Functions of Several Complex Variables*" (Translations of Amer. Math. Soc. Vol. 6, 1963), applied his theory to harmonic analysis on unitary groups, deepening the well-known Peter-Weyl Theorem and initiating the research on harmonic analysis on classical groups, compact Lie groups and compact homogeneous spaces.

In the late 1950's, a systematic research program based on Hua's work was successfully carried out by Sheng Gong ([1]-[8]), which studied harmonic analysis on unitary groups. Gong developed the concepts, definitions and methods which form the foun dation of harmonic analysis on classical groups, compact Lie groups and compact homogeneous spaces in China. His ideas and methods has greatly influnced the research on these topics in China.

This research , initiated by Loo-keng Hua and set on a foundation by Sheng Gong, was suspended during the cultural revolution. In the late 1970's, the research regained its strength in China.

The basic idea is that unitary groups are the Silov boundaries of the classical domains of type one, and both the rotation groups and unitary symplectic groups are the Silov boundaries of the real classical domains of type one and the classical domains of quaternions of type one, respectively. With this point of view and under the guidance of Sheng Gong, Shikun Wang, Daozhen Dong, Zhuqi He and Guangxiao Chen and others systematically studied harmonic analysis on the rotation groups and on the unitary symplectic groups.

After completion of some basic reseach on harmonic analysis on classical groups. it is nature to study harmonic analysis on compact Lie groups. Under the great inspiration of Sheng Gong's works, Shixiong Li, Xuean Zheng, Dashan Fan and others have made a series of studies on this topic ([8],[17]-36], [43]-49]).

Compact Lie groups, the surface of the unit spfere in n-dimensional Euclidean space and compact symmetric spaces are special compact homogeneous spaces. Their geometric properties are determined by their isometric transformation groups. And their tangential spaces have certain algebric structures. From this point view, Xuean

M. Cheng et al. (eds.), Harmonic Analysis in China, 266–286.

Zheng, Shanzhen Lu and Daozhen Dong(37]-[42]) have done research on harmonic analysis on compact symmetric spaces.

At present, research on harmonic analysis in these directions is being carried on vigorously in China.

Most important results on harmonic analysis on classical groups in China were collected and published in the book of Sheng Gong ([1]) with the title "*Harmonic Analysis on Classical Groups*" (Springer-Verlag, 1991). We regret that, in order to save space, we must omit all the results on this classical topic here.

The purpose of this article is to introduce some important results on harmonic analysis on compact Lie groups and compact homogeneous spaces in China.

PART I HARMONIC ANALYSIS ON COMPACT LIE GROUPS

§1. Introduction.

Let G be a compact Lie group with Lie algebra g. Let H be a Cartan subalgebra of g, and $\widehat{G}$ be the unitary dual of G. For every λ in $\widehat{G}$, choose an irreducible unitary representation U_λ of G in λ with $U_\lambda(x) = (u_{ij}^\lambda(x))$ for x in G. Write d_λ for the representation dimension of U_λ and let $\chi_\lambda(x) = trU_\lambda(x) = \Sigma_{k=1}^{d_\lambda} u_{kk}^\lambda(x)$ be the character of U_λ, then

$$\{e_{ij}^\lambda(x) \equiv d_\lambda^{\frac{1}{2}} u_{ij}^\lambda(x), 1 \le i,j \le d_\lambda, \lambda \in \widehat{G}\} \tag{1.1}$$

is a complete orthonormal function system of $L^2(G)$.

Let $C^\infty(G)$ be the collection of all C^∞ functions on G. Let $S'(G)$ be the collection of all continous linear functionals on $C^\infty(G)$ with $f : C^\infty(G) \to C, \phi \in C^\infty(G) \to < f, \phi > \in C$ for f in $S'(G)$. And let $T : G \to G, x \to x^{-1}$. Set

$$\widehat{f}_\lambda = < f, U_\lambda \circ T > = (\widehat{f}_{ij}^\lambda), 1 \le i,j \le d_\lambda \tag{1.2}$$

with $\widehat{f}_{ij}^\lambda = < f, \overline{u_{ji}^\lambda} >$. In particular, if f is in $L(G)$, then (1.2) becomes

$$\widehat{f}_\lambda = \int_G f(x) U_\lambda(x^{-1}) dx \tag{1.3}$$

with dx being the probability Haar measure on G. Then f in $S'(G)$ has the followong Fouier expansion

$$\begin{aligned} f &\overset{S'}{=} \sum_{\lambda \in \widehat{G}} d_\lambda tr(\widehat{f}_\lambda U_\lambda(.)) \\ &\overset{S'}{=} \sum_{\lambda \in \widehat{G}} d_\lambda \sum_{i,j=1}^{d_\lambda} \widehat{f}_{i,j}^\lambda u_{ij}^\lambda(.) \\ &\overset{S'}{=} \sum_{\lambda \in \widehat{G}} \sum_{i,j=1}^{d_\lambda} a_{ij}^\lambda(f) e_{ij}^\lambda(.) \\ &\overset{S'}{=} \sum_{\lambda \in \widehat{G}} d_\lambda \chi_\lambda * f \end{aligned} \tag{1.4}$$

In (1.4), $a_{ij}^{\lambda}(f) = d_{\lambda}^{\frac{1}{2}} \hat{f}_{ij}^{\lambda}$ is called the Fourier coefficient of f with respect to the function $e_{ji}^{\lambda}(x)$ in the system in (1.1) and the convergence in (1.4) is the sense in $S'(G)$.

It is easy to see that f is in $S'(G)$ if and only if there exists a positive number A such that

$$\left|\hat{f}_{ij}^{\lambda}\right| \leq A(1 + |\lambda|)^{A} \tag{1.5}$$

for all λ in $\hat{G}$ and $1 \leq i, j \leq d_{\lambda}$.

§2. Fourier Coefficients of Integrable Functions on Compact Lie Groups.

The Riemann-Lebesgue lemma concerning, Fourier coefficients in the classical theory of Fourier series is well-known. For compact Lie group, the picture is very different from the classical one. We have

Theorem 2.1. *([17]) Let G be a non-commutative compact Lie group (i.e. G is not a n-dimensional torus). Then for f being in $L^p(G), p \geq 2$ or f being in $C(G)$, the Riemann-Lebesgue lemma is also true that is*

$$\overline{\lim_{|\lambda| \to +\infty}} \sup_{1 \leq i,j \leq d_{\lambda}} \left|a_{ij}^{\lambda}(f)\right| = 0. \tag{2.1}$$

and (2.1) cannot be improved. If f is in $L^p(G)$ for $1 \leq p < 2$, there exists some f in $L^p(G)$, such that

$$\overline{\lim_{|\lambda| \to +\infty}} \sup_{1 \leq i,j \leq d_{\lambda}} \left|a_{ij}^{\lambda}(f)\right| = +\infty, \tag{2.2}$$

where $a_{ij}^{\lambda}(f)$ is as in (1.4).

More precisely, we have

Theorem 2.2. *([17]) Let G be a compact Lie group, f be in $L^p(G)$ for $1 \leq p \leq 2$. Then the Fourier coefficients $\{a_{ij}^{\lambda}(f)\}$ of f have the following asymptotic estimates*

$$a_{ij}^{\lambda}(f) = o(d_{\lambda}^{\frac{1}{p} - \frac{1}{2}}), \ (|\lambda| \to +\infty), \tag{2.3}$$

and the estimates in (2.3) can not be improved.

In particular, we have

Theorem 2.3. *([17]) Let $\{\epsilon_{ij}, 1 \leq i, j \leq d_{\lambda}, \lambda \in \hat{G}\}$ be a set of positive number satisfying*

$$\lim_{|\lambda| \to \infty} \sup_{1 \leq i,j \leq d_{\lambda}} \epsilon_{ij}^{\lambda} = 0. \tag{2.4}$$

Then there exists a f in $L^p(G)$ with $1 \leq p \leq 2$ such that

$$\overline{\lim_{|\lambda| \to +\infty}} \sup_{1 \leq i,j \leq d_{\lambda}} \left|a_{ij}^{\lambda}(f)\right| (\epsilon_{ij}^{\lambda} d_{\lambda}^{\frac{1}{p} - \frac{1}{2}})^{-1} = 1. \tag{2.5}$$

In particular, there exists a f in $L^p(G)$ with $1 \leq p \leq 2$ such that

$$\overline{\lim_{|\lambda| \to +\infty}} \sup_{1 \leq i,j \leq d_{\lambda}} \left|a_{ij}^{\lambda}(f)\right| (\epsilon_{ij}^{\lambda} |\lambda|^{m(\frac{1}{p} - \frac{1}{2})})^{-1} = 1, \tag{2.6}$$

where $m = \frac{1}{2}(dimG - rankG), 1 \leq p \leq 2$.

Theorem 2.4. *([17]) If G is a commutative compact Lie group, then $d_\lambda = 1$ for $\lambda \in \widehat{G}$ and $m = \frac{1}{2}(dimG - rankG) = 0$, and Theorem 2.2 or Theorem 2.3 becomes the classical Riemann-Lebesgue lemma. If G is a non-commutative compact Lie group, then Theorem 2.2 or Theorem 2.3 gives sharp estimates for asymptotic behaviors of Fourier coefficients in $L^p(G)$.*

Let $C^k(G)$ be all functions on G which are continuously k times differentible, $C^{k,\alpha}(G), 0 < \alpha < 1$, be all functions on $C^k(G)$ whose k-th order partial derivatives are in Lip α, $L_p^k(G)$ be all functions in $L_p(G)$ which have up to k times partial derivatives in L_p norm, $L_p^{\lambda,\alpha}(G)$ be all functions in $L_p^k(G)$, whose k-th order partial derivatives $D^\beta f$ satisfy

$$\sup_{|\beta|=k} |y|^{-\alpha} \Big(\int_G \left| D^\beta f(xy) - D^\beta f(x) \right|^p dx \Big)^{\frac{1}{p}} < +\infty.$$

Theorem 2.5. *([18],[23],[26],[28]) If $f \in L_p^k(G), 1 < p \leq 2$, in particular, if $f \in C^k(G) \subset L_2^k(G)$, then, for $p' = \frac{p}{p-1}$*

$$\sum_{\lambda \in \widehat{G}} d_\lambda^{1-\frac{1}{2}p'} \sum_{1 \leq i,j \leq d_\lambda} \left| a_{ij}^\lambda(f)(1+|\lambda|)^k \right|^{p'} \leq A^{p'} < +\infty.$$

Furthermore, if $f \in L_p^k(G), 1 < p \leq 2$ or $f \in L_1^k(G), k = 0, 2, 4, \cdots$, then

$$\sup_{1 \leq i,j \leq d_\lambda} \left| a_{ij}^\lambda(f) \right| (1+|\lambda|)^k = o(d_\lambda^{\frac{1}{p}-\frac{1}{2}}), (|\lambda| \to +\infty).$$

If $f \in L_p^{k,\alpha}(G)$ or $f \in L_1^k(G), k = 1, 3, 5, \cdots$, then

$$\sup_{1 \leq i,j \leq d_\lambda} \left| a_{ij}^\lambda(f) \right| (1+|\lambda|)^{k+\alpha} = O(d_\lambda^{\frac{1}{p}-\frac{1}{2}}), (|\lambda| \to +\infty).$$

§3 Poisson Summation Formula and Summations by Spherical Means and Cubical Means.

Let $L_I^k(H)$ be all functions in $L(H)$ having up to k-th order partial derivatives and being invariant under the Weyl group, $\delta(h) = \Pi_{\alpha>0}(2i \sin \frac{1}{2}(\alpha, h)), P(h) = \Pi_{\alpha>0}(\alpha, h), \delta = \frac{1}{2}\sum_{\alpha>0} \alpha, D = P(\frac{\partial}{\partial h}), Q = \{a \in H, \exp a = unit\ element\ of\ G\}$, exp be the exponential mapping of g onto G. Then we have the following Poisson summation formula.

Theorem 3.1. *([21], [35]) Let $\Phi \in L_I^k(H), k = \frac{1}{2}(dimG - rankG)$,*

$$\varphi(h) = (2\pi)^{-\frac{1}{2}rankG} \int_H \Phi(Y) e^{-i(h,Y)} dY, \tag{3.1}$$

where $(\cdot,\cdot)$ is the invariant inner product on g. Then there exists a integrable central function $\tilde{\Phi}(x)$ on G, such that

$$\tilde{\Phi}(x) \sim \sum_{\lambda\in\widehat{G}} \varphi(\lambda+\delta)d_\lambda\chi_\lambda(x), \tag{3.2}$$

and for $x, y \in G, x = y \exp hy^{-1}, h \in H$,

$$\tilde{\Phi}(x) = \tilde{\Phi}(\exp h) = A_G \sum_{a\in Q} \delta(h+a)^{-1} D\Phi)(h+a). \tag{3.3}$$

where A_G depends only on G.

In [23],[35], we gave an another Poisson summation formula, here we omit it.

Theorem 3.2. *([34], [35]) Let φ and Φ be as in Theorem 3.1, $\Phi_R(h) = R^{-rankG}$ $\Phi(Rh), \varphi_R(h) = \varphi(h/R), \widetilde{\Phi}_R(x)$ be as in (3.3) and $\varphi(0) = 1$. Set*

$$S_R^\varphi(f,x) = f * \widetilde{\Phi}_R(x). \tag{3.4}$$

Then for $f \in L^p(G), p \geq 1$, the following conclusions are valid:

1) $\| S_R^\varphi(f,\cdot) \|_p \leq A(G,\varphi,R) \| f \|_p$;

2) If $P(h)(D\Phi)(h)$ is integrable on H, then

$$\sup_{R>0}\{\| S_R^\varphi(f) \|_p\} \leq A(G,\varphi) \| f \|_p,$$

and

$$\lim_{R\to\infty} \| S_R^\varphi(f) - f \|_p = 0;$$

3) Let $B(t) = ess.\sup_{|h|>t} \left|P(h)^{-1}(D\Phi)(h)\right|$. If $B(|X|)$ is integrable on g, then

a) $S_R^\varphi(f,x)$ converges to $f(x)$ a.e. as $R \to +\infty$,

b) If f is continuous, then $S_R^\varphi(f,x)$ converges to $f(x)$ uniformly,

c) $mes.\{x \in G \,|\sup_{R>0}\{S_R^\varphi(f,x)|\} > \alpha\} \leq A(G,\varphi)\alpha^{-1} \| f \|_1$;

4) If $\int_1^\infty t^{\frac{1}{2}(dimG+rankG)-1} \sup_{|h|\geq t} |\varphi(h)|\, dt < +\infty$, then the Fourier series of $S_R^\varphi(f,x)$ converges to $S_R^\varphi(f,x)$ almost everywhere.

Following Sheng Gong and S.Bochuet, we set

$$H_s^\varphi(t) = t^{-1}\int_0^\infty \varphi(u/t)u^{s+1}J_s(u)du, \tag{3.5}$$

$$W_s^\varphi(t) = H_s^\varphi(t)/t^{2s+1}, \tag{3.6}$$

where $J_s(u)$ is the Bessel function of order s.

As a radial function on H, the Fourier transform of $W_{(q-1)/2}^\varphi(|h|)$ is $\varphi(|h|)$. In this case, we write the kernel function as $K_R^\varphi(x)$ instead of $\widetilde{\Phi}_R(x)$.

Theorem 3.3. *([21],[34],[35]) If $H_s^\varphi \in L_1^{\frac{1}{2}(n-q)}(0,+\infty)$, then, for $x = y\exp hy^{-1}$,*

$$\begin{aligned} K_R^\Phi(x) &= K_R^\Phi(\exp h) \\ &= (-1)^{\frac{1}{2}(n-q)} A_G R^n \sum_{a\in Q} \delta(h+a)^{-1} P(h+a) W_{(n-2)/2}^\varphi(R\,|h+a|) \end{aligned} \tag{3.7}$$

and

$$K_R^\Phi(x) \sim \sum_{\lambda\in\widehat{G}} \varphi(|\lambda+\delta|/R) d_\lambda \chi_\lambda(x), \tag{3.8}$$

where $n = dimG$, $q = rankG$.

The more interesting examples of $\varphi(t)$ are the following

1) $\varphi(t) = (1-t^\alpha)_+^\delta = \begin{cases} (1-t^\alpha)^\delta, & 0\le t\le 1 \\ 0, & t>1, \end{cases}$ with $\mathrm{Re}\alpha > 0$, $\mathrm{Re}\delta > 0$.

2) $\varphi(t) = e^{-t^\alpha}$, $t \ge 0$, $\mathrm{Re}\delta > 0$.

Write $W_s^{\alpha,\delta}(|h|)$, $W_s^{A,\delta}(|h|)$ instead of $W_s^\varphi(|h|)$ corresponding to $\varphi(t) = (1-t^\alpha)_+^\delta$ and $\varphi(t) = e^{-t^\alpha}$, respectively, and write $W_s^\delta(|h|)$ instead of $W_s^{2,\delta}(|h|)$. The kernel functions in (3.7) corresponding $W_s^{\alpha,\delta}(|h|)$, $W_s^\delta(|h|)$ and $W_s^{A,\alpha}(|h|)$ are $K_R^{\alpha,\delta}(x)$, $K_R^\delta(x)$ and $K_R^{A,\alpha}(x)$. Then we have

Theorem 3.4. *([32],[33]) $W_s^{\alpha,\delta}(|h|)$ and $K_R^{\alpha,\delta}(x)$ have the following decompositions*

$$W_R^{\alpha,\delta}(|h|) = (\frac{\alpha}{2})^\delta W_s^\delta(|h|) + \sum_{k=1}^N a_k W_s^{\delta+k}(|h|) + r(|h|), \tag{3.9}$$

$$K_R^{\alpha,\delta}(x) = (\frac{\alpha}{2})^\delta K_R^\delta(x) + \sum_{k=1}^N a_k K_R^{\delta+k}(x) + \tilde{r}(x), \tag{3.10}$$

satisfying

$$|r(|h|)| \le c_0(1+|h|)^{-|\alpha|-2s-2}, \tag{3.11}$$

$$\left|(\frac{d}{dt})^k r(t)\right| \le c_k(1+t)^{-|\alpha|-2s-2-k},\ k = 1,2,\cdots,N+[Re\delta]. \tag{3.12}$$

Theorem 3.5. *([34]) $W_s^{A,\alpha}(|h|)$ has the following estimate*

$$\left|(\frac{\partial}{\partial h})^\beta W_s^{A,\alpha}(|h|)\right| \le C_\beta(1+|h|)^{-|\alpha|-2s-2-|\beta|}. \tag{3.13}$$

Corollary 3.6. *([21],[34],[35]) If $Re\alpha > 0$ and $Re\delta > \frac{1}{2}(dimG-1)$, then*

$$S_R^{\alpha,\delta}(f,x) = f * K_R^{\alpha,\delta}(x)$$

and

$$S_R^{A,\delta}(f,x) = f * K_R^{A,\delta}(x)$$

satisfy Theorem 3.2.

Let $k_N(\theta) = \Sigma_{m=-\infty}^{\infty} \mu N_m e^{im\theta}$ with $\lim_{\mu\to\infty} \mu N_m = 1$ and $\theta \in R$. For $x \in G$, $x = y \exp hy^{-1}$, $y \in G$, $h \in H$, let $h = (h_1, \cdots, h_q)$ be the regular coordinnate of h. Set

$$K_N(x) = K_N(\exp h) = \prod_{j=1}^{q} (k_N((h_j))_j^n / B_N, \tag{3.14}$$

such that

$$\int_G K_N(x)dx = 1,$$

where $|k_N(\theta)| \leq AN(1+N\,|\theta|)^{-1-\alpha}$, $\alpha > 0$, $n_j = dim\widetilde{g}/rank\widetilde{g}$ for any simple compact Lie algebra or commutative compact Lie algebra and $n_k = k$ for $u_k = (A_{k-1})_u \oplus H_1$.

Theorem 3.7. *([25]) There exists critical index α_0 with $0 \leq \alpha_0 \leq 1$, such that for $\alpha > \alpha_0$:*

1) $f * K_N(x)$ converges to $f(x)$ for $f \in L(G)$ almost everywhere;

2) $\lim_{n\to\infty} \| f * K_N - f \|_p = 0$, for $f \in L^p(G), p \geq 1$;

3) $f * K_N(x)$ converges to $f(x)$ uniformly for $f \in C(G)$;

4) $mes.\{x \in G, \sup_N |f * K_N(x)| > \lambda\} \leq A_G \lambda^{-1} \| f \|_1$.

If $\alpha \leq \alpha_0$, there exists a F in $L(G)$ such that 1) of this theorem can not be valid for the F.

To prove Theorem 3.7, the following result is needed. Let $b = (b_1, \cdots, b_q) \in H$, $V_b = \{h = (h_1, \cdots, h_q) \in H, |h_j| \leq b_j\}$, $G(rV_b) = \{t \exp ht^{-1}, t \in G, h \in rV_b\}$, it is easy to see that $G(rV_b)$ is not a geodesic convex set. Set

$$f_b^*(x) = \sup_{r>0}\{\int_{G(rv_b)} |f(xy)|\, dy/mes.G(rV_b)\}.$$

Then we have

Theorem 3.8. *([26]) If $f \in L^p(G), p > 1$, then*

$$\| f_b^* \|_p \leq A_p \| f \|_p,$$

if $f \in L(G)$, then

$$mes.\{x \in G, f_b^*)(x) > \lambda\} \leq A_1 \lambda^{-1} \| f \|_1,$$

where $A_p (p \geq 1)$ depends only on G and p.

§4 Convergence of Fourier Series, Appoximation.

Let ω b a closed bounded domain in H with $0 \in \omega$, $N\omega = \{h = NY \in H, Y \in \omega\}$. Set

$$S_{N_\omega}(f, x) = \sum_{\lambda+\delta \in N_\omega \cap (\widehat{G}+\delta)} d_\lambda tr(\widehat{f}_\lambda U_\lambda(x)),\ N = 1, 2, \cdots . \tag{4.1}$$

Then (4.1) is the partial sum of Fourier series of f in $L(G)$ related to ω. The kernel functions are

$$D_{N_\omega}(x) = \sum_{\lambda+\delta\in N_\omega\cap(\widehat{G}+\delta)} d_\lambda \chi_\lambda(x), \; N = 1, 2, \cdots . \tag{4.2}$$

Let ω_0 be the cube $\{h = (h_1, \cdots, h_q) \in H, |h_j| \leq 1, \; j = 1, 2, \cdots, q, \; q = dimH,$ $(h_1, \cdots, h_q)$ is the regular coordinate of $h\}$. Write $S_N(f, x)$ and $D_N(x)$ instead of $S_{N_{\omega_0}}(f, x)$ and $D_{N_{\omega_0}}(x)$. Then $D_N(x)$ are called Dirichlet kernels.

Theorem 4.1. *([34]) Let $H_+ = \{h \in H, \; (\alpha_j, h) \geq 0, \; j = 1, 2, \cdots, q, \; \alpha_1, \cdots, \alpha_q$ are the simple roots$\}, \widehat{\omega} = \cup_{\sigma\in W}\sigma(\omega \cap H_+)$, W be the Weyl group, $\widehat{G}_1 = (\delta + \cup_{\sigma\in W}\sigma(\widehat{G})) \cap H_+$, $(\widetilde{\omega}, \widehat{G}) = \widetilde{\omega} \cap (\cup_{\sigma\in W}\sigma(\widehat{G}_1))$, and let*

$$d_{N_{\widetilde{\omega}}}(h) = \sum_{\lambda+\delta\in N_{\widetilde{\omega}}\cap(\widetilde{\omega},\widehat{G})} e^{i(\lambda+\delta,h)}. \tag{4.3}$$

Then, for $x = y \exp hy^{-1}, x, y \in G, h \in H$,

$$\begin{aligned} D_{N_\omega}(x) &= D_{N_{\widetilde{\omega}}}(x) \\ &= \delta(h)^{-1} P(i\delta)^{-1} P(\frac{\partial}{\partial h})\{d_{N_{\widetilde{\omega}}}(h)\}. \end{aligned} \tag{4.4}$$

$$\begin{aligned} D_N(x) &= D_{N_{\widetilde{\omega}_0}}(x) \\ &= \delta(h)^{-1} P(i\delta)^{-1} P(\frac{\partial}{\partial h})\{d_N(h)\}. \end{aligned} \tag{4.5}$$

The Lebesgue constants are

$$\rho_{N_\omega}(G) = \int_G |D_{N_\omega}(x)| \, dx, \tag{4.5}$$

$$\rho_N(G) = \int_G |D_N(x)| \, dx. \tag{4.6}$$

And by $\rho_N(G) = A(N)$, we mean $\rho_N(G) = A(N) + o(A(N))$. We have

Theorem 4.2. *([34],[20],[22],[24]) Let $(A_k)_u, (B_k)_u, (C_k)_u, (D_k)_u, (G_2)_u, (F_4)_u,$ $(E_6)_u, (E_7)_u, (E_8)_u$ be all simple compact Lie algebras, H_k be k-dimensional commutative compact Lie algebra. Then Lie algebra g of G has the following decomposition*

$$\begin{aligned} g = &\sum_{j=1}^{r_A}(A_{k_j})_u \oplus \sum_{j=1}^{r_B}(B_{k_j})_u \oplus \sum_{j=1}^{r_C}(C_{k_j})_u \oplus \sum_{j=1}^{r_D}(D_{k_j})_u \oplus \\ &\oplus \sum_{j=1}^{r_2}(G_2)_u \oplus \sum_{j=1}^{r_4}(F_4)_u \oplus \sum_{j=1}^{r_6}(E_6)_u \oplus \sum_{j=1}^{r_7}(E_4)_u \oplus \sum_{j=1}^{r_8}(E_8)_u \oplus H_k. \end{aligned} \tag{4.7}$$

Let g include $r_1(A_1)'_u s$, for $\widetilde{g}$ being a simple Lie algebra, let $\chi(\widetilde{g}) = 1$ if $\dim\widetilde{g} \not\equiv 0$ mod 3 and $\chi(\widetilde{g}) = 0$ if $\dim\widetilde{g} \equiv 0$ mod 3, and let $\overline{\chi}(\widetilde{g}) = 1 - \chi(\widetilde{g})$. Let

$$\begin{aligned} a = &\sum_{j=1}^{r_A}(dim(A_{k_j})_u + \chi((A_{k_j})_u)) + \sum_{j=1}^{r_B}(dim(B_{k_j})_u - \chi((B_{k_j})_u)) \\ &+ \sum_{j=1}^{r_C}(dim(C_{k_j})_u - \chi((C_{k_j})_u)) + \sum_{j=1}^{r_D}(dim(D_{k_j})_u - \chi((D_{k_j})_u)) \\ &+ r_2(1 + dim((G_2)_u)), \end{aligned}$$

$$\begin{aligned} b = &\sum_{j=1}^{r_4}(dim(F_4)_u - rank(F_4)_u) + \sum_{j=1}^{r_6}(dim(E_6)_u - rank(E_6)_u) \\ &+ \sum_{j=1}^{r_7}(dim(E_7)_u - rank(E_7)_u) + \sum_{j=1}^{r_8}(dim(E_8)_u - rank(E_8)_u) \\ &- 3r_4 - 3r_6 - 4r_7 - 9r_8, \end{aligned}$$

$$c = \sum_{j=1}^{r_B}\chi((B_{k_j})_u) + \sum_{j=1}^{r_C}\chi((C_{k_j})_u) + \sum_{j=1}^{r_D}\chi((D_{k_j})_u),$$

$$d = r_2 + 2r_6 + 3r_7 + r_8 + \max\{k - \sum_{j=1}^{r_A}\chi((A_{k_j})_u), \sum_{j=1}^{r_A}\overline{\chi}((A_{k_j})_u) - r_1\}.$$

Then

$$\rho_N(G) \doteq A_G N^{a+b}(\log N)^{c+d},$$

where A_G depends only on G.

Theorem 4.3. *([20]) Let ω be any closed bounded domain in H. Then there esists $A(G,\omega) > 0$, such that*

$$\rho_{N_\omega}(G) \geq A(G,\omega)\rho_N(G),$$

where $A(G,\omega)$ depends only on G and ω.

Theorem 4.4. *([20],[24]) Let $f \in C^{a+b,\omega}(G)$, and $\omega(\frac{1}{N})(\log N)^{c+d} \to 0$ for $N \to \infty$. Then $S_N(f,x)$ converges to $f(x)$ uniformly.*

Theorems 4.2 and 4.3 imply that if $1 \leq p < 2$ and G is not commutative, then there exists a f in $L^p(G)$ such that

$$\overline{\lim_{N\to\infty}} \parallel S_{N(\omega)}(f,\cdot) - f \parallel_p \neq 0.$$

Now we consider absolute convergence of Fourier series in the sense of

$$\sum_{\lambda\in\widehat{G}} d_\lambda \sum_{1\leq i,j\leq d_\lambda} \left|\widehat{f}_{ij}^\lambda u_{ij}^\lambda(x)\right| < +\infty. \tag{4.8}$$

Obviously, (4.8) implies

$$\sum_{\lambda\in\widehat{G}} d_\lambda \left|\chi_\lambda(x) * f\right| < +\infty. \tag{4.9}$$

Theorem 4.5. *([18]) If $f \in L_2^{k,\alpha}$ with $k+\alpha > \frac{1}{2}\dim G$, in particular, if $f \in C^{k,\alpha}(G)$ then*

$$\sum_{\lambda\in\widehat{G}} d_\lambda \sum_{1\le i,j\le d_\lambda} \left|\widehat{f}_{ij}^{\lambda} u_{ij}^{\lambda}(x)\right| \le A(G,k,\alpha) \parallel f \parallel_{2,k,\alpha},$$

therefore the Fourier series of $f(x)$ converges to $f(x)$ uniformly and absolutoly in the sence of (4.8) and (4.9).

Let

$$E_R(f) = \inf_{b_{ij}^{\lambda}\in C} \parallel f - \sum_{|\lambda+\delta|\le R} b_{ij}^{\lambda} u_{ij}^{\lambda}(\cdot) \parallel_\infty,$$

$$E_R(f)_p = \inf_{b_{ij}^{\lambda}\in C} \parallel f - \sum_{|\lambda+\delta|\le R} b_{ij}^{\lambda} u_{ij}^{\lambda}(\cdot) \parallel_p .$$

Then we have

Theorem 4.6. *([26]) Let $f \in C^{k,w}(G)$, w be the modulus of continuity, $\alpha \ge k+1, \delta > \frac{1}{2}(\dim G - 1)$. Then*

$$\left|S_R^{\alpha,\delta}(f,g) - f(g)\right| \le A(G,\alpha,\delta) \parallel f \parallel_{k,w} R^{-k} w(\frac{1}{R}),$$

therefore

$$E_R(f) \le A(G,\alpha,\delta) \parallel f \parallel_{k,w} R^{-k} w(\frac{1}{R}).$$

Conversely, if $\Sigma_{N=1}^{\infty} N^{k-1} E_N(f) < +\infty$, then $f \in C^{k,w}(G)$, where

$$w(f^{(k)}, \frac{1}{R}) \le A \sum_{N>\frac{1}{2}R} N^{k-1} E_N(f).$$

Theorem 4.7. *([23]) If $f \in L_p^{k,w}(G)$, then for $\alpha \ge k+1, \delta > \frac{1}{2}(\dim G - 1)$,*

$$\parallel S_R^{\alpha,\delta}(f,\cdot) - f \parallel_p \le A(G,k,\alpha) \parallel f \parallel_{p,k,w} R^{-k} w(\frac{1}{R}),$$

therefore

$$E_R(f)_p \le A(G,k,\alpha) \parallel f \parallel_{p,k,w} R^{-k} w(\frac{1}{R}).$$

Conversely if $\Sigma_{N=1}^{\infty} N^{k-1} E_N(f)_p < +\infty$, then $f \in L_p^{k,w}(G)$, where

$$w(\frac{1}{R}) \le A \Sigma_{N>\frac{1}{2}R} N^{k-1} E_N(f)_p.$$

§5 Riesz and Bessel Potentials, Riesz ans Bessel Transforms and Fractional Derivatives.

Let $\triangle$ be the Laplacian of G, $Re\alpha > 0$. Set $I_\alpha = (-\triangle)^{-\frac{1}{2}\alpha}, \widetilde{I}_\alpha = (|\delta|^2 I - \triangle)^{-\frac{1}{2}\alpha}$ with δ being the half of the sum of all positive roots, I_α is the Riesz potential of order $\alpha, \widetilde{I}_\alpha$ is the Bessel potential, I^α and $\widetilde{I}^\alpha$ are the fractional derivatives for $\alpha < 1$.

Let $\mu_\lambda, \lambda \in \widehat{G}$ be the characteristic values of Laplacian $\triangle$ satisfying $\triangle U_\lambda(x) = -\mu_\lambda U_\lambda(x)$. Then

$$I_\alpha f \overset{S'}{=} \sum_{\lambda \neq 0} \mu_\lambda^{-\frac{1}{2}\alpha} d_\lambda \chi_\lambda * f, \tag{5.1}$$

$$\widetilde{I}_\alpha f \overset{S'}{=} \sum_{\lambda \neq 0} |\lambda + \delta|^{-\alpha} d_\lambda \chi_\lambda * f, \tag{5.2}$$

$$I^\alpha f \overset{S'}{=} \sum_{\lambda \in \widehat{G}} \mu_\lambda^{\frac{1}{2}\alpha} d_\lambda \chi_\lambda * f, \tag{5.3}$$

$$\widetilde{I}_\alpha f \overset{S'}{=} \sum_{\lambda \in \widehat{G}} |\lambda + \delta|^{\alpha} d_\lambda \chi_\lambda * f. \tag{5.4}$$

Let $X_1, \cdots, X_n$ be a orthonormal basis of g. Let $\widetilde{X}_j$ be the left invariant vector field generated by X_j for $j = 1, 2, \cdots, n$, $n = \dim G$. The Riesz transforms are $R_j = \widetilde{X}_j(-\triangle)^{-\frac{1}{2}}, j = 1, 2, \cdots, n$ and $\widetilde{R}_j = \widetilde{X}_j(|\delta|^2 I - \triangle)^{\frac{1}{2}}$ are the Bessel transforms.

Theorem 5.1. *([28],[40]) There are functions $I_\alpha(x)$ and $\widetilde{I}_\alpha(x)$ on G, such that*

1) $I_\alpha(x)$ and $\widetilde{I}_\alpha(x)$ are in $L^p(G), p > 1$

2) $I_\alpha(x) \sim \Sigma_{\lambda \neq 0} \mu_\lambda^{\frac{1}{2}\alpha} d_\lambda \chi_\lambda(x)$ and $\widetilde{I}_\alpha(x) \sim \Sigma_{\lambda \neq 0} |\lambda + \delta|^{-\alpha} d_\lambda \chi_\lambda$

3) I_α and $\widetilde{I}_\alpha(x)$ are infinitely differentiable on $G \backslash \{e\}$, e is the unit element of G, and

$$I_\alpha(x) = \widetilde{I}_\alpha(x) + \sum_{p=1}^{m} b_k \widetilde{I}_{\alpha+2k}(x) + r(x),$$

where $r(x)$ is $m - n$ times differentiable, m is any positive integer, $n = \dim G$.

4) Let $X \in g, X = x_1 X_1 + \cdots + x_n X_n$, $\{V, \psi\}$ be a coordinate neighborhood with $e \in V \subset G, \psi : V \to g$, $\psi(e) = (0, 0, \cdots, 0)$. Then the main part of $I_\alpha(x)$ or $\widetilde{I}_\alpha(x)$ is $A_G |x|^{\alpha - n}$, with $|x| = \exp|X| = (x_1^2 + \cdots + x_n^2)^{\frac{1}{2}}$ for $x = \exp X, X \in g$ and $\alpha - n \neq 0, 2, 4, \cdots$, or $A_G |x|^{\alpha-n} \log|x|^{-1}$ for $\alpha - n = 0, 2, 4, \cdots$.

Theorem 5.2. *([28],[40]) Let $\widetilde{K}_j(x)$ be the kernel function of Riesz transform R_j for $j = 1, 2, \cdots, n$, $\widetilde{K}_j(x)$ be the kernel function of Bessel transform of $\widetilde{R}_j$. Then*

1) $K_j(x) = \widetilde{X}_j I_1(x), \widetilde{K}_j(x) = \widetilde{X}_j \widetilde{I}_1(x)$

2) $K_j(x)$ and $\widetilde{K}_j(x)$ are infinitely differentiable on $G \backslash \{e\}$,

3) For $f \in L^p(G), p \geq 1$, then

$$R_j f(x) = P.V.K_j(\cdot) * f(x), \ \ \widetilde{R}_j f(x) = P.V.\widetilde{K}_j(\cdot) * f$$

4) The main part of $K_j(x)$ *and* $\widetilde{K}_j(x)$ *is*

$$A_G x_j \left|x\right|^{-n}$$

5) For $p > 0$. R_j *and* $\widetilde{R}_j$ *are* $H^p(G)$ *bounded.*

Theorem 5.3. *([40]) There exist functions* $I^\alpha(x)$ *and* $\widetilde{I}^\alpha(x)$ *on* G*, such that*

1) $(I^\alpha f)(x) = \int_G (f(xy^{-1}) - f(x)) I^\alpha(y)dy + B(G,\alpha)f(x)$,

$$(\widetilde{I}^\alpha f)(x) = \int_G (f(xy^{-1}) - f(x))\widetilde{I}^\alpha(y)dy + B(G,\alpha)f(x),$$

with $0 < Re\alpha < 1$, f *being differentiable;*

2) $I^\alpha(x) = \widetilde{I}^\alpha(x) + \Sigma_{k=1}^m b_R \widetilde{I_{2k-\alpha}}(x) + r(x)$ *with* $r(x)$ *being* $2m - n - 1$ *times differentiable,* m *being any positive integer;*

3) The main part of $I^\alpha x$ *or* $\widetilde{I^\alpha}(x)$ *is*

$$A(G,\alpha)\left|x\right|^{-n-\alpha},$$

and $I^\alpha(x)$ *or* $\widetilde{I^\alpha}(x)$ *is infinitely differentiable on* $G\backslash\{e\}$.

§6 Hormander Type Multiplier Theorem.

Under the inner product $(\cdot,\cdot)$, H is a q-dimensional Euclidean space R^q with $q = \mathrm{rank}G$. Let ϕ be a essential bounded central function on $H = R^q, \phi$ satisfying Hormander condition of order k with k being positive integer is defined by

$$\left|(\frac{\partial}{\partial n})^\alpha \phi(h)\right| \le A(h)\left|h\right|^{-|\alpha|}, \text{for } 0 \le |\alpha| \le k, \tag{6.1}$$

$$\int_{r<|h|<2r} A(h)^2 dh \le A_0 r^q, \text{for } 0 < r, \tag{6.2}$$

and we simply call ϕ satisfies H_k-condition.

We call ϕ satisfies strong H_α-condition for $\alpha > 0$ if

$$|I^\alpha \phi(h)| \le A(h)\left|h\right|^{-\alpha}, \tag{6.3}$$

and $A(h)$ satisfies (6.2).

We call ϕ satisfies weak H_α-condition for $\alpha > 0$ if

1) The inequality

$$\left|\Delta_Y^k \phi(h)\right| \le A(h)\left|Y\right|^\alpha \left|h\right|^{-\alpha} \tag{6.4}$$

holds for $k > \alpha > 0$.

2) $A(h)$ stisfies (6.2), where

$$\Delta_Y \phi(h) = \phi(h-Y) - \phi(h), \Delta_Y^k \phi = \Delta_Y(\Delta_Y^{k-1}\phi).$$

With the above three conditions, the following lemmas are valid.

Lemma 6.1. *([30]) Let ϕ be a essential bounded function on H. Then ϕ satisfying H_k-condition implies ϕ satisfying strong H_k-condition or conversely.*

Lemma 6.2. *([36]) ϕ satisfying strong H_α-condition implies ϕ satisfying weak H_α-condition. Conversely, ϕ satisfying weak H_α-condition implies ϕ satisfying strong $H_{\alpha-\epsilon}$-condition for $0 < \epsilon < \alpha$ but not uniformly, that is*

$$\left|I^{\alpha-\epsilon}\phi(h)\right| \leq A_\epsilon(h)\,|h|^{\epsilon-\alpha}, \tag{6.5}$$

and

$$\int_{r\leq|h|\leq 2r} A_\epsilon(h)^2 dh \leq \epsilon^{-1}A_0 r^q. \tag{6.6}$$

Lemma 6.3. *([36]) ϕ satisfying strong H_α-condition implies ϕ satisfying strong H_β-condition for $0 < \beta < \alpha$ uniformly. Similarly, ϕ satisfying weak H_α-condition implies ϕ satisfying weak H_β-condition for $0 < \beta < \alpha$ uniformly.*

Lemma 6.4. *([36]) Let ϕ satisfy H_k-condition and for all $|\alpha| = k$, $|h|^k\,(\frac{\partial}{\partial h})^\alpha$ satisfy weak H_b-condition. Then ϕ satisfies weak H_β-condition for $0 < \beta \leq k + b$. Similarly, if ϕ satisfies strong H_α-condition and $|h|^\alpha I^\alpha\phi(h)$ satisfies weak H_b-condition, then ϕ satisfies weak H_β-condition for $0 < \beta \leq \alpha + b$. Conversely if ϕ satisfies weak H_α-condition, then for $0 < \beta < \alpha, |h|^\beta I^\beta\phi(h)$ satisfies strong H_b-condition for $0 < b < \alpha - \beta$.*

For compact Lie group G, without loss of generality, we only consider ϕ on H to be continuous. If ϕ is a slowly incresing function, then ϕ defines a multiplier operator T_ϕ by

$$T_\phi \overset{S'}{=} \sum_{\lambda\in\widehat{G},\lambda\neq 0} \phi(\lambda+\delta)d_\lambda\chi_\lambda * f \overset{S'}{\equiv} {\sum}' \phi(\lambda+\delta)d_\lambda\chi_\lambda * f. \tag{6.7}$$

If the ϕ in (6.7) is radial, then ϕ can define another multiplier operator $\widetilde{T_\phi}$ as

$$\widetilde{T_\phi} \overset{S'}{=} {\sum}' \phi(\mu_\lambda)d_\lambda\chi_\lambda * f, \tag{6.8}$$

where $-\mu_\lambda$ is the characteristis value of Laplacian on G.

Theorem 6.5. *([36]) Let ϕ satisfy weak H_α-condition for $\alpha = \frac{1}{2}n + b, b > 0, n = dimG$. Then T_ϕ is $H^p(G)$ bounded for $\frac{n}{n+b} < p \leq 1$.*

Theorem 6.6. *([36]) Let ϕ be radial and satisfy weak H_α-condition for $\alpha = \frac{1}{2}n+b = k+\beta, 0 < b, 0 < \beta \leq 1$ and let ϕ be in $C^{k,\beta}(H\setminus\{o\})$. Then both T_ϕ and $\widetilde{T_\phi}$ are $H^p(G)$ bounded for $\frac{n}{n+b} < p \leq 1$.*

Part II HARMONIC ANALYSIS ON COMPACT HOMOGENEOUS SPACES

Let M be a connected compact Riemannian manifold. Let G be the connected component of isomeric transfoumation group of M and G be transitively acting on M. For $o \in M$, let K be the isotropy group of G at o, then $G/K \cong M$ is isometric isomorphism. Then M is called a compact homogeneous space.

Every compact Lie group G and unit sphere S_{n-1} in n-dimensional Euclidean space R^n are special compact homogeneous spaces. Hence, if a theorem is valid for compact homogeneous spaces, then it is valid for compact Lie groups and spheres naturally.

§7 Introduction.

Let M be as above, $x \in G, x : M \to M, m \in M \to x \cdot m \in M$. For any m on M, there exists an x in G such that $m = x \cdot o = xk \cdot o, k \in K$. Let $\widehat{G}$ be as in Part I, for $\lambda \in \widehat{G}$ choose a unitary irreducible representation U_λ of G in λ, and set $I_\lambda = \int_K U_\lambda(k)dk$. Then the unitary dual $\widehat{M}$ of M is

$$\widehat{M} = \{\lambda \in \widehat{G}, I_\lambda \neq \text{null matrix}\}.$$

Let $\chi_\lambda(m) = tr(U_\lambda(m)I_\lambda) = tr(U_\lambda(x)I_\lambda)$ for $m = x \cdot 0, S'(M)$ be the linear continuous functionals on $C^\infty(M)$. Then $f \in S'(M)$ has the following Fourier expansion

$$\begin{aligned} f &\overset{S'}{=} \sum_{\lambda \in \widehat{M}} d_\lambda \chi_\lambda * f \\ &\overset{S'}{=} \sum_{\lambda \in \widehat{M}} d_\lambda tr(\widehat{f}_\lambda U_\lambda) \end{aligned} \tag{7.1}$$

where $\widehat{f}_\lambda = < f, I_\lambda \overline{U'_\lambda} >$. If f is integrable on M, then

$$\widehat{f}_\lambda = I_\lambda \int_M f(m) U_\lambda(m^{-1}) dm \tag{7.2}$$

with $U_\lambda(m^{-1}) = U_\lambda(x^{-1})$ for $m = x \cdot 0$.

Lemma 7.1. *([37]) $f \in S'(M)$ if and only if there exists a positive number A such that*

$$tr(\widehat{f}_\lambda \overline{\widehat{f'_\lambda}}) \leq A(1 + |\lambda|)^A. \tag{7.3}$$

Let $g = \mathcal{P} \oplus \mathcal{K}$ be the orthogonal direct sum with $\mathcal{K}$ being the Lie algebra of K. Set

$$P = \{x = \exp X \in G, X \in \mathcal{P}\}.$$

Then P is diffeomorphic to M, hence for any $n \in M$ there exists a unique p in P, such that $n = p \cdot 0, n^{-1} = p^{-1} \cdot 0$. Write $n^{-1} \cdot m = p^{-1} \cdot m$ for $n = p \cdot 0 \in M, m \in M$.

Theorem 7.2. *([37]) Let f_1, f_2 be as in $L(M)$ and set*

$$f_1 * f_2(m) = \int_K \int_M f_1(k \cdot n^{-1}m) f_2(n) dn dk. \tag{7.4}$$

Then $\widehat{f_1 * f_2}_\lambda = \widehat{f}_{1\lambda}\widehat{f}_{2\lambda}$. If f_1 is a central function, then

$$f_1 * f_2(m) = \int_M f_1(n^{-1}m) f_2(n) dn. \tag{7.4}$$

$f \in S'(G)$ is central if and only if

$$f \stackrel{S'}{=} \sum_{\lambda \in \widehat{M}} a_\lambda d_\lambda \chi_\lambda, \ |a_\lambda| \leq A(1 + |\lambda|)^A. \tag{7.5}$$

All central functional in S' is denoted by $S'_I(G)$. Obviously, $\chi_\lambda(m)$ is a central function, and if M is the sphere S_{n-1}, then $d_\lambda \chi_\lambda(n^{-1}m)$ is a zonal harmonic.

Let $X_1, \cdots, X_l, l = \dim M$ be an orthonormal basis of $\mathcal{P}, \widetilde{X_1}, \cdots, \widetilde{X_l}$ be vector fields on M difined by

$$(\widetilde{X_j} f)(m) = \frac{d}{dt} f(m \cdot \exp tX_j \cdot 0)|_{t=0}. \tag{7.5}$$

Then the Laplacian Δ on M is

$$(\Delta f)(m) = \sum_{j=1}^{l} (\frac{d}{dt})^2 f(m \cdot \exp tX_j \cdot o)|_{t=0} \tag{7.6}$$

satisfying

$$\Delta U_\alpha(m) I_\lambda = -\mu_\lambda U_\lambda I_\lambda,$$
$$\mu_\lambda = |\lambda + \delta|^2 - |\delta|^2,$$

for λ in $\widehat{M}$ and δ being as above.

§8 Poisson Kernels, Abel Summation..

The Poisson kernels on M is

$$P(t, m) = \sum_{\lambda \in \widehat{G}} e^{-t\mu_\lambda^{\frac{1}{2}}} d_\lambda \chi_\lambda(m), t > 0. \tag{8.1}$$

It satisfies

$$(\frac{\partial^2}{\partial t^2} + \Delta) P(t, m) = 0. \tag{8.2}$$

For f in $S'(G)$, set

$$P_t(f)(m) = P(t, \cdot) * f(m), \ t > 0. \tag{8.3}$$

Then $P_t(f)(m) \in C^\infty(M)$.

In order to study the operator family $\{P_t, \ t > 0\}$, we must study the operator family $\{S_t^A, \ t > 0\}$ defined by

$$S_t^A(f)(m) = A(t, \cdot) * f(m), \tag{8.4}$$

$$A(t, m) = \sum_{\lambda \in \widehat{M}} e^{-t|\lambda + \delta|} d_\lambda \chi_\lambda(m), \ t > 0. \tag{8.5}$$

Theorem 8.1. *([37]) Let P_t and S_t^A be as above. Then, for $0 < t < 1$,*

$$P_t = S_t^A + S_t^A \circ (\sum_{k=1}^{N} b_k(t) t \widetilde{I}_{2k-1} + t b_{n+1}(t) \widetilde{r}_N(t)),$$

where, $b_k(t)$ is bounded and infinitely differentiable on (0,1),and operator $\widetilde{r}_N(t)$ has an $N - l - 1$ times defferentiable kernel function on M, and I_{2k-1} is the Bessel potential on M.

For $f \in S'(M)$, let

$$\begin{aligned} P_+^*(f)(m) &= \sup_{t>0}\{|P_t(f)(m)|\} \\ S_+^{A*}(f)(m) &= \sup_{t>0}\{|P_t(f)(m)|\}. \end{aligned} \tag{8.6}$$

Then we have

Theorem 8.2. *([37]) S_+^{A*} is type of (p,p) with $p > 1$ and weak type of (1.1). Therefore P_+^* is of type (p,p) with $p > 1$ and weak type (1,1).*

Theorem 8.3. *([37]) If $f \in L^p(M)$ for $p \geq 1$, then*
1) $\lim_{t\to 0} \| S_t^A(f) - f \|_p = 0$,
2) $\lim_{t\to 0} \| P_t(f) - f \|_p = 0$,
3) $\lim_{t\to 0} S_t^A(f)(m) = f(m)$ *a.e.,*
4) $\lim_{t\to 0} P_t^A(f)(m) = f(m)$ *a.e.*

Theorem 8.4. *([38]) Let $\alpha > 0$, set*

$$P_{\alpha\nabla}^*(f)(m) = \sup_{|n^{-1}m|<\alpha t} |P_t(f)(n)|, \tag{8.7}$$

$$S_{\alpha\nabla}^A(f)(m) = \sup_{|n^{-1}m|<\alpha t} \left|S_t^A(f)(n)\right|. \tag{8.8}$$

Then $S_{\alpha\nabla}^A$ and $P_{\alpha\nabla}^*$ are of type (p,p) with $p > 1$ and weak type (1,1).

Theorem 8.5. *([38]) $P(t,m)$ are pisitive kernels. For $f \in S'(M)$, if $P_\nabla^*(f)$ is bounded at all the points of a subset $S \subset M$ of positive measure, then $P_t(f)(m)$ has nontangential limits at almost every point of S.*

§9 H^p Space.

The $H^p(M)$ for $0 < p \leq 1$ defined in [39] consists of all $f \in S'(M)$ satisfying

$$P_\nabla^*(f) \in L^p(M), 0 < p \leq 1. \tag{9.1}$$

The (p,q,s) atom is defined as follows

Definition 9.1. *([39]) Let $0 < p \le 1 \le +\infty$ and $p < q, l = dimM, s = [l(1-\frac{1}{p})]$ be a non-negative integer. If $a(m) \in L^q(M)$ satisfying*

1) supp $a \subset B(m_0, r)$, where $B(m_0, r)$ is a geodesic sphere of radius r and centered at m_0;

2) $\| A \|_q \le |B|^{\frac{1}{q}-\frac{1}{p}}$ where $|B| = mes.B(m_0, r)$;

3) $\int_M a(m)P_k(\pi(m)I_\pi)dm = 0, 0 \le k \le s$, where π is a faithful representation of $G, I_\pi = \int_k \pi(k)dk$. $P_k(\cdot)$ is any polynomial of degree k of real variables of $\pi(m)$; then $a(\cdot)$ is a (p, q, s) atom.

4) A exceptional atom $a(\cdot)$ is a C^∞-function on M satisfying $\| a(\cdot) \|_\infty \le 1$.

Definition 9.2. *([39]) The atomic Hardy space $H_a^{p,q,s}(M)$ is defined as follows*

$$H_a^{p,q,s}(M) = \{f \in s'(M), f \overset{s'}{=} \sum \lambda_R a_R, \| f \cdot \|_{H_a^{p,q,s}} = \inf(\sum_R |\lambda_R|^p)^{\frac{1}{p}} < +\infty\}. \quad (9.2)$$

In (9.2), every a_k is a (p, q, s) atom or a exceptional atom.

Let $S_I(H)$ be the class of all those central C^∞ function Φ on H, such that

$$\sup_{h \in H} \left|(1+|h|)^b (\frac{\partial}{\partial h})^\beta (h)\right| < +\infty$$

for all positive b and all q-tupls $\beta = (\beta_1, \cdots, \beta_q)$ of nonnegative integers. Set $\Phi_t(h) = t^{-q}\Phi(t^{-1}h)$. Then, for $m = x \cdot o$,

$$\widetilde{\Phi_t}(m) = A_G \sum_{a \in Q} \int_K \Delta(h(xk) + a)^{-1}(D\Phi_t)(hxk) + a)dk \quad (9.3)$$

is a central C^∞ function on M. Let $C_I^\infty(M)$ be the class of all central C^∞ functions on M, then $S_I(H) \to C_I^\infty(M), \Phi \in S_I(H) \to \widetilde{\Phi} = \widetilde{\Phi_1}$ in (9.3) is surjective.

Definition 9.3. *([30]) Let r be a positive integer, $l = dimM$,*

$$K_r = \{\widetilde{\Phi} \in C_I^\infty(M), \ \| \widetilde{\Phi} \|_{k_r} = \sup_{h \in H, |\beta| \le r} (1+|h|)^{r+l} \left|(\frac{\partial}{\partial h})^\beta \Phi(h)\right| \le 1\},$$

$$\widetilde{\Phi_\nabla^*}(f)(m) = \sup_{|n^{-1}m| \le t} \left|\widetilde{\Phi_t} * f(m)\right|,$$

$$f_r^*(m) = \sup_{\widetilde{\Phi} \in K_r} \{\widetilde{\Phi_\nabla^*}(f)(m)\}.$$

Then $f_r^*(m)$ is called the Grand maximal function of f in $S'(M)$ related to K_r.

There exists a function $\psi \in L(1,\infty) \cap C^\infty(1,\infty)$ such that

$$\widetilde{\Phi_t}(f)(m) = \int_1^\infty \psi(s) S_{ts}^A(f)(m)ds \quad (9.4)$$

satisfying $\widetilde{\Phi_t}$ is generated by $\phi \in S_I(H)$ according to (9.3). Let

$$\widetilde{\varphi}_t^\lambda(f)(m) = \sup \left| f * \widetilde{\phi}_t(n)(\frac{t}{|n^{-1}m| + t})^\lambda \right|.$$

Then we have

Theorem 9.4. *([39]) Let $S^A_\nabla(f)$ and $P^*_\nabla(f)$ be as above. Then there exists $A_p > 0$, such that*

1) $\| S^A_\nabla(f) \|_p \leq A_p \| P^*_\nabla(f) \|_p$,

2) $P^*_\nabla(f) \|_p \leq A_p \| S^A_\nabla(f) \|_p$.

Theorem 9.5. *([39]) If $\lambda > l/p$, then*

$$\| \widetilde{\varphi}^\lambda_T(f) \|_p \leq C_p \| \widetilde{\varphi}^*_\nabla(f) \|_p .$$

Theorem 9.6. *([39]) Let $r \geq \lambda + 1, \lambda > l/p$. Then, for $f \in H^p(M)$, the following inequality is valid*

$$f^*_r(m) \leq c_p \widetilde{\varphi}^\lambda_T(f)(m).$$

Theorem 9.7. *([30]) Let $p > 0, r > (l/p) + 1$. Then $f^*_r \in L^p(M)$ implies $S^A_\nabla(f) \in L^p(M)$. Furthermore,*

$$S^A_\nabla(f)(m) \leq c f^*_r(m).$$

Theorem 9.8. *([39]) Let*

$$\widetilde{\varphi^*_+}(f)(m) = \sup_{t>0} \left| \widetilde{\phi}_t(f)(m) \right| .$$

Then

$$\| \widetilde{\varphi^*_\nabla}(f)(m) \|_p \leq C_p \| \widetilde{\phi^*_+}(f) \|_p .$$

Since (9.4) implies

$$\| \widetilde{\varphi^*_+}(f) \|_p \leq \| \psi \|_1 \| S^A_+(f) \|_p \leq \| \psi \|_1 \| S^A_\nabla(f) \|_p, \tag{9.5}$$

and the following inequalities are valid

$$\begin{aligned} \| P^*_+(f) \|_p &\leq A_p \| S^A_+(f) \|_p, \\ \| S^A_+(f) \|_p &\leq A_p \| P^*_+(f) \|_p . \end{aligned} \tag{9.6}$$

Furthermore, Theorem 9.4 implies

$$P^*_+(f)(m) = \sup_{0<t<\epsilon} |P_t(f)(m)| . \tag{9.7}$$

We have

Theorem 9.9. *([39]) The following conclusions are equivalent to $f \in S'(M)$:*

1) $\sup_{0<t<\epsilon} |P_t(f)(m)| \in L^p(M)$,

2) $\sup_{0<t<\epsilon} |S_t(f)(m)| \in L^p(M)$,

3) $S^A_+(f) \in L^p(M)$,

4) $P^*_+(f) \in L^p(M)$,

5) $S^A_\nabla(f) \in L^p(M)$,

6) $P^*_\nabla(f) \in L^p(M)$,

7) $\widetilde{\varphi}^*_\nabla(f) \in L^p(M)$,

8) $\widetilde{\varphi}^*_\nabla(f) \in L^p(M)$,

9) $\widetilde{\varphi}^*_T(f) \in L^p(M), \lambda \in l/p$,

10) $f^*_r \in L^p(M), r > (l/p) + 1$.

Theorem 9.10. *([39]) Let $H^p(M)$ and $H_a^{p,q,s}(M)$ be as above. Then*

1) $H^p(M) = H_a^{p,q,s}(M)$;

2) There exist positive constants A_p and B_p, such that

$$B_p \parallel P_{\bigtriangledown}^*(f) \parallel_p \leq \parallel f \parallel_{H_a^{p,q,s}} \leq A_p \parallel P_{\bigtriangledown}^*(f) \parallel_p .$$

§10 Some Important operators.

In order to completes the proof of Theorem 9.4, we need the following important operators.

1) Riesz potential I_α for $Re\alpha > 0, I_\alpha = (-\Delta)^{\frac{1}{2}\alpha}$.

$$I_\alpha f \overset{S'}{=} \sum_{\lambda \in \widehat{M}, \lambda \neq 0} \mu_\lambda^{\frac{1}{2}\alpha} d_\lambda \chi_\lambda * f, \ f \in S'(M);$$

2) Bessel potential $\widetilde{I_\alpha}$

$$\widetilde{I_\alpha} f \overset{S'}{=} \sum_{\lambda \in \widehat{M}, \lambda \neq 0} |\lambda + \delta|^{-\alpha} d_\lambda \chi_\lambda * f;$$

3) Riesz and Bessel transformations:

$$R_j = \widetilde{X}_j(-\Delta)^{\frac{1}{2}}, \ \widetilde{R}_j = \widetilde{X}_j(|\delta|^2 I - \Delta)^{-\frac{1}{2}}, j = 1, 2, \cdots, l;$$

4) Fractional dervatives $I^\alpha = (-\Delta)^{\frac{1}{2}\alpha}$ and $\widetilde{I}^\alpha = (|\delta|^2 I - \Delta)^{\frac{1}{2}\alpha}$ for $Re\alpha > 0$.

Theorem 10.1. *([40]) Let the kernel functions of I_α and $\widetilde{I}_\alpha$ be $I_\alpha(m)$ and $\widetilde{I}_\alpha(m)$, respectively. Then*

1) $\widetilde{I}_\alpha(m)$ is a C^∞ function on $M \backslash \{o\}$, and the main part of $\widetilde{I}_\alpha(m)$ at o is $A_M^\alpha |m|^{\alpha - l}$ for $\alpha - l \neq 0, 2, 4, \cdots$, where $|m|$ is the Riemannian distance between m and o, and is $A_M^\alpha |m|^{\alpha - l} \log |m|$ for $\alpha - l = 0, 2, 4, \cdots$, $l = dimM$.

2) $I_\alpha = \widetilde{I}_\alpha + \Sigma_{k=1}^N a_k \widetilde{I_{\alpha_2 k}} + r_{N_\alpha}$, where $r_{N_\alpha}(m)$ is at least $N - l$ times differentiable.

3) Let $\widetilde{I}_\alpha(m) = A_M^\alpha |m|^{\alpha - l} + b_\alpha(m)$ or $\widetilde{I}_\alpha(m) = A_M^\alpha |m|^{\alpha - l} \log |m| + b_\alpha(m)$. Then, $b_\alpha(m)$ is at least $N - l$ times differentiable for $|m| < \epsilon$.

Theorem 10.2. *([40]) I_α and $\widetilde{I}_\alpha$ are H^p bounded for $0 < p \leq 1$ and L^p bounded for $p \geq 1$.*

Theorem 10.3. *([40]) The kernel functions for R_j and $\widetilde{R}_j$ are $K_j(m)$ and $\widetilde{K}_j(m)$ such that*

1) $K_j(m) = \widetilde{X}_j I_1(m)$, $\widetilde{K}_j(m) = \widetilde{X}_j \widetilde{I}_1(m)$;

2) the main part of $\widetilde{K}_j$ and K_j at o is $A_M x_j |x|^{-n-1}$ where $m = \exp X$, $X = \Sigma x_j X_j$, $x = (x_1, \cdots, x_l)$.

3) R_j and $\widetilde{R}_j$ are $L^p(M)$ bounded for $p > 1$ and $H^p(M)$ bounded for $0 < p \leq 1$ and also $(H^p(M), L^p(M))$ bounded for $0 < p \leq 1$.

Theorem 10.4. *([40]) Let $I^\alpha(m)$ and $\widetilde{I}^\alpha(m)$ be the kernel functions of I^α and $\widetilde{I}^\alpha$ for $0 < Re\alpha < 1$. Then*

1) $\widetilde{I}^\alpha f(m) = \int_M (f(mn) - f(m))\widetilde{I}^\alpha(n)dn + B_\alpha f$, for f in $C^1(M)$;

2) $I^\alpha(m)$ is a C^∞ function on $M\setminus\{o\}$, and the main part of $I^\alpha(m)$ at 0 is $|m|^{-l-\alpha}$;

3) $I^\alpha = \widetilde{I}^\alpha + \Sigma_{k=1}^N a_k \widetilde{I}_{2k-\alpha} + r_N^\alpha$, where $r_N^\alpha(m)$ is at least $N-l-1$ times differentiable on M.

§11 The dual of $H^p(M)$, Molecular Theory.

For the dual space of $H^p(M)$, we have

Theorem 11.1. *The dual space of $H^1(M)$ is $BMO(M)$ and the dual space of $H_a^{p,q,s}(M)$ for $0 < p < 1$ is the space $L_{\frac{1}{p}-1,q,s}(M)$.*

The (p, q, s, ϵ) molecular centred at m_0 is defined as follows

Definition 11.2. *([39],[41]) Let $0 < p \le 1 < q \le +\infty$, $p < q, l = dimM$, $s \ge [l(\frac{1}{p}-1)]$ be a nonnegative integer, $\epsilon > \max\{l/n, \frac{1}{p}-1\}$, $a = 1-\frac{1}{p}+\epsilon$, $b = 1-\frac{1}{q}+\epsilon$, $M \in L^q(M)$ is called a (p, q, s, ϵ) molecular centered at m_0 if*

1) $\mathcal{N}_q(M) = \| M \|_q^{a/b} \| M \cdot |m_0^{-1}m|^{nb} \|_q^{1-\frac{a}{b}} < +\infty$,

2) $\int_M M(m)P_k(\pi(m)I_\lambda)dm = 0$ for $0 \le k \le s$,

3) A C^∞ function on M is called an exceptional molecular and in this case $\mathcal{N}_q(M) = \| M \|_\infty$.

Theorem 11.3. *If M is a (p, q, s, ϵ) molecular then $M \in H_a^{p,q,s}(M)$ and*

$$\| M \|_{H_a^{p,q,s}} \le c\mathcal{N}_q(M).$$

REFERENCES

1. Gong Sheng, *Harmonic Analysis on Classical Groups*, Springer Verlag. 1991.
2. ———, Acta Math Sinica **10** (1960), 239–260.
3. ———, ibid **12** (1962), 17–31.
4. ———, ibid **13** (1963), 152–161.
5. ———, ibid **13** (1963), 323–331.
6. ———, ibid **13** (1965), 305–325.
7. ———, J.of University of Science and Technology of China **9** (1979), 25–30.
8. Gong Sheng, Li Shixiong and Zheng Xueau, *Procedings of Analysis Conference*, Singapore 1986, (1988), pp. 69–113, North-Holland.
9. Zhong Jiaging, J. of China Univ. of Sci. & Tech. **9** (1979), 31–43.
10. Wong Shikun and Dong Daozheng, Chinese Ann. of Math **4A** **(2)** (1983), 195–206.
11. ———, ibid **4A** **(3)** (1983), 369–378.
12. ———, ibid **4A** **(5)** (1983), 547–556.
13. He Zuqi and Chen Guangxiao, J. of Math. Res. & Expo. **1** **(1)** (1981), 29–41.
14. ———, ibid **3** **(1)** (1983), 97–100.
15. ———, ibid **3** **(2)** (1983), 23–26.
16. ———, ibid **3** **(3)** (1983), 51–54.
17. Zheng Xuean, Acta Math, Sinica **35** **(1)** (1992), 20–32.
18. ———, J. of Anhui Univ. **3** (1992), 1–7.
19. ———, J. of Univ. of Sci. & Tech. of China **22** **(3)** (1992), 303–307.
20. ———, J. of Math. Res. & Expo. **12** **(2)** (1992), 235–252.

21. ———, Northeastern Math. J. **7** **(3)** (1991), 295–301.
22. ———, J. of Anhui Univ. **2** (1990), 199–203.
23. ———, Chin. Adv. in Math. **19** **(2)** (1990 199–203).
24. ———, Chin. Ann. of Math. **10A** **(4)** (1989), 407–451.
25. ———, Acta Math, Sinica **31** **(4)** (1988), 443–447.
26. ———, Chin. Adv. in Math. **16** **(1)** (1987), 61–66.
27. ———, J. of Anhui Univ. (Mathematics) (1987), 96–99.
28. ———, Chin. Sci. Bull. **32** (1987), 1657–1663.
29. ———, J. of Anhui Univ. (Mathematics) (1985), 8–11.
30. ———, Chin. Ann. of Math. **6A** **(2)** (1985), 237–241.
31. ———, Chin. Sci. Bull. **30** (1985), 1758.
32. ———, ibid **29** (1984), 1342.
33. ———, ibid **29** (1984), 767.
34. ———, Chin. Adv. in Math. **13** **(2)** (1984), 103– 118.
35. ———, Northeastern Math. J. **5** **(3)** (1989), 301–308.
36. ———, *Hormander's Maltipler Theorem on compact Lie Groups (preprint).*
37. Zheng Xuean and Lu Shanzhen, *Harmonic analysis on Compact Homogeneous Spaces (I)*, Chin. Adv. in Math..
38. Lu Shanzhen and Zheng Xuean, *Harmonic analysis on Compact Homogeneous Spaces (II)*, J. of Beijing Normal Univ. **28** **(3)** (1992), 265–275.
39. ———, *Harmonic analysis on Compact Homogeneous Spaces (III)*, ibid **28** **(3)** (1992), 276–286.
40. Zheng Xuean, *Harmonic analysis on Compact Homogeneous Spaces (IV)*, Acta Math. Sinica (to appear).
41. Dong Daozheng and Zheng Xuean, *Harmonic Analysis on Compact Homogeneous Spaces–Mulecular Theory of H^p Spaces*, J. of Henan Univ. **22** **(2)** (1992), 1–7.
42. ———, *Harmonic analysis on Compact Homogeneous Spaces – The dual Spaces of H^p Spaces*, Chin. Ann. of Math. (to appear).
43. Fan Dashan, Chin. Ann. of Math. **6A** **(4)** (1985).
44. ———, Chin. Sci. Bull. **24** **(17)** (1984).
45. ———, Acta Math. Sinica **29** **(5)** (1986).
46. ———, J. of Math. Res. & Expo. **8** **(1)** (1988).
47. ———, Chin. Ann. of Math. **13A** **(3)** (1992), 281–288.
48. Xu Zengfu, J. of Math. Res. & Expo. **7** **(2)** (1987).
49. Zhao Xeshen, Chin. Ann. of Math. **8A** **(6)**.

Harmonic Analysis on Bounded Symmetric Domains

KEHE ZHU

Department of Mathematics and Statics
State University of New York

1. Introduction

It is perhaps fair to say that the most influential work of modern Chinese harmonic analysts is the book "*Harmonic Analysis of Functions of Several Complex Variables in the Classical Domains*" by the late Professor L. K. Hua [21]. Although the book was written over 30 years ago, it is still a very useful reference for information about Cartan domains. In this article we survey a few topics of analysis on bounded symmetric domains which were studied after the appearance of Hua's classic.

The topics chosen in this article are unavoidably incomplete and biased; I have only chosen topics which I have had some working knowledge of.

2. Symmetries and Analytic Invariants

Let Ω be a bounded symmetric domain in $\mathbf{C}^n$. Throughout the paper we assume Ω is in its standard Harish-Chandra realization, so that Ω is circular, convex, and contains the origin of $\mathbf{C}^n$. We also assume that Ω is irreducible, although most results we present here are valid in the reducible case as well.

2.1. Cartan's classification. The irreducible bounded symmetric domains were completely classified (up to a biholomorphic map) by E. Cartan. There are only six types of bounded symmetric domains, also called Cartan domains. The following is a complete list of them.

Type I: The space of all $n \times m$ complex matrices z with $z^*z < I_m$, where n and m are positive integers with $n \leq m$.

Type II: The space of all $n \times n$ anti-symmetric complex matrices z with $z^*z < I_n$, where n is any integer greater than 4.

Type III: The space of all $n \times n$ symmetric complex matrices z with $z^*z < I_n$, where n is any integer greater than 1.

Research supported by the US National Science Foundation

1991 Mathematics Subject Classification: 32M15, 43A85, 32A37

M. Cheng et al. (eds.), Harmonic Analysis in China, 287–307.

Type IV: The space of all $z \in \mathbf{C}^n$ with

$$\left[\sum_{k=1}^{n} |z_k|^2\right]^2 - \left|\sum_{k=1}^{n} z_k^2\right|^2 < \left[1 - \sum_{k=1}^{n} |z_k|^2\right]^2,$$

where n is any integer greater than 4. Such a domain is also called a Lie ball.

Type V: This type consists of a single domain in complex dimension 16.

Type VI: This type consists of a single domain in complex dimension 27.

Domains of types I–IV are called classical domains; and domains of types V and VI are called exceptional domains.

2.2. Let $\partial\Omega$ be the full (or topological) boundary of Ω. The Shilov (or distinguished) boundary of Ω will be denoted by $b\Omega$. We use dv to denote the normalized volume measure on Ω, and $d\sigma$ the normalized surface measure on $b\Omega$.

The following function spaces will be frequently used in our later analysis. The space of holomorphic functions on Ω will be denoted by $H(\Omega)$. Let $\mathbf{C}_0(\Omega)$ be the space of complex-valued continuous functions on Ω which can be uniformly approximated by continuous functions with compact support. The space of continuous functions on Ω which can be continuously extended to $\partial\Omega$ is denoted by $\mathbf{C}(\overline{\Omega})$.

2.3. Symmetries. For every $a \in \Omega$ there exists a unique holomorphic mapping $\varphi_a : \Omega \to \Omega$ satisfying the following conditions:

(1) $\varphi_a \circ \varphi_a(z) = z$ for all $z \in \Omega$.

(2) $\varphi_a(0) = a$ and $\varphi_a(a) = 0$.

(3) φ_a has a unique fixed point.

The mappings φ_a are called symmetries of Ω; they are involutive automorphisms of the domain. It is easy to see that the unique fixed point of φ_a is the geodesic mid-point between 0 and a in the Bergman metric. See [20][55] for other properties of symmetries.

When $\Omega = \mathbf{D}$, the open unit disk in the complex plane $\mathbf{C}$, the symmetries φ_a are given by

$$\varphi_a(z) = \frac{a - z}{1 - \bar{a}z}, \qquad a, z \in \mathbf{D}.$$

An explict description of the symmetries for the open unit ball B_n in $\mathbf{C}^n$ can be found in [31]. In the case of matrix domains of types I–III, the symmetries are given by

$$\varphi_a(z) = (I - aa^*)^{-1/2}(a - z)(I - a^*z)^{-1}(I - a^*a)^{1/2};$$

see [22]. The symmetries are also explicitly written down in [47] in the case of the so-called tube type domains.

2.4. The automorphism group. Let $\text{Aut}(\Omega)$ be the set of all biholomorphic maps from Ω onto Ω. It is clear that $\text{Aut}(\Omega)$ is a group with the group operation being composition; it is called the automorphism group of Ω. This is a real analytic Lie group.

It is easy to show that for every φ in $\text{Aut}(\Omega)$ there exists a point a in Ω and a unitary transfomation on $\mathbf{C}^n$ such that $\varphi(z) = U\varphi_a(z)$ for all $z \in \Omega$. If we write points in $\mathbf{C}^n$ as column vectors and identify U with the unitary matrix corresponding to the natural basis of $\mathbf{C}^n$, then Uw, $w \in \mathbf{C}^n$, becomes matrix multiplication. In the case of the open unit ball, every unitary transformation of $\mathbf{C}^n$ is in Aut (Ω). However, if the rank of Ω is greater than 1, there are unitaries of $\mathbf{C}^n$ which do not belong to Aut (Ω). The set of all unitaries in Aut (Ω) will be denoted by K. It is clear that K is a subgroup of Aut (Ω).

2.5. Analytic invariants. Every irreducible bounded symmetric domain is uniquely determined (up to a biholomorphic map) by three analytic invariants: $r = r(\Omega)$, $a = a(\Omega)$, and $b = b(\Omega)$; see [24]. These invariants are all nonnegative integers. The number r is called the rank of Ω, which is of course positive. For most of our applications it will not be necessary to know the precise definition of these invariants. Another invariant of Ω, called the genus of Ω, is defined by

$$N = N(\Omega) = a(r-1) + b + 2.$$

It is clear that N is an integer greater than or equal to 2.

The ranks of the six types of Cartan domains listed earlier are (in the corresponding order): n, $[n/2]$, n, 2, 2, and 3. The genera of those domains are $m+n$, $2(n-1)$, $n+1$, n, 12, and 18, respectively.

When $b = b(\Omega) = 0$, the domain Ω will be called a tube type domain. See [4] for other background material on bounded symmetric domains.

2.6. Polar coordinates. Recall that r is the rank of Ω. There exist r points $e_1, \cdots, e_r$ in $\partial\Omega$ such that every point z in Ω admits a representation

$$z = U(t_1 e_1 + \cdots + t_n e_n), \qquad U \in K, 0 \le t_k < 1, k = 1, \cdots, r.$$

This is called the polar representation of z. If f is a nonnegative or Lebesgue integrable function on Ω, then f can be integrated in polar coordinates as follows:

$$\int_\Omega f(z)\, dv(z) = c \int_0^1 \cdots \int_0^1 \int_K f(Ut)\, dU \prod_{k=1}^r t_k^{2b+1} \prod_{j<k} |t_j^2 - t_k^2|^a \, dt_1 \cdots dt_r,$$

where $t = t_1 e_1 + \cdots + t_r e_r$, dU is the normalized Haar measure on K, and c is an appropriate constant. See [16] for more details on polar coordinates.

3. Reproducing Kernels

The bulk of Hua's book [21] is devoted to the explict computation of the Bergman, Szegö, and Poisson kernels on the classical domains. These computations are important in that they enable us to extract very precise information about kernel functions and other properties of the classical domains. However, when the results are too specific, they sometimes prevent people from getting good insight into certain general problems. In this section we present a theory of reproducing kernels on general bounded symmetric domains (not just the classical ones). The results here are specific enough for applications and yet general enough for one to work in a more uniform way.

3.1. Homogeneous expansions. For every holomorphic function $f(z)$ on Ω we can write

$$f(z) = \sum_{n=0}^{\infty} f_n(z), \qquad z \in \Omega,$$

where $f_n(z)$ is a homogeneous polynomial of degree n. This decomposition is unique and is called the homogeneous expansion of f. For a more refined and more useful homogeneous type expansion see [16].

It follows from the symmetry of Ω that if $n \neq m$, and if f_n and f_m are homogeneous polynomials of degrees n and m, respectively, then

$$\int_\Omega f_n(z)\overline{f_m(z)}\,dv(z) = 0,$$

and

$$\int_{b\Omega} f_n(\zeta)\overline{f_m(\zeta)}\,d\sigma(\zeta) = 0.$$

In particular, if f is holomorphic and integrable with respect to volume measure, then

$$f(0) = \int_\Omega f(z)\,dv(z).$$

This simple formula is usually what one needs to compute the Bergman kernel of Ω.

3.2. The Bergman kernel. Let $L^2_a(\Omega)$ be the closed subspace of $L^2(\Omega, dv)$ consisting of holomorphic functions. It is called the Bergman space of Ω. Being a closed subspace of a Hilbert space, the Bergman space is a Hilbert space by itself with the induced inner product:

$$\langle f, g\rangle = \int_\Omega f(z)\overline{g(z)}\,dv(z).$$

Using the mean-value property of holomorphic functions, we can easily prove that each point evaluation is a bounded linear functional on $L^2_a(\Omega)$. By Riesz representation, for every $z \in \Omega$ there exists a unique function K_z in $L^2_a(\Omega)$ such that

$$f(z) = \langle f, K_z\rangle, \qquad f \in L^2_a(\Omega).$$

The function

$$K(z, w) = \overline{K_z(w)}, \qquad z, w \in \Omega,$$

is called the Bergman kernel of Ω. Elementary functional and complex analysis show that the Bergman kernel has the following additional properties:

(1) $K(z, w)$ is holomorphic in z and conjugate holomorphic in w.
(2) $K(z, w) = \overline{K(w, z)}$ for all z and w in Ω.
(3) $K(z, z) > 0$ for every $z \in \Omega$.

(4) $K(z,w)$ is uniquely determined by its diagonal values $K(z,z)$.
(5) If $\{e_n\}$ is any orthonormal basis for $L^2_a(\Omega)$, then

$$K(z,w)=\sum_{n=1}^{\infty}e_n(z)\overline{e_n(w)},\qquad z,w\in\Omega,$$

with the series convergent uniformly on compact subsets of $\Omega\times\Omega$.
(6) If $\varphi\in\mathrm{Aut}\,(\Omega)$, then

$$J_\varphi(z)K(\varphi(z),\varphi(w))\overline{J_\varphi(w)}=K(z,w)$$

for all z and w in Ω, where $J_\varphi(z)$ is the complex Jacobian determinant of φ at z.

3.3. The Bergman kernel via symmetries. Recall that for every volume integrable holomorphic function f in Ω we have

$$f(0)=\int_\Omega f(w)\,dv(w).$$

This along with the uniqueness in the Riesz representation theorem shows that $K(z,0)=1$ for all $z\in\Omega$. Replacing z by $\varphi_w(z)$ and applying the invariance of the Bergman kernel (property (6) of the last subsection), we obtain

$$K(z,w)=J_w(z)\overline{J_w(w)},\qquad z,w\in\Omega,$$

where $J_w(z)$ is the complex Jacobian determinant of the symmetry φ_w at z. In the case where the symmetries can be explicitly written down, such as the open unit ball in $\mathbf{C}^n$, the formula above enables us to compute the Bergman kernel.

For every $a\in\Omega$ let

$$k_a(z)=\frac{K(z,a)}{\sqrt{K(a,a)}},\qquad z\in\Omega.$$

It is clear that each k_a is a unit vector in $L^2_a(\Omega)$. The vectors k_a are called the normalized reproducing kernels of Ω. By the formula of $K(z,w)$ in the proceding paragraph we have

$$k_a(z)=J_a(z)\frac{\overline{J_a(a)}}{|J_a(a)|},\qquad a,z\in\Omega.$$

The quotien above is clearly unimodulus. Actually, more is true.

Theorem. [55] *For all a and z in Ω we have $k_a(z)=(-1)^nJ_a(z)$, where n is the complex dimension of Ω.*

Since the real Jacobian determinant of an automorphism φ at the point z is equal to $|J_\varphi(z)|^2$, we arrive at the following change of variables formula.

Corollary. *If f is positive or volume integrable on Ω, then for every $a\in\Omega$ we have*

$$\int_\Omega f(z)\,dv(z)=\int_\Omega f\circ\varphi_a(z)|k_a(z)|^2\,dv(z)$$

and (after replacing f by $f \circ \varphi_a$)

$$\int_\Omega f \circ \varphi_a(z)\, dv(z) = \int_\Omega f(z) |k_a(z)|^2\, dv(z).$$

Another consequence of the relationship between the Bergman kernel and the set of symmetries of Ω is that the kernel function $K(z, w)$ never vanishes on $\Omega \times \Omega$. This is because the complex Jacobian determinant of an automorphism never vanishes (by the chain rule).

Since each φ_a is involutive, we have $J_a(\varphi_a(z))J_a(z) = 1$ for all a and z in Ω. It follows that $k_a(\varphi_a(z))k_a(z) = 1$ for all a and z in Ω.

3.4. A Formula of Faraut and Koranyi. The genus of Ω plays an essential role in analysis on bounded symmetric domains. The following result of Faraut and Koranyi describes the structure of the Bergman kernel $K(z, w)$ of Ω in terms of its genus. This result is not as precise as the formulas found in Hua's book for the classical domains. But it holds for a general bounded symmetric domain and is specific enough for most applications.

Theorem. [16] *The Bergman kernel of Ω is given by*

$$K(z, w) = \frac{1}{h(z, w)^N}, \qquad z, w \in \Omega,$$

where N is the genus of Ω, and $h(z, w)$ is the unique polynomial in z and $\overline{w}$ satisfying

$$h(t, t) = \prod_{k=1}^{r} (1 - t_k^2), \qquad t = \sum_{k=1}^{r} t_k e_k,$$

in polar coordinates.

Note that in the case $\Omega = B_n$, the open unit ball in $\mathbf{C}^n$, we have $N = n + 1$ and $h(z, w) = 1 - \langle z, w\rangle$. Here $\langle z, w\rangle$ is the natural inner product in $\mathbf{C}^n$.

3.5. Some weighted reproducing kernels. For any complex number α consider the measure

$$dv_\alpha(z) = h(z, z)^\alpha\, dv(z).$$

It is known that the measure dv_α is finite if and only if $\operatorname{Re} \alpha > -1$; see [25] for example. When this happens, we also normalize dv_α so that it has total mass 1.

The Möbius invariance of the Bergman kernel implies that the measure dv_α is inviariant under the action of unitaries in Aut (Ω). It follows that for every $\operatorname{Re} \alpha > -1$ and every bounded holomorphic function f on Ω we have

$$f(0) = \int_\Omega f(w)\, dv_\alpha(w) = c_\alpha \int_\Omega f(w)\, h(w, w)^\alpha\, dv(w).$$

Replacing f by $f \circ \varphi_z$, making a change of variables, and using the transformation law for the Bergman kernel, we obtain

$$f(z) = c_\alpha h(z,z)^{N+\alpha} \int_\Omega f(w) \frac{h(w,w)^\alpha \, dv(w)}{|h(z,w)|^{2(N+\alpha)}}.$$

Replacing f by $fh(\,\cdot\, z)^{N+\alpha}$ we deduce

$$f(z) = c_\alpha \int_\Omega f(w) \frac{h(w,w)^\alpha \, dv(w)}{h(z,w)^{N+\alpha}} = \int_\Omega \frac{f(w)\, dv_\alpha(w)}{h(z,w)^{N+\alpha}}, \qquad z \in \Omega.$$

Since this holds for all bounded holomorphic functions f, an easy limit argument shows that it also holds for holomorphic functions in $L^1(\Omega, dv_\alpha)$. This proves the following.

Theorem. [34] *Suppose* $\operatorname{Re}\alpha > -1$. *Then the reproducing kernel of the weighted Bergman space*

$$L^2_a(dv_\alpha) = L^2(\Omega, dv_\alpha) \cap H(\Omega)$$

is given by

$$K_\alpha(z,w) = \frac{1}{h(z,w)^{N+\alpha}}, \qquad z, w \in \Omega.$$

Equivalently, the orthogonal projection P_α *from* $L^2(\Omega, dv_\alpha)$ *onto* $L^2_a(dv_\alpha)$ *is given by*

$$P_\alpha f(z) = \int_\Omega \frac{f(w)\, dv_\alpha(w)}{h(z,w)^{N+\alpha}}, \qquad f \in L^2(\Omega, dv_\alpha).$$

For $p > 0$ and $\operatorname{Re}\alpha > -1$ we shall also consider the weighted Bergman spaces

$$L^p_a(dv_\alpha) = L^p(\Omega, dv_\alpha) \cap H(\Omega).$$

When $\alpha = 0$ we write $L^p_a(\Omega)$ instead of $L^p_a(dv_0)$. Also in this case we use P instead of P_0.

3.6. Forelli-Rudin type theorems. For many problems on bounded symmetric domains it is necessary to estimate the kernel functions. We find the following result of Faraut and Koranyi useful.

Theorem. [16] *For* $t > -1$ *and* c *real let*

$$I_{t,c}(z) = \int_\Omega \frac{h(w,w)^t \, dv(w)}{|h(z,w)|^{N+t+c}}, \qquad z \in \Omega.$$

Then $I_{t,c}(z)$ *is bounded in* z *if* $c < -a(r-1)/2$; *and* $I_{t,c}(z) \sim h(z,z)^{-c}$ *if* $c > a(r-1)/2$.

Note that when $r = 1$, so that Ω is biholomorphic to the open unit ball in $\mathbf{C}^n$, then $I_{t,0}(z) \sim -\log(1-|z|^2)$; see [18]. In the higher rank case, the behavior of $I_{t,c}(z)$ is related to certain hypergeometric functions in the case $|c| \le a(r-1)/2$. See [48] for details.

Using the theorem above and the classical Schur's test for boundedness of integral operators (see [54] for example) we can prove the following.

Theorem. [60] *For A and B real let $T_{A,B}$ be the integral operator defined by*

$$T_{A,B}f(z) = h(z,z)^A \int_\Omega \frac{h(w,w)^B}{|h(z,w)|^{N+A+B}} f(w)\,dv(w), \qquad z \in \Omega.$$

If α is real, $1 \le p \le +\infty$, and

$$-A + \frac{a(r-1)}{2} < \frac{\alpha + 1 + \frac{a(r-1)}{2}}{p} < B + 1,$$

then the operator $T_{A,B}$ is bounded on $L^p(\Omega, dv_\alpha)$.

In the rank 1 case, the converse of the above theorem also holds. Recall that the Bergman projection is the operator P defined by

$$Pf(z) = \int_\Omega K(z,w) f(w)\,dv(w), \qquad z \in \Omega.$$

As a consequence of the theorem above we conclude that the Bergman projection is bounded on $L^p(\Omega, dv)$ if $r = 1$ and $1 < p < +\infty$. If $r = 2$ and $a = 1$ (such a domain is necessarily of type II), then the Bergman projection is bounded on $L^p(\Omega, dv)$ when $3/2 < p < 3$. The (un)boundedness of the Bergman projection on $L^p(\Omega, dv)$ is a tricky issue in general; see [11].

4. Fractional Radial Differential Operators

4.1. Motivation. Recall that $\mathbf{D}$ is the open unit disk in $\mathbf{C}$. If $\operatorname{Re}\beta > -1$ and f is a bounded analytic function in $\mathbf{D}$, then by **3.5**

$$f(z) = \int_{\mathbf{D}} \frac{f(w)\,dv_\beta(w)}{(1 - z\overline{w})^{2+\beta}}, \qquad z \in \mathbf{D}.$$

Differentiating under the integral sign, we obtain

$$f^{(k)}(z) = c_{\beta,k} \int_{\mathbf{D}} \frac{\overline{w}^k f(w)\,dv_\beta(w)}{(1 - z\overline{w})^{2+\beta+k}}, \qquad z \in \mathbf{D}.$$

This suggests that we consider operators of the following form on the space $H(\mathbf{D})$:

$$f \mapsto g(z) = \int_{\mathbf{D}} \frac{f(w)\,dv_\beta(w)}{(1 - z\overline{w})^{2+\beta+\alpha}}.$$

Such operators constitute a class of fractional differential operators on $H(\mathbf{D})$. It is easy to check that if α is a positive integer, then the operator above is indeed a differential operator with polynomial coefficients. This idea can be generalized to bounded symmetric domains.

4.2. Existence and uniqueness. Recall that $H(\Omega)$ is the linear space of all holomorphic functions in Ω. We equip $H(\Omega)$ with the topology of uniform convergence on compact sets.

Theorem. [61] *Suppose α and β are complex numbers with real parts greater than -1. Then there exists a unique continuous linear operator $D^{\alpha,\beta} : H(\Omega) \to H(\Omega)$ such that*

$$D_z^{\alpha,\beta}\left[h(z,w)^{-(N+\beta)}\right] = h(z,w)^{-(N+\alpha+\beta)}$$

for all z and w in Ω.

We call the operators $D^{\alpha,\beta}$ fractional radial differential operators. They have the following additional properties:
(1) $D^{\alpha,\beta} f_s(z) = D^{\alpha,\beta} f(sz)$ for all $z \in \Omega$, $s \in (0,1)$, and $f \in H(\Omega)$. Here f_s is the function on Ω defined by $f_s(w) = f(sw)$, $w \in \Omega$.
(2) If α is a positive integer, then $D^{\alpha,\beta}$ is a linear partial differential operator with polynomial coefficients of order αr, where r is the rank of Ω.
(3) For every $f \in H(\Omega)$ and $z \in \Omega$ we have

$$D^{\alpha,\beta} f(z) = c_\beta \lim_{s\to 1^-} \int_\Omega \frac{h(w,w)^\beta}{h(z,w)^{N+\beta+\alpha}} f_s(w)\, dv(w),$$

where c_β is the normalizing constant in dv_β. The existence of the limit is always guaranteed.

4.3. Invertibility. On a bounded symmetric domain there are many ways of defining fractional differential operators. One of the advantages for using the type here is the existence of an inverse.

Theorem. [61] *Suppose α and β are complex numbers with real parts greater than -1. Then $D^{\alpha,\beta} : H(\Omega) \to H(\Omega)$ is invertible for* $\operatorname{Re}(\alpha+\beta) > -1$.

We denote the inverse of $D^{\alpha,\beta}$ by $D_{\alpha,\beta}$. The operators $D_{\alpha,\beta}$ can be thought of as fractional (radial) integral operators. They have the following additional properties:
(1) Each $D_{\alpha,\beta}$ is continuous on $H(\Omega)$.
(2) $D_{\alpha,\beta} f_s(z) = D_{\alpha,\beta} f(sz)$ for all $z \in \Omega$, $s \in (0,1)$, and $f \in H(\Omega)$.
(3) For every $f \in H(\Omega)$ and $z \in \Omega$ we have

$$D_{\alpha,\beta} f(z) = c_{\alpha+\beta} \lim_{s\to 1^-} \int_\Omega \frac{h(w,w)^{\alpha+\beta}}{h(z,w)^{N+\beta}} f_s(w)\, dv(w),$$

where $c_{\alpha+\beta}$ is the normalizing constant in $dv_{\alpha+\beta}$. Again the existence of the limit is always guaranteed.

4.4. Invariance. Another advantage of the fractional differential operators introduced above is its invariance under the action of the unitaries in $\operatorname{Aut}(\Omega)$. See [45] for a disscussion on the algebra of invariant differential operators on Ω. Note that the operators disscussed here include the operators defined in [44][45][49] (where α is assumed to be a positive integer) thanks to the uniqueness result. Note also that when α is a positive integer, the operators $D^{\alpha,\beta}$ are explictly written down in [45] as partial differential operators with polynomial coefficients.

5. Some Banach Spaces of Holomorphic Functions

Several types of spaces of holomorphic functions have been under investigation in recent years. These include the Hardy spaces, the Bergman spaces, the Besov spaces, and the Bloch spaces. The following is a brief account of these spaces.

5.1. Hardy spaces. Recall that $b\Omega$ is the Shilov boundary of Ω. For $0 < p < +\infty$ the Hardy space $H^p(\Omega)$ consists of holomorphic functions f in Ω such that

$$\|f\| = \sup_{0<t<1} \left[\int_{b\Omega} |f(t\zeta)|^p \, d\sigma(\zeta) \right]^{\frac{1}{p}} < +\infty.$$

The Hardy space $H^\infty(\Omega)$ consists of all bounded holomorphic functions on Ω.

It is well known that every function in $H^p(\Omega)$ has radial limits almost everywhere on the distinguished boundary $b\Omega$. It follows that each $H^p(\Omega)$ can be thought of as a closed subspace of $L^p(b\Omega, d\sigma)$. The Szegö projection is then the orthogonal projection from $L^2(b\Omega, d\sigma)$ onto $H^2(\Omega)$. In the rank one case, it is well known that the Szegö projection maps $L^p(b\Omega, d\sigma)$ onto $H^p(\Omega)$ if and only if $1 < p < +\infty$. Although many papers have been written about Hardy spaces on bounded symmetric domains, we find the following result most surprising.

Theorem. [11][23] *Let Ω be an irreducible tube-type bounded symmetric domain of rank greater than 1. Then the Szegö projection is not bounded on $L^p(b\Omega, d\sigma)$ unless $p = 2$.*

Because of the unboundedness of the Szegö projection on L^p spaces, the duality problem for Hardy spaces is still quite open, even in the usually easy case $1 < p < +\infty$ (except $p = 2$ of course).

Note that by the usual duality of L^p spaces and the Hahn-Banach extension theorem the dual space of $H^p(\Omega)$, $1 \le p < +\infty$, can be identified with $QL^q(\Omega, d\sigma)$ under the integral pairing on $b\Omega$ with measure $d\sigma$, where Q is the Szegö projection. Thus we just have to identify the spaces $QL^q(\Omega, d\sigma)$ intrinsically.

5.2. Weighted Bergman spaces. Recall that for $\operatorname{Re}\alpha > -1$ and $0 < p < +\infty$ the weighted Bergman space $L^p_a(dv_\alpha)$ is the closed subspace of $L^p(\Omega, dv_\alpha)$ consisting of holomorphic functions. The orthogonal projection from $L^2(\Omega, dv_\alpha)$ onto $L^2_a(dv_\alpha)$ is denoted by P_α.

Again, using the usual duality of L^p spaces and the Hahn-Banach extension theorem, we can show that the dual space of $L^p_a(dv_\alpha)$, $1 \le p < +\infty$, can be identified with the space $P_\alpha L^q(\Omega, dv_\alpha)$ under the integral pairing on Ω with measure dv_α. In the rank 1 case, we always have $P_\alpha L^q(\Omega, dv_\alpha) = L^q_a(dv_\alpha)$ for $1 < q < +\infty$; and $P_\alpha L^\infty(\Omega)$ is equal to the Bloch space. Thus the duality problem in this case is well understood. However, when the rank of Ω is greater than 1, the situation becomes much more complicated. We do not have an easy way of identifying the spaces $P_\alpha L^q(\Omega, dv_\alpha)$ intrinsically. Examples are known in which $P_\alpha L^q(\Omega, dv_\alpha) = L^q_a(dv_\alpha)$; and examples are known when these two spaces are not the same. See [11].

When q is close to 1, namely, $1 < q < 1+2/[a(r-1)]$, we can show that $PL^q(\Omega, dv)$ is equal to the space of holomorphic functions f with $h(z,z)^N D^N f(z)$ belonging to $L^q(\Omega, dv)$, where $D^N = D^{N,0}$ is the differential operator discussed in the last section. We do not know whether or not this is true for other $p \geq 1$. However, we believe the situation in the weighted case should be similar to the unweighted case.

Although the duality problem for (weighted or unweighted) Bergman spaces is still pretty much open in the range $p > 1$, recently we have been able to settle the case $0 < p \leq 1$ completely.

Let $B_\infty(\Omega)$ be the space of holomorphic functions f in Ω such that

$$\|f\| = \sup\left\{h(z,z)^N |D^N f(z)| : z \in \Omega\right\} < +\infty.$$

Then $B_\infty(\Omega)$ is a Banach space with the norm above. It is a special Besov space; see **5.3**. This space is also discussed and used in [44][51].

Theorem. [61] *Suppose $0 < p \leq 1$, $\alpha > -1$, and $\beta = (N+\alpha)/p - N$. Then the dual space of $L^p_a(dv_\alpha)$ can be identified with $B_\infty(\Omega)$ under the following integral pairing:*

$$\langle f, g\rangle = \lim_{t\to 1^-} \int_\Omega f(tz)\overline{g(z)}\, dv_\beta(z), \quad f \in L^p_a(dv_\alpha),\ g \in B_\infty(\Omega).$$

The case $p = 1$ has been studied by several authors, including [10][44][51]. We expect a similar result to hold for Hardy spaces of Ω with small exponent. But we have been unable to prove such a result, mainly because of the lack of a sharp Hardy-Littlewood type embedding theorem for Hardy spaces. The duality problem for Hardy spaces H^p, $1 < p < 1$, on various special domains has been studied in a number of papers, including see [15][19][27][32][59].

5.3. Besov spaces. Despite the lack of a satisfactory theory for either the Bergman spaces or the Hardy spaces, a relatively complete theory for Besov spaces has been developed recently in [60][61]. Recall that $D^N = D^{N,0}$ is the differential operator defined in **4.2**; it is a partial differential operator of order rN with polynomial coefficients. We define the holomorphic Besov spaces on Ω in terms of this differential operator.

For $1 \leq p \leq +\infty$ let $B_p(\Omega)$ be the space of holomorphic functions f in Ω such that

$$\|f\|_p = \left[\int_\Omega \left|h(z,z)^N D^N f(z)\right|^p d\lambda(z)\right]^{\frac{1}{p}} < +\infty,$$

where

$$d\lambda(z) = K(z,z)\, dv(z)$$

is the Möbius invariant measure on Ω. Note that when $p = +\infty$, the norm above should be understood as the sup-norm.

The Besov spaces $B_p(\Omega)$ obey the usual duality relations. Thus if $1 \le p < +\infty$ then the dual of $B_p(\Omega)$ is $B_q(\Omega)$. Let $B_0(\Omega)$ be the closure of the set of polynomials in $B_\infty(\Omega)$. Then the dual of $B_0(\Omega)$ is $B_1(\Omega)$. Note that a holomorphic function f in Ω belongs to $B_0(\Omega)$ if and only if the function $h(z,z)^N D^N f(z)$ is in $\mathbf{C}_0(\Omega)$.

The Besov spaces also obey the usual interpolation relations. Thus

$$\left[B_{p_0}(\Omega), B_{p_1}(\Omega)\right]_\theta = B_p(\Omega)$$

whenever $1 \le p_0 < p < p_1 \le +\infty$ with

$$\frac{1}{p} = \frac{1-\theta}{p_0} + \frac{\theta}{p_1}, \qquad \theta \in (0,1).$$

We used the operator D^N to define the holomorphic Besov spaces $B_p(\Omega)$. This is only for the sake of simplicity; we can characterize the spaces $B_p(\Omega)$ in terms of other fractional differential operators. For example, if $1 \le p < +\infty$, $\beta > -1$, and

$$\alpha > \frac{N-1}{p} + \frac{a(r-1)}{2}\left(1 - \frac{1}{p}\right),$$

then a function f in $H(\Omega)$ belongs to $B_p(\Omega)$ if and only if the function $h(z,z)^\alpha D^{\alpha,\beta} f(z)$ is in $L^p(\Omega, d\lambda)$.

We can also characterize the Besov spaces $B_p(\Omega)$ in terms of weighted Bergman projections. Specifically, if $\alpha > -1$ and $1 \le p \le +\infty$ then $B_p(\Omega) = P_\alpha L^p(\Omega, d\lambda)$. Similarly, we have $B_0(\Omega) = P_\alpha \mathbf{C}_0(\Omega)$.

We have already seen the application of the Besov space $B_\infty(\Omega)$ to the duality problem for weighted Bergman spaces with exponent less than or equal to 1. The other main application of this theory of Besov spaces is in the study of little Hankel operators on weighted Bergman spaces of Ω. See [60][61].

5.4. Bloch spaces. The theory of Bloch spaces on bounded symmetric domains is developed in [35][36]. Recall that $K(z,z)$ is the Bergman kernel of Ω. For every $z \in \Omega$ we call the matrix

$$B(z) = \frac{1}{2}\left(\frac{\partial^2}{\partial z_i \partial \bar{z}_j} \log K(z,z)\right)$$

the Bergman matrix at the point z. It is well known that each $B(z)$ is positive definite. Given a function $f \in H(\Omega)$ define

$$Q_f(z) = \sup\left\{\frac{\left|\sum_{k=1}^n w_k \frac{\partial f}{\partial z_k}(z)\right|}{\sqrt{\langle B(z)w, w\rangle}} : w \in \mathbf{C}^n, |w| = 1\right\}, \qquad z \in \Omega.$$

The quantity $Q_f(z)$ is the Bergman length of the complex gradient of f at z. The Bloch space of Ω, denoted $\mathcal{B}(\Omega)$, is then the space of holomorphic functions f in Ω such that

$$\|f\| = \sup_{z\in\Omega} Q_f(z) < +\infty.$$

It is easy to see that the above is a complete semi-norm on $\mathcal{B}(\Omega)$. Furthermore, the space $\mathcal{B}(\Omega)$ is invariant under the action of Aut (Ω) with $\|f \circ \varphi\| = \|f\|$ for every $f \in \mathcal{B}(\Omega)$ and $\varphi \in$ Aut (Ω). The following theorem gives several characterizations of the Bloch space.

Theorem. [35] *Suppose f is holomorphic in Ω. Then the following conditions are equivalent.*

(1) $f \in \mathcal{B}(\Omega)$.
(2) There exists a constant $C > 0$ such that $|f(z) - f(w)| \le C\beta(z, w)$ for all $z, w \in \Omega$, where $\beta(z, w)$ is the Bergman distance between z and w.
(3) The function $\widetilde{\Delta}(|f|^2)(z)$ is bounded in Ω, where $\widetilde{\Delta}$ is the invariant Laplacian on Ω.
(4) There exists a constant $C > 0$ such that $\left|\nabla(f \circ \varphi_z)(0)\right| \le C$ for all $z \in \Omega$, where $\nabla f(z)$ is the complex gradient of f at z.

Note that when the rank of Ω is 1, namely, when Ω is the open unit ball in $\mathbf{C}^n$, then f is in the Bloch space if and only if $(1 - |z|^2)|\nabla f(z)|$ is bounded in Ω. Also in this case the Bloch space is equal to the image of $L^\infty(\Omega)$ under any of the weighted Bergman projections P_α, $\alpha > -1$. Thus $\mathcal{B}(\Omega) = B_\infty(\Omega)$ if Ω is the open unit ball. Unfortunately, this is no longer true when the rank of Ω is greater than 1; see [51]. Nevertheless, we always have the following inclusion: $\mathcal{B}(\Omega) \subset B_\infty(\Omega)$, with equality holding if and only if Ω is the open unit ball in $\mathbf{C}^n$.

The little Bloch space of Ω, denoted $\mathcal{B}_0(\Omega)$, is defined to be the closure in $\mathcal{B}(\Omega)$ of the set of polynomials [36]. This definition is a little artificial. However, we have no other choice in defining the little Bloch space. The most natural approach would be to take one of the conditions (except (1)) in the preceding theorem and to replace the boundedness assumption by a little "*oh*" condition. The trouble is we only obtain constant functions in this way when the rank of Ω is greater than 1.

With the little Bloch space defined in the preceding paragraph we always have $\mathcal{B}_0(\Omega) \subset B_0(\Omega)$; equality holds if and only if Ω is the open unit ball; see [51].

The Bloch space characterizes holomorphic functions f such that the so-called big Hankel operator on the Bergman space with symbol $\bar{f}$ is bounded [12], while the space $B_\infty(\Omega)$ characterizes holomorphic functions f such that the so-called little Hankel operator on the Bergman space with symbol $\bar{f}$ is bounded [60].

6. Möbius Invariant Function Spaces

What makes a bounded symmetric domain special is essentially its automorphism group. Thus it is natural to look at the interaction between the automorphism group and the norm when one studies a Banach space of holomorphic functions on Ω. The natural action of Aut (Ω) on a function space is via composition.

6.1. Möbius invariant Banach spaces. Let X be a Banach space of holomorphic functions in Ω. We say that X is Möbius invariant if X is invariant under the natural action of Aut (Ω), $\|f \circ \varphi\| = \|f\|$ for all $f \in X$ and $\varphi \in$ Aut (Ω), and the action of K on X is continuous.

In many applications it is necessary to consider semi-norms instead of norms. If in the definition of a norm $\| \ \|$ we allow $\|x\| = 0$ for some nonzero x, then the resulting function $\| \ \|$ is called a semi-norm. A semi-Banach space is then a complex linear space together with a semi-norm which is complete (modulo the subspace whose elements have zero semi-norm). We say that a semi-Banach space is Möbius invariant if it is invariant under the natural action of Aut(Ω), $\|f \circ \varphi\| = \|f\|$ for all $f \in X$ and $\varphi \in \mathrm{Aut}(\Omega)$, and the action of K on X is continuous.

We already mentioned in **5.4** that the Bloch space is Möbius invariant. It is an open problem to find a Möbius invariant (semi-)norm on each of the Besov spaces $B_p(\Omega)$. In the case of the disk, the semi-norm defined by

$$\|f\| = \left[\int_{\mathbf{D}} (1 - |z|^2)^p |f'(z)|^p \frac{dA(z)}{(1 - |z|^2)^2}\right]^{\frac{1}{p}}$$

is clearly Möbius invariant on $B_p(\mathbf{D})$. Partial results for the open unit ball are obtained in [29].

6.2. Extremal invariant semi-Banach spaces. Among relatively nice semi-Banach spaces of holomorphic functions in Ω there exists a maximal one and a minimal one. The maximal space is the Bloch space; see [37]. In the case of the unit disk the minimal space is the Besov space $B_1(\mathbf{D})$, which can be shown to consist of analytic functions f which admit the following representation:

$$f(z) = \sum_{k=1}^{\infty} a_k \varphi_{w_k}(z), \qquad z \in \Omega,$$

where $\{a_k\} \in l^1$ and $\{w_k\} \subset \mathbf{D}$. The Möbius invariant norm on $B_1(\mathbf{D})$ is then given by $\|f\| = \inf \sum_k |a_k|$. See [9].

6.3. Möbius invariant Hilbert spaces. Although there exist many Möbius invariant semi-Banach spaces on a given bounded symmetric domain Ω, there aren't that many Möbius invariant semi-Hilbert spaces of holomorphic functions on Ω. In fact, there is no nontrivial Hilbert space of holomorphic functions whose norm is invariant under Aut(Ω). As for semi-Hilbert spaces, the following theorem says that there is essentially just one such space on each domain.

Theorem. [6][7] *For every bounded symmetric domain Ω there exists one and only one (up to a constant multiple of the semi-inner product) "natural" semi-Hilbert space of holomorphic functions on Ω.*

In the case of the open unit disk the Möbius invariant semi-Hilbert space is the Dirichlet space consisting of analytic functions f on $\mathbf{D}$ such that

$$\|f\|^2 = \int_{\mathbf{D}} |f'(z)|^2 \, dv(z) < +\infty;$$

see [5]. In the case of the open unit ball B_n in $\mathbf{C}^n$ the Möbius invariant semi-Hilbert space is the space of holomorphic functions $f(z) = \sum a_\alpha z^\alpha$ such that

$$\|f\|^2 = \sum |\alpha| \frac{\alpha!}{|\alpha|!} |a_\alpha|^2 < +\infty;$$

see [28][57]. The description of the Möbius invariant semi-Hilbert space in the general case is given in [2][3][46].

6.4. Weighted actions of Aut (Ω). Recall that for $\varphi \in \operatorname{Aut}(\Omega)$ we use $J_\varphi(z)$ to denote the complex Jacobian determinant of the mapping φ at the point z. Since φ is invertible, $J_\varphi(z)$ is nonvanishing. Thus for every complex number τ the principal branch of $J_\varphi^\tau(z) = \left[J_\varphi(z)\right]^\tau$ is well defined on Ω. We let U_φ^τ be the operator on $H(\Omega)$ defined by

$$U_\varphi^\tau f(z) = f \circ \varphi(z) J_\varphi^{\tau/N}(z), \qquad z \in \Omega, \ f \in H(\Omega).$$

It is shown in [6][7] that for every τ lying in a certain subset of $[0, +\infty)$ (the so-called Wallach set of Ω) there exists one and only one (up to constant multiple of the inner product) "natural" Hilbert space of holomorphic functions which is invariant under the actions of U_φ^τ, $\varphi \in \operatorname{Aut}(\Omega)$. It is also shown in [6][7] for every τ in a certain discrete subset of the Wallach set there exists one and only one (up to constant multiple of the semi-inner product) nontrivial "natural" semi-Hilbert (but not Hilbert) space of holomorphic functions which is invariant under the actions of U_φ^τ, $\varphi \in \operatorname{Aut}(\Omega)$.

7. The Berezin Transform

In classical harmonic analysis on the circle the Poisson transform plays a very prominent role. This is especially so for problems related to Hardy spaces. It has been realized in recent years that there is a Bergman space version of the Poisson transform, namely, the Berezin transform.

Recall that the unit vectors k_z, $z \in \Omega$, in $L_a^2(\Omega)$ are called normalized reproducing kernels of Ω. Let f be a Lebesgue integrable function on Ω. The Berezin transform of f is the function

$$\widetilde{f}(z) = \int_\Omega f \circ \varphi_z(w)\, dv(w) = \int_\Omega f(w)\, |k_z(w)|^2\, dv(w), \qquad z \in \Omega.$$

It is easy to see that the Berezin transform is invariant under the action of the automorphism group of Ω. Thus $\widetilde{f \circ \varphi}(z) = \widetilde{f} \circ \varphi(z)$ for all $f \in L^1(\Omega, dv)$, $\varphi \in \operatorname{Aut}(\Omega)$, and $z \in \Omega$.

7.1. Fixed points. It is clear from the symmetry of Ω and the mean-value property of holomorphic functions that the Berezin transform fixes holomorphic functions in $L^1(\Omega, dv)$. It follows that the Berezin transform fixes complex pluriharmonic functions in $L^1(\Omega, dv)$. We find the following result about fixed points of the Berezin transform very surprising.

Theorem. [1] *Suppose f is Lebesgue integrable on the open unit ball B_n of $\mathbf{C}^n$. If $1 \le n \le 11$, then $\widetilde{f} = f$ if and only if f is $\mathcal{M}$-harmonic. This does not hold for any $n \ge 12$.*

Recall that a function f on Ω is called $\mathcal{M}$-harmonic if it is second differentiable and annihilated by the invariant Laplacian of Ω. In particular, pluriharmonic functions are $\mathcal{M}$-harmonic.

7.2. Algebraic properties. It is obvious that the Berezin transform is linear, preserves complex conjugation, and preserves positivity. It is natural to ask if the Berezin transform is one-to-one.

Assume f is Lebesgue integrable. Consider the function

$$F(z,w) = \int_\Omega f(u)K(z,u)K(u,w)\,dv(u), \qquad z,w \in \Omega.$$

It is clear that F is holomorphic in z and conjugate holomorphic in w. Moreover, $\widetilde{f} = 0$ on Ω if and only if F vanishes on the diagonal of $\Omega \times \Omega$. By a well-known theorem in several complex variables (see Exercise 3 on page 326 of [26] for example), F vanishes on the diagonal of $\Omega \times \Omega$ if and only if F is identically zero on $\Omega \times \Omega$. Since functions of the form $K(z,\cdot)K(\cdot\, w)$ $(z,w \in \Omega)$ span the space of all continuous functions on $\overline{\Omega}$ by the Stone-Weierstrass theorem, we conclude that $\widetilde{f} = 0$ implies $f = 0$. This proves the following folk theorem.

Theorem. *The Berezin transform is one-to-one on $L^1(\Omega, dv)$.*

7.3. BMO in the Bergman metric. It is well known that the boundary BMO theory for higher rank bounded symmetric domains is hard; see [14][17] for the case of the bidisk. However, there exists a nice theory of BMO in terms of the Bergman metric on Ω. The best application of the Berezin transform so far is perhaps in the development of this theory.

Recall that $\beta(z,w)$ is the Bergman distance between z and w. For $z \in \Omega$ and $R > 0$ let $D(z,R) = \{w \in \Omega : \beta(z,w) < R\}$ be the Bergman metric ball centered at z with radius R. Several useful properties of $D(z,R)$ are obtained in [12][13]. For a locally integrable function f on Ω we let $\widehat{f}_R(z)$ be the average of f over $D(z,R)$ with respect to the unweighted volume measure dv.

For $R > 0$ the space $BMO_R(\Omega)$ consists of locally integrable functions f such that

$$\|f\| = \sup_{z\in\Omega}\left[\frac{1}{v(D(z,R))}\int_{D(z,R)}\left|f(w) - \widehat{f}_R(z)\right|^2 dv(w)\right]^{\frac{1}{2}} < +\infty.$$

It was shown in [12] that the space $BMO_R(\Omega)$ is actually independent of R. We shall write $BMO_\partial(\Omega) = BMO_R(\Omega)$. A companion space $VMO_\partial(\Omega)$ can be introduced in the obvious way.

Just as the classical BMO on the circle can be described using the Poisson integral (Garsia's lemma), the space $BMO_\partial(\Omega)$ can be characterized by the Berezin transform.

Theorem. [12] *A function $f \in L^2(\Omega, dv)$ belongs to $BMO_\partial(\Omega)$ if and only if*

$$\|f\|_{BMO_\partial} = \sup_{z\in\Omega} \left[\widetilde{|f|^2}(z) - |\widetilde{f}(z)|^2\right]^{\frac{1}{2}} < +\infty.$$

Furthermore, for $f \in BMO_\partial(\Omega)$ we have

$$|\widetilde{f}(z) - \widetilde{f}(w)| \le 2\sqrt{2}\, \|f\|_{BMO_\partial}\beta(z,w)$$

for all z and w in Ω.

It follows easily from the theorem above that $BMO_\partial(\Omega) \cap H(\Omega) = \mathcal{B}(\Omega)$, the Bloch space of Ω. However, we do not always have $VMO_\partial(\Omega) \cap H(\Omega) = \mathcal{B}_0(\Omega)$; we have this only when Ω is the open unit ball. In fact, $VMO_\partial(\Omega) \cap H(\Omega)$ consists of just constant functions when the rank of Ω is greater than one. See [12].

7.4. Applications to operator theory. The space $BMO_\partial(\Omega)$ is used in [12] to characterize bounded Hankel operators on the Bergman space of Ω. Similarly, the space $VMO_\partial(\Omega)$ can be used to characterize compact Hankel operators on the Bergman space. The application of the Berezin transform to the study of Schatten class Hankel operators on the Bergman space can be found in [8][50][56].

The Berezin transform has also been found to be a very useful tool in the theory of Toeplitz operators on the Bergman space. See [12][52][54].

8. Some Extremal Problems

We include a few extremal problems related to the Bergman kernel and the Bergman metric. The solution of the first problem also uses the standard variational argument.

8.1. Theorem. *Suppose $1 \le p < +\infty$ and $z \in \Omega$. Then the extremal problem*

$$\sup\{\operatorname{Re} f(z) : \|f\|_p \le 1, f \in L^p_a(\Omega)\}$$

has a unique solution; this solution is given by

$$f_z(w) = [k_z(w)]^{\frac{2}{p}}, \qquad w \in \Omega.$$

The existence and uniqueness of the solution follow from standard functional analysis. To find the optimal solution we first apply the variational method to show that if $f_z(w)$ is the optimal solution then it must satisfy the following condition:

$$\frac{f(z)}{f_z(z)} = \int_\Omega |f_z(w)|^{p-2}\overline{f_z(w)}f(w)\,dv(w),$$

where f is any function in $L_a^p(\Omega)$. Furthermore, general functional analysis shows that there is at most one function f_z satisfying the condition above. A direct calculation, with the nonvanishing property of the Bergman kernel, shows that the function

$$f_z(w) = [k_z(w)]^{\frac{2}{p}}, \qquad w \in \Omega,$$

has the desired property.

Similar results can be proven for Hardy spaces as well as weighted Bergman spaces on Ω. Also, the same argument works on other types of domains as long as the reproducing kernel is nonvanishing.

In the case of Bergman spaces on bounded symmetric domains, it is likely that Theorem 8.1 can also be proved by first showing the result at 0 and then applying the symmetries of Ω. See [43] for the case of the open unit ball.

8.2. Recall that the Bloch space $\mathcal{B}(\Omega)$ consists of holomorphic functions f in Ω with

$$\|f\|_{\mathcal{B}} = \sup\{Q_f(z) : z \in \Omega\} < +\infty.$$

A holomorphic function f in Ω belongs to $\mathcal{B}(\Omega)$ if and only if $|f(z) - f(w)| \le C\beta(z,w)$ for some constant $C > 0$ and all $z, w \in \Omega$. The following result improves upon this relationship between the Bloch space and the Bergman metric.

Theorem. [58] *For all z and w in a rank one domain Ω we have*

$$\beta(z,w) = \sup\{|f(z) - f(w)| : \|f\|_{\mathcal{B}} \le 1, f \in \mathcal{B}(\Omega)\}.$$

We suspect the result above is true for all bounded symmetric domains. But we are unable to come up with a proof, or a reference, or a counter-example. However, we can still prove the following variant of the above result.

8.3. Theorem. [53] *For every bounded symmetric domain Ω there exists a constant $C > 0$ such that*

$$C^{-1}\beta(z,w) \le \sup\left\{|\widetilde{f}(z) - \widetilde{f}(w)| : \|f\|_{BMO_\partial} \le 1\right\} \le C\beta(z,w)$$

for all z and w in Ω.

This result was used in [53] to characterize the pointwise multipliers of the spaces $BMO_\partial(\Omega)$ and $VMO_\partial(\Omega)$. Multipliers of the Bloch space of the ball B_n are also characterized in [53]; the result on the disk was proved earlier by J. Arazy and M. Anderson, but they never published it. We have been unable to characterize the multipliers of the Bloch space for an arbitrary bounded symmetric domain.

References

1. P. Ahern, M. Flores, and W. Rudin, An invariant volume-mean-value property, *J. Funct. Anal.* **111** (1993), 380-397.
2. J. Arazy, Realization of the invariant inner products on the highest quotients of the composition series, *Ark. Mat.* **30** (1992), 1-24.
3. J. Arazy, Integral formulas for the invariant inner products in spaces of analytic functions on the unit ball, *Function Spaces*, 9-23, Lecture Notes in Pure and Appl. Math. **136**, Dekker, New York, 1992.
4. J. Arazy, A survey of invariant Hilbert spaces of analytic functions on bounded symmetric domains, preprint.
5. J. Arazy and S. Fisher, The uniqueness of the Dirichlet space among Möbius invariant Hilbert spaces, *Ill. J. Math.* **29** (1985), 449-462.
6. J. Arazy and S. Fisher, Weighted actions of the Möbius group and their invariant Hilbert spaces, *Math. Publication Series (II)* **23**, University of Haifa, Israel, 1989.
7. J. Arazy and S. Fisher, Invariant Hilbert spaces of analytic functions on bounded symmetric domains, *Topics in Operator Theory: Ernst D. Hellinger memorial volume*, 67-91, *Operator Theory Adv. Appl.* **48**, Birkhäuser, Basel, 1990.
8. J. Arazy, S. Fisher, S. Janson, and J. Peetre, Membership of Hankel operators on the ball in unitary ideals, *J. London Math. Soc.* **43** (1991), 485-508.
9. J. Arazy, S. Fisher, and J. Peetre, Möbius invariant function spaces, *J. Reine Angew. Math.* **363** (1985), 110-145.
10. D. Békollé, The dual of the Bergman space A^1 in symmetric Siegel domains of type II, *Trans. Amer. Math.* **296** (1986), 607-619.
11. D. Békollé and A. Bonami, Estimates for the Bergman and Szegö projections in two symmetric domains of $\mathbf{C}^n$, preprint.
12. D. Békollé, C. Berger, L. Coburn, and K. Zhu, BMO in the Bergman metric on bounded symmetric domains, *J. Funct. Anal.* **93** (1990), 310-350.
13. C. Berger, L. Coburn, and K. Zhu, Function theory in Cartan domains and the Berezin-Toeplitz symbol calculus, *Amer. J. Math.* **110** (1988), 921-953.
14. S.-Y. A. Chang, Carleson measure on the bi-disk, *Ann. Math.* **109** (1979), 613-620.
15. P. Duren, B. Romberg, and A. Shields, Linear functionals on H^p spaces with $0 < p < 1$, *J. Reine Angew. Math.* **238** (1969), 32-60.
16. J. Faraut and A. Koranyi, Function spaces and reproducing kernels on bounded symmetric domains, *J. Funct. Anal.* **88** (1990), 64-89.
17. R. Fefferman, Bounded mean oscillation on the polydisk, *Ann. Math.* **110** (1979), 395-406.
18. F. Forelli and W. Rudin, Projections on spaces of holomorphic functions in balls, *Indiana Univ. Math. J.* **24** (1974), 593-602.
19. A. Frazier, The dual of H^p of the polydisk for $0 < p < 1$, *Duke Math. J.* **39** (1972), 369-379.
20. S. Helgason, *Differential Geometry, Lie Groups, and Symmetric Spaces*, Academic Press, New York, 1978.

21. L. K. Hua, *Harmonic Analysis of Functions of Several Complex Variables in the Classical Domains*, Amer. Math. Soc. Transl. Ser. **6**, Amer. Math. Soc., Providence, RI, 1963.
22. J. Isidro and L. Stacho, *Holomorphic Automorphism Groups in Banach Spaces: An Elementary Introduction*, North Holland Math. Studies **105**, Elsevier, New York, 1985.
23. B. Jöricke, Continuity of the Cauchy projection in Hölder norms for classical domains, *Math. Nachr.* **113** (1983), 227-244.
24. A. Koranyi, Analytic invariants of bounded symmetric domains, *Proc. Amer. Math. Soc.* **19** (1968), 279-284.
25. A. Koranyi, The volume of symmetric domains, the Koecher Gamma function, and an integral of Selberg, *Studia Sci. Math. Hungarica* **17** (1982), 129-133.
26. S. Krantz, *Function Theory of Several Complex Variables*, John Wiley & Sons, 1982.
27. J. Mitchell and K. Hahn, Representation of linear functionals in H^p spaces over bounded symmetric domains, *J. Math. Anal. Appl.* **56** (1976), 379-396.
28. J. Peetre, Möbius invariant function spaces in several variables, preprint, 1982.
29. M. Peloso, Möbius invariant spaces on the unit ball, *Michigan Math. J.* **39** (1992), 509-536.
30. L. Rubel and R. Timoney, An extremal property of the Bloch space, *Proc. Amer. Math. Soc.* **75** (1979), 45-50.
31. W. Rudin, *Function Theory in the Unit Ball of* $\mathbf{C}^n$, Springer-Verlag, New York, 1980.
32. J. Shapiro, Mackey topologies, reproducing kernels, and diagonal maps on the Hardy and Bergman spaces, *Duke Math. J.* **43** (1976), 187-202.
33. J. Shi, On the rate of growth of the means M_p of holomorphic and pluriharmonic functions on bounded symmetric domains in $\mathbf{C}^n$, *J. Math. Anal. Appl.* **126** (1987), 161-175.
34. M. Stoll, Mean value theorems for harmonic and holomorphic functions on bounded symmetric domains, *J. Reine Angew. Math.* **283** (1977), 191-198.
35. R. Timoney, Bloch functions in several complex variables, I, *Bull. London Math. Soc.* **12** (1980), 241-267.
36. R. Timoney, Bloch functions in several complex variables, II, *J. Reine Angew. Math.* **319** (1980), 1-22.
37. R. Timoney, Maximal invariant spaces of analytic functions, *Indiana Univ. Math. J.* **31** (1982), 651-663.
38. H. Upmeier, Jordan algebras and harmonic analysis on symmetric spaces, *Amer. J. Math.* **108** (1986), 1-25.
39. H. Upmeier, Toeplitz C^*-algebras on bounded symmetric domains, *Ann. Math.* **119** (1984), 549-576.
40. H. Upmeier, *Jordan algebras in analysis, operator theory, and quantum mechanics*, CBMS Regional Conference Series in Math. **67**, Amer. Math. Soc., 1987.
41. H. Upmeier, Fredholm indices for Toeplitz operators on bounded symmetric domains, *Amer. J. Math.* **110** (1988), 811-832.
42. S. Vagi, Harmonic Analysis in Cartan and Siegel Domains, *Studies in Harmonic Analysis* (J. M. Ash, Ed.), MAA Studies in Math. **13**, New York, 1976.

43. D. Vukotić, A sharp estimate for A^p_α functions in $\mathbf{C}^n$, *Proc. Amer. Math. Soc.* **117** (1993), 753-756.
44. Z. Yan, Duality and differential operators on the Bergman spaces of bounded symmetric domains, *J. Funct. Anal.* **105** (1992), 171-186.
45. Z. Yan, Invariant differential operators and holomorphic function spaces, preprint.
46. Z. Yan, Integral formulas for invariant inner products on Hilbert spaces of holomorphic functions, preprint.
47. Z. Yan, Möbius transformations on bounded symmetric domains of tube type,preprint.
48. Z. Yan, A class of generalized hypergeometric functions in several variables, *Canad. J. Math.* **44** (1992), 1317-1338.
49. Z. Yan, Differential operators and function spaces, *Several Complex Variables in China*, 121-142, Contemp. Math. **142**, Amer. Math. Soc., Providence, 1993.
50. D. Zheng, Schatten class Hankel operators on Bergman spaces, *Integral Equations and Operator Theory* **13** (1990), 442-459.
51. K. Zhu, Duality and Hankel operators on the Bergman spaces of bounded symmetric domains, *J. Funct. Anal.* **81** (1988), 260-278.
52. K. Zhu, Positive Toeplitz operators on weighted Bergman spaces of bounded symmetric domains, *J. Operator Theory* **20** (1988), 329-357.
53. K. Zhu, Multipliers of BMO in the Bergman metric with applications to Toeplitz operators, *J. Funct. Anal.* **87** (1989), 31-50.
54. K. Zhu, *Operator Theory in Function Spaces*, Marcel Dekker, New York, 1990.
55. K. Zhu, On certain unitary and composition operators, *Proc. Symp. Pure Math.* **51** (Part II), 371-385, Amer. Math. Soc., Providence, 1990.
56. K. Zhu, Schatten class Hankel operators on the Bergman space of the unit ball, *Amer. J. Math.* **113** (1991), 147-167.
57. K. Zhu, Möbius invariant Hilbert spaces of holomorphic functions in the unit ball of $\mathbf{C}^n$, *Trans. Amer. Math. Soc.* **323** (1991), 823-842.
58. K. Zhu, Distances induced by Banach spaces of holomorphic functions on complex domains, to appear in *J. London Math. Soc.*.
59. K. Zhu, Bergman and Hardy spaces with small exponents, *Pacific J. Math.* **162** (1994), 189-199.
60. K. Zhu, Holomorphic Besov spaces on bounded symmetric domains, to appear in *Quarterly J. Math.*,
61. K. Zhu, Holomorphic Besov spaces on bounded symmetric domains, II, preprint.

Other *Mathematics and Its Applications* titles of interest:

A. van der Burgh and J. Simonis (eds.): *Topics in Engineering Mathematics*. 1992, 266 pp. ISBN 0-7923-2005-3

F. Neuman: *Global Properties of Linear Ordinary Differential Equations*. 1992, 320 pp. ISBN 0-7923-1269-4

A. Dvurecenskij: *Gleason's Theorem and its Applications*. 1992, 334 pp. ISBN 0-7923-1990-7

D.S. Mitrinovic, J.E. Pecaric and A.M. Fink: *Classical and New Inequalities in Analysis*. 1992, 740 pp. ISBN 0-7923-2064-6

H.M. Hapaev: *Averaging in Stability Theory*. 1992, 280 pp. ISBN 0-7923-1581-2

S. Gindinkin and L.R. Volevich: *The Method of Newton's Polyhedron in the Theory of PDE's*. 1992, 276 pp. ISBN 0-7923-2037-9

Yu.A. Mitropolsky, A.M. Samoilenko and D.I. Martinyuk: *Systems of Evolution Equations with Periodic and Quasiperiodic Coefficients*. 1992, 280 pp. ISBN 0-7923-2054-9

I.T. Kiguradze and T.A. Chanturia: *Asymptotic Properties of Solutions of Nonautonomous Ordinary Differential Equations*. 1992, 332 pp. ISBN 0-7923-2059-X

V.L. Kocic and G. Ladas: *Global Behavior of Nonlinear Difference Equations of Higher Order with Applications*. 1993, 228 pp. ISBN 0-7923-2286-X

S. Levendorskii: *Degenerate Elliptic Equations*. 1993, 445 pp. ISBN 0-7923-2305-X

D. Mitrinovic and J.D. Kečkić: *The Cauchy Method of Residues, Volume 2*. Theory and Applications. 1993, 202 pp. ISBN 0-7923-2311-8

R.P. Agarwal and P.J.Y Wong: *Error Inequalities in Polynomial Interpolation and Their Applications*. 1993, 376 pp. ISBN 0-7923-2337-8

A.G. Butkovskiy and L.M. Pustyl'nikov (eds.): *Characteristics of Distributed-Parameter Systems*. 1993, 386 pp. ISBN 0-7923-2499-4

B. Sternin and V. Shatalov: *Differential Equations on Complex Manifolds*. 1994, 504 pp. ISBN 0-7923-2710-1

S.B. Yakubovich and Y.F. Luchko: *The Hypergeometric Approach to Integral Transforms and Convolutions*. 1994, 324 pp. ISBN 0-7923-2856-6

C. Gu, X. Ding and C.-C. Yang: *Partial Differential Equations in China*. 1994, 181 pp. ISBN 0-7923-2857-4

V.G. Kravchenko and G.S. Litvinchuk: *Introduction to the Theory of Singular Integral Operators with Shift*. 1994, 288 pp. ISBN 0-7923-2864-7

A. Cuyt (ed.): *Nonlinear Numerical Methods and Rational Approximation II*. 1994, 446 pp. ISBN 0-7923-2967-8

Other *Mathematics and Its Applications* titles of interest:

G. Gaeta: *Nonlinear Symmetries and Nonlinear Equations.* 1994, 258 pp.
ISBN 0-7923-3048-X

V.A. Vassiliev: *Ramified Integrals, Singularities and Lacunas.* 1995, 289 pp.
ISBN 0-7923-3193-1

N.Ja. Vilenkin and A.U. Klimyk: *Representation of Lie Groups and Special Functions.* Recent Advances. 1995, 497 pp. ISBN 0-7923-3210-5

Yu. A. Mitropolsky and A.K. Lopatin: *Nonlinear Mechanics, Groups and Symmetry.* 1995, 388 pp. ISBN 0-7923-3339-X

R.P. Agarwal and P.Y.H. Pang: *Opial Inequalities with Applications in Differential and Difference Equations.* 1995, 393 pp. ISBN 0-7923-3365-9

A.G. Kusraev and S.S. Kutateladze: *Subdifferentials: Theory and Applications.* 1995, 408 pp. ISBN 0-7923-3389-6

M. Cheng, D.-G. Deng, S. Gong and C.-C. Yang (eds.): *Harmonic Analysis in China.* 1995, 318 pp. ISBN 0-7923-3566-X

M.S. Livšic, N. Kravitsky, A.S. Markus and V. Vinnikov: *Theory of Commuting Nonselfadjoint Operators.* 1995, 314 pp ISBN 0-7923-3588-0